Reinhard Selten (Ed.)

Game Equilibrium Models III

Strategic Bargaining

With Contributions by

W. Albers, E. Bennett, W. Güth, H. Haller, J. D. Laing
A. Okada, R. Selten, E. van Damme, M. Wooders

With 29 Figures

Springer-Verlag

Berlin Heidelberg New York
London Paris Tokyo
Hong Kong Barcelona
Budapest

Professor Dr. Reinhard Selten
Institut für Gesellschaft- und
Wirtschaftswissenschaften
der Universität Bonn
Wirtschaftstheoretische Abteilung I
Adenauerallee 24-42
D-5300 Bonn 1, FRG

ISBN 978-3-642-08110-1

Printed in Germany

2142/7130-543210

Preface to the Series "Game Equilibrium Models"

The four volumes of the series "Game Equilibrium Models" are the result of a research year at the Center for Interdisciplinary Research of the University of Bielefeld, Germany. The German name of this center is Zentrum für interdisziplinäre Forschung, but everybody who is familiar with this unique institution refers to it by the official acronym *ZiF*.

In the time from October 1, 1987, to September 30, 1988, the ZiF was the home of the interdisciplinary research group which produced the papers in the four volumes of this series. Participants coming from many parts of the world lived in the guest apartments of the ZiF for the whole time or part of it and worked on a common project. The name of the project was "Game Theory in the Behavioral Sciences". It occurred to me only later that "Game Equilibrium Models" - the title of the series - even more appropriately indicates the unifying theme of the research group.

Among the participants were economists, biologists, mathematicians, political scientists, psychologists and a philosopher. A lively interaction resulted from the mix of disciplines. The common methodological basis of non-cooperative theory was the shared culture which facilitated communication across disciplines. The intense exchange of ideas coming from different fields had a profound influence on the thinking of many among the participants.

It was not easy to find a coherent way to group the papers into the four volumes and to find appropriate titles for the books. These and other difficult decisions have been made by an editorial committee consisting of Wulf Albers, Rudolf Avenhaus, Eric van Damme, Werner Güth, Peter Hammerstein, Ronald Harstad, Franz Weissing, and myself.

In the behalf of the whole research group I want to thank all those who helped to make the research year possible. We owe special thanks to the staff of the ZiF and in particular to Mrs. Lilo Jegerlehner for her technical help in the preparation of the four volumes.

Finally, I want to express my gratitude to all those who assisted me in the organizational and editorial work, especially to Franz Weissing whose efforts were indispensable.

Bielefeld/Bonn, January 1991

Reinhard Selten

Contents

Contributors

Wulf Albers, Institut für mathematische Wirtschaftsforschung (IMW), Universität Bielefeld, D-4800 Bielefeld 1, FRG

Elaine Bennett, Bunche Hall, UCLA, Los Angeles, CA 90024, USA

Eric van Damme, CentER for Economic Research, Hogeschoollaan 225, P.O. box 90153, 5000 LE Tilburg, The Netherlands

Werner Güth, Johann Wolfgang Goethe-Universität Frankfurt/M., Fachbereich Wirtschaftswissenschaften, Mertonstraße 17, D-6000 Frankfurt am Main 11, FRG

Hans Haller, Department of Economics, Virginia Polytechnic Institute, Blacksburg, VA 24061, USA

James D. Laing, Department of Decision Sciences, University of Pennsylvania, 1311 Steinberg Hall - Dietrich Hall, Philadelphia, PA 19104-6366, USA

Akira Okada, Graduate School of Policy Science, Saitama University, Urawa 338, Japan

Reinhard Selten, Institut für Gesellschafts- und Wirtschaftswissenschaften der Universität Bonn, wirtschaftstheoretische Abteilung 1, Adenauerallee 24-42, D-5300 Bonn 1, FRG

Myrna Wooders, Department of Economics, University of Toronto, 150 St. George Street, Toronto, Canada M5S 1A1

Introduction to the Series "Game Equilibrium Models"

Game equilibrium models are descriptions of interactive decision situations by games in extensive or normal form. The analysis of such models is based on the equilibrium point concept, often refined by additional requirements like subgame perfectness. The series consists of four volumes:

I: Evolution and Game Dynamics
II: Methods, Morals and Markets
III: Strategic Bargaining
IV: Social and Political Interaction.

The game equilibrium models presented in these books deal with a wide variety of topics. Just one example from each of the volumes may provide an illustration: Egg trading in hermaphrodite fish (*Friedman and Hammerstein* in Volume I), the social organization of irrigation systems (*Weissing and Ostrom* in Volume II), wage bargaining (*Haller* in Volume III), and beheading games in mediaeval literature (*O'Neill* in Volume IV).

Non-cooperative game theory is a useful research tool not only in economics and the social sciences, but also in biology. Game theory has been created as a theory of conflict and cooperation among rational individuals. For a long time strong rationality assumptions seemed to be indispensable at the foundations of game theory. In this respect, biological applications have changed our perspectives. Game equilibrium may be reached as the result of processes of natural selection without any rational deliberation. Therefore, game equilibrium models can contribute to the explanation of behavioral and structural features of animals and plants.

The interpretation of game equilibrium as the result of evolution raises the question of dynamic stability with respect to evolutionary processes. Similar problems also arise in theories of game learning. The first volume contains three papers on game dynamics. Two of them are concerned with the dynamic foundations of evolutionary game theory and the third one explores stability in a model of anticipatory learning. The remaining papers in the first volume present evolutionary game equilibrium models ranging from abstract investigations of phenomena like bluffing or group-based altruism to the examination of concrete systems observed in nature like "competition avoidance in a dragonfly mating system". Not only theoretical clarifications of the foundations of evolutionary game theory and related research can be found in **Evolution and Game Dynamics**, but also exciting new biological applications.

The title of the second volume, **Methods, Morals, and Markets**, points to several areas of research which attract the interest mainly of economists, but also of political scientists, mathematicians and philosophers. The first paper is a sophisticated mathematical contribution which applies new tools to basic questions of non-cooperative game theory. The word "method" mainly refers to this paper, but to some extent also to the next three contributions, which discuss basic conceptual problems in the interpretation of game equilibrium. Two papers relate to the philosophical notion of the social contract and its exploration with the help of game theoretical models. This work in concerned with "morals", a theme which is also touched by a paper on irrigation institutions. The remaining four papers of the second volume explore game equilibrium models of markets; two of these contributions are experimental and compare theoretical solutions with experimental data.

The third volume on **Strategic Bargaining** collects ten papers on game equilibrium models of bargaining. All these papers look at bargaining situations as non-cooperative games. Unlike in cooperative game theory, cooperation is not taken for granted, but explained as an outcome of equilibrium analysis. General models of two-person and n-person bargaining are explored, sometimes enriched by institutional detail like the availability of long-term contracts. Other papers explore bargaining in special contexts like wage negotiations. Two contributions concern spatial games; one of these contributions is experimental.

The exploration of strategic models of bargaining is an active field of research which attracts the attention of many game theorists and economists. The ten papers in the third volume contribute to the progress in this field.

The fourth volume on **Social and Political Interaction** mainly presents game equilibrium models in the area of political science. Three of the papers concern topics in other fields: the distribution of foreign language skills, altruism as a social dilemma (an experimental paper) and beheading games in mediaeval literature. Five contributions to the area of international relations deal with game theoretical models of the balance of power, of alliance formation, and of an issue in armament policy. An investigation of inspection problems like those arising in connection with the non-proliferation treaty also touches the area of international relations. Other papers on problems of political science deal with the game theoretical resolution of the Condorcet paradox by equilibrium selection, the modelling of political pressure exerted by firms on the government and the draft resistance problem.

The main emphasis is on biology in Volume I, on economics in Volumes II and III, and on political science in Volume IV. This is the result of an attempt to group the great variety of papers resulting from a year long interdisciplinary research project in a reasonably coherent way. However, not only biologists, but also economists and a psychologist have contributed to Volume I. Similarly, not only economists and mathematicians, but also political scientists as well as a biologist and a psychologist are among the authors of Volumes II and III. All four volumes are the result of the cooperation of researchers from many disciplines united by a common interest in game equilibrium models within and beyond the borders of their fields.

Bielefeld/Bonn, January 1991 Reinhard Selten

Introduction to Volume III: "Strategic Bargaining"

A topic that received a lot of attention from the research group throughout the year the group was together at the ZiF was that of strategic bargaining. The inspiration came from papers by Binmore (1985), Rubinstein (1982) and Selten (1981) which were carefully studied and critically discussed at the time the research project started. All the papers collected in this volume, except the one by *Selten and Güth*, deal with bargaining under conditions of complete information. The papers of *Okada* and *Haller* consider bilateral bargaining problems. The *Selten/Wooders* paper deals with bargaining in a market context in which the number of participants varies over time. The remaining papers consider coalitional bargaining problems with a fixed number of players and they can be viewed as pursuing the "Nash program" of investigating whether concepts from cooperative game theory can be implemented by means of noncooperative bargaining procedures. The *Bennett/van Damme* paper is restricted to games with transferable utility, while *Bennett* considers general NTU games. *Laing* and *Albers/Laing* consider bargaining in a spatial context where the problem is which location to choose. *Laing* provides a theoretical analysis whereas *Albers/Laing* present experimental results obtained in this setting.

In the first paper, *Okada* investigates bargaining in the context of a supergame. Given is a 2-person normal form game G with a unique Nash equilibrium point which has to be played infinitely often. It is assumed that both players discount future payoffs with a common discount factor δ. In contrast to the traditional repeated game analysis, Okada assumes that binding contracts may be signed. In each period in which no contract is in force, a player, say i, is randomly selected to propose a contract. A contract specifies a correlated strategy pair as well as for how long this pair should be played. If the opponent (player $j = -i$) accepts the proposal, it becomes binding; if j rejects, j may make a proposal himself. If the latter proposal is accepted, it is enforced, otherwise there is disagreement in this period and players are forced to play G noncooperatively once. In the next period, however, negotiations continue according to the specified procedure. Okada's model differs from related papers in the literature: Both the disagreement point and the length of the contract are endogeneously determined. Okada shows that there is a unique stationary subgame perfect equilibrium (SPE) and that, if the discount factor δ tends to 1, the equilibrium payoffs converge to the Nash cooperative bargaining solution of G when the Nash equilibrium of G is taken as the threat point. Okada also shows that the result no longer holds if only short term contracts can be signed.

In his second paper, *Okada* considers a slightly different model: Player do not discount the future (they want to maximize their per period payoffs) and, in each period, player 1 is the first to make a proposal. The main result in this paper is that if the memory of the players is incomplete (i.e., there exists a time t such that the players' actions at time $t' \geq t$ cannot depend on what happened before time t), the set of outcomes generated by stationary SPE coincides with the Pareto efficient individually rational outcomes of G. With complete memory, however, also inefficient outcomes can result in equilibrium.

Haller studies bargaining between capital and labor about how to divide a per period surplus of S. The bargaining procedure is a variant of Rubinstein's alternating offer model with both players using the same discount rate δ. In each period, one of the players makes a proposal concerning which fraction α of S capital should get and the other player can accept or reject. If the proposal α is accepted, capital gets αS in each of the following periods, while labor gets $(1-\alpha)S$ per period. If there is no agreement, labor can choose to go on strike for one period, or to be paid according to the conditions of the status quo contract in this period. (The latter yields labor $(1-\alpha)S$.) Hence, we have a bargaining problem with a status quo point as well as a threat point. (Apart from the random moves, Haller's game is a special case of Okada's first model.) Haller shows that there is a unique stationary SPE. In this equilibrium labor never strikes and gets paid the same as in the old contract. However, there also exists a nonstationary SPE. In this equilibrium labor threatens to strike, capital reacts by offering labor a higher share of the surplus, an offer that labor accepts.

When there are more than two players, one not only has to address the question of how the gains from cooperation will be divided, but also which coalition will be formed. There are four papers in this volume that address these questions for the case with a fixed number of players and all four papers (*Bennett*, *Laing* and *Albers/Laing*) use the so-called proposal making model to address them. The proposal making model is a bargaining procedure that was introduced in Selten's 1981 paper on characteristic function bargaining. Loosely the procedure may be described as follows. When there are no offers on the table, a player who has the initiative may propose a coalition together with a payoff division for that coalition. When an offer is already on the table, the player who has to move may accept or reject it. If he rejects it, it disappears from the table and he may make a new proposal. If all players involved in a proposal accept it, then the proposal is enforced. In Selten's 1981 paper it was shown that, for a certain class of TU games, there is a one-to-one correspondence between the stationary SPE payoffs of the proposal making model and the so-called semistable demand vectors. The latter is a solution concept from cooperative game theory that was introduced by W. Albers. Intuitively, a semistable demand vector is a vector of aspiration levels (one for each player) with the property that no player can raise his aspiration level, but no player is forced to lower it either.

Laing and *Bennett* investigate whether Selten's result can be extended to the NTU case. Bennett shows that indeed the one-to-one correspondence continues to hold for NTU games in which players that are not in the winning coalition receive payoff zero. She also proves a "folk theorem": Any proposal that is individually rational can be sustained by means of a (possibly) nonstationary SPE. Laing studies the proposal making model in a spatial context. The simplest situation is as follows. There is an odd number of players who have to make a collective decision of which point to choose in the Euclidean plane. If the majority agrees on a point, this point becomes the collective choice. Each player has his own bliss point and he prefers to have the chosen point to be as close as possible to his bliss point. Note that in this setup, the payoff of a player who is not in the winning coalition depends upon which agreement is reached by that coalition. Laing shows that the stationary SPE induce a subset of the semistable demand vectors. *Albers and Laing* report on experimental results obtained in this context. The experiments involved five players and the simple absolute majority rule. Also the experimental bargaining procedure differed from the proposal making model as it also incorporated a verification phase. (Basically the latter allowed a proposer to have second thoughts and to withdraw the proposal.) Albers and Laing find that in initial rounds players start with prominent (e.g. symmetric) proposals but that, as the game progresses and players gain experience, people begin to appreciate more fully the competitive structure of the game and progress towards a stable outcome.

Bennett and van Damme use a completely different noncooperative bargaining procedure, which they call the demand commitment model. In this model play proceeds by each player in turn setting a demand, i.e. a price for which he is willing to participate in any coalition. (Such a demand may again be interpreted as an aspiration level.) Having set his price, a player i can form a coalition S (with $i \in S$) if he wants, provided that all coalition partners in S have already committed themselves and the coalition can afford all demands. If a player does not want (or cannot) form a coalition, he selects a player that is not yet committed. If no coalition is formed, all players have payoff zero. Bennett/van Damme investigate the special case of apex games: There is one major player and n minor ones and the only profitable coalitions consist of the major player and a minor one, or of all the minor players together. Bennett/van Damme show that there is a unique SPE if the major player starts the game, but that there are multiple SPE when a minor player starts. They then refine the SPE concept, to what they call credible SPE, by forcing players to break ties in a way that ex ante profits them most. The main result then is that this refined concept implements the set of stable demand vectors.

Selten and Wooders study bargaining in thin markets. In each of countably many periods new participants become active in the market while participants that have formed a group leave. In each period the players play a game that is very much like the proposal making model, except that now a proposal (in the Selten/Wooders terminology: a firm) specifies wages for a certain number of players and that several proposals may be on the table simultaneously. If a player does not want to accept any proposal and also does not want to make a proposal himself, he simply waits for (hopefully) a better deal in the next period. Selten/Wooders show that a SPE exists and that, except in special circumstances, in equilibrium participants get no more than their competitive payoffs even with positive waiting costs. Hence, even if markets are thin the first mover in a period typically does not have an advantage since the constant inflow into the market exerts a competitive pressure.

Finally, *Selten and Güth* consider a 2-person bargaining game with incomplete information. A buyer is interested to buy a work of art, but this may be a fake or an original, and he has no way to find it out before it is too late. It is, however, common knowledge that the seller knows whether it is a fake or not. Bargaining is supposed to proceed by players simultaneously announcing a price offer and trade takes place if and only if the offers coincide. Since the game has many equilibria, the equilibrium selection theory of Harsanyi and Selten is applied to arrive at a unique answer. It is shown that the solution may be one of three types depending on the relation between w (the probability of a fake) and h (the buyer's value of an original): (i) If w is high, there is no trade, (ii) for intermediate values of w there is trade at a price equal to the buyer's value, and (iii) if w is low, trade takes place at an intermediate price which, however, is such that the buyer suffers more from the risk of art forgery than does the honest art dealer.

Bielefeld/Tilburg, January 1991 Eric van Damme

References:

Binmore, K. (1985). Bargaining and Coalitions. In: Alvin E. Roth (ed.), Game Theoretic Models of Bargaining. Cambridge, MA: Cambridge University Press.

Rubinstein, A. (1982). Perfect Equilibrium in a Bargaining Model. Econometrica **50**: 97-108.

Selten, R. (1981). A Noncooperative Model of Characteristic Function Bargaining. In: V. Böhm and H.H. Nachtkamp (eds.): Essays in Game Theory and Mathematical Economics in Honor of Oskar Morgenstern. pp 131-151.

A NONCOOPERATIVE APPROACH TO THE NASH BARGAINING PROBLEM

by

Akira Okada

Abstract: We present a noncooperative repeated bargaining model for a two-person game in normal form in which the Nash bargaining solution can be implemented as a unique stationary subgame perfect equilibrium point in the limiting case that the discount factor of payoffs goes to 1. Our bargaining game is based on the supergame model in a way that two players are allowed to negotiate for a long-term contract on actions before they select their own actions independently. The disagreement point of the Nash bargaining solution is determined to be a Nash equilibrium point of the normal form game. If short-term contracts only are enforceable, the Nash bargaining solution can not be implemented.

1. Introduction

The purpose of this paper is to consider the two-person bargaining problem initiated by Nash (1950, 1953) in the framework of the theory of a noncooperative game. We present a noncooperative repeated bargaining model for a two-person game in normal form by introducing the possibility of negotiations for actions into the supergame model. It is studied under what situations a noncooperative equilibrium point leads to the Nash bargaining solution in this repeated bargaining model.

In our earlier work (Okada, 1988a), we considered a question arising from the Folk Theorem of the supergame: Does a noncooperative equilibrium point necessarily lead to a Pareto efficient outcome in a supergame if binding agreements on actions are possible among players? In order to answer this question, we presented a two-person repeated bargaining game based on the supergame model in which players can negotiate for actions at every period before they choose their own actions independently and furthermore they can make a long-term contract on actions. Once they reach an agreement, it is assumed to be enforced.

The rule of our bargaining game was as follows. At period 1, player 1 first proposes an action pair and a term of contract. Then, player 2 either accepts or rejects player 1's proposal. If player 2 accepts it, they will take the agreed action pair for as many periods as they agreed. After the contract ends, they will negotiate again for a new contract. If player 2 rejects it, he can make a counterproposal to player 1. In turn, player 1 either accepts or rejects it. If he rejects it, their negotiations at period 1 break down and they must select their own actions independently. Afterwards, the game proceeds to the next period 2 and the same negotiation process will be repeated over infinitely many periods. We employed the limit of average payoffs as the players' preferences in our repeated bargaining game and proved that the set of stationary subgame perfect equilibrium outcomes in the bargaining game is identical to that of Pareto efficient and individually rational outcomes in the component game. We also pointed out that nonstationary subgame perfect equilibrium points may lead to Pareto inefficient outcomes even under the strong institutional assumption that any long-term contract on actions are enforceable.

From the viewpoint of a solution for the bargaining problem, a Pareto efficient and individually rational outcome is not sufficient because there are usually many outcomes satisfying both conditions. The basic question addressed in the bargaining theory is to which outcome rational players will agree in the set of Pareto efficient and individually rational outcomes.

In this paper, we will extend our earlier analysis and investigate under what situations a noncooperative equilibrium point of our bargaining model leads to a unique outcome and, if this happens, what the outcome of negotiations is. For this purpose, we will change the rule of our bargaining game described above in a way that players discount their future payoffs by some common discount factor. It turns out that in the case of discounted payoffs the equilibrium outcome significantly depends on who is the first proposer at every period. Therefore, we will assume that at every period each of two players is selected as the first proposer by some random mechanism with equal probability.

We will prove that for any discount factor there exists a unique stationary subgame perfect equilibrium outcome in our bargaining game. The equilibrium point prescribes the following behavior of players in negotiations. If negotiations break down, they select some Nash equilibrium point in the component game. This Nash equilibrium point turns out to be the disagreement point of the Nash bargaining solution. In the equilibrium point, the payoff pair which each player demands if he has the opportunity to make the last proposal within one period, called his equilibrium demand pair, depends on the discount factor as follows. When the discount factor is very small, each player demands his maximum payoff in the set of feasible individually rational payoff pairs. The feasible payoff pair yielding this maximum payoff is called his maximum point. When the discount factor goes beyond some critical level, the player whose maximum point has the smaller Nash product begins to decrease his demand. If the discount factor becomes greater, the other player also begins to decrease his demand and the two players' equilibrium demand pairs have the identical Nash products. Generally, their equilibrium demand pairs are different and thus their (normalized) discounted equilibrium payoff pair is not Pareto efficient. However, as the discount factor goes to 1, the equilibrium demand pairs of players and their (normalized) discounted payoff pair all converge to the Nash bargaining solution of the component game.

In the literature, there are several important works on the noncooperative implementation of the Nash bargaining solution. Among those are the works by Rubinstein (1982), Binmore (1980) and Moulin (1982). The bargaining model studied in this paper is closely related to Moulin's model. Moulin considered the following n-person bargaining model. At the beginning of every period, one player is selected randomly out of n players. This player proposes a lottery over the set of finite alternatives. Then all other players either accept or reject this proposal. If everyone accepts it, this is enforced as the final outcome. If at least one player rejects it, a random mechanism stops negotiations with probability ε and a (predetermined) disagreement point is enforced. With probability $1-\varepsilon$, the next period is played under the same rule. Moulin proved that the set of all stationary subgame perfect equilibrium payoff vectors converges to the Nash bargaining solution in the Hausdorff metric as the stopping probability ε goes to 0.

Our bargaining model differs from Moulin's model (and also from Rubinstein/Binmore's model) in the following two aspects. First, our model is constructed from a game in normal form and the disagreement point of bargaining is not given outside the model. The rule of our bargaining game requires two players to play the component game noncooperatively in case of the breakdown of negotiations which happens after they have rejected each other's proposals. This makes it possible that the disagreement point is endogenously determined to be a Nash equilibrium point of the component game. Secondly, although any long-term contract on actions is assumed to be enforceable, the term of contract is a strategic variable in our model. Therefore, by imposing the upperlimit of the term of contract, we can examine how the term of enforceable contracts influences the outcome of negotiations. We show that the Nash bargaining solution cannot be attained as a stationary subgame perfect equilibrium point if only short-term contracts, i.e., contracts of term 1 are enforceable in our bargaining game. In a unique stationary subgame perfect equilibrium point of such a modified game, players demand their maximum points regardless of the discount factor. Therefore, the (normalized) discounted payoff pair of players is generally Pareto inefficient in the equilibrium point. We can conclude from these results that the Nash bargaining solution can be implemented as a noncooperative equilibrium point in our bargaining model only in the limiting case that the discount factor of players goes to 1 and that there is no upperlimit of the term of enforceable contracts.

The paper is organized as follows. Section 2 provides some preliminaries about a two-person game in normal form. The Nash bargaining solution of a two-person cooperative game is defined. Section 3 presents our noncooperative bargaining model. The concept of a stationary subgame perfect equilibrium point (briefly called an equilibrium point) is introduced. Section 4 analyzes the equilibrium point of our bargaining model. The main results of the paper are proved. With the help of an example of the bargaining problem for dividing a cake of size 1, we point out the difference between negotiation outcomes in our model and Rubinstein's model. Section 5 investigates how the equilibrium point of our bargaining game changes if short-term contracts only are assumed to be enforceable. Finally, Section 6 provides concluding remarks.

2. Two-Person Cooperative Games

Let $G = (S_1,S_2;u_1,u_2)$ be a two-person game in normal form where S_i $(i = 1,2)$ is the set of pure strategies for player i, and u_i his payoff function. In this paper, we assume for $i = 1,2$ that S_i is a finite set and u_i is a real-valued function on the Cartesian product $S = S_1 \times S_2$. We further assume that the payoff function u_i for player i $(= 1,2)$ represents the von Neumann-Morgenstern utility function.

A mixed strategy q_i for player i is a probability distribution on S_i. Let Q_i denote the set of all mixed strategies for player i. The Cartesian product $Q = Q_1 \times Q_2$ is the set of all mixed strategy pairs for both players. For a mixed strategy pair $q = (q_1,q_2) \in Q$, the expected payoff for player i is given by

$$Eu_i(q_1,q_2) = \sum_{s_1} \sum_{s_2} u_i(s_1,s_2)\, q_1(s_1)\, q_2(s_2) \tag{2.1}$$

where $q_i(s_i)$, $i = 1,2$, is the probability that q_i assigns to s_i. We will use the notation $u_i(q_1,q_2)$ for simplicity instead of $Eu_i(q_1,q_2)$ whenever no confusion arises.

Definition 2.1. A mixed strategy pair $q^* = (q_1^*,q_2^*)$ is said to be a *Nash equilibrium point* of G iff

$$u_i(q^*) \geq u_i(q^*/q_i), \quad \forall\ q_i \in Q_i, \quad i = 1,2$$

where q^*/q_i is the strategy pair which results from q^* if q_i^* is replaced with q_i.

A correlated (mixed) strategy p in G is a probability distribution on $S = S_1 \times S_2$. Let P denote the set of all correlated strategies in G. For a correlated strategy p, the expected payoff for player i $(= 1,2)$, simply denoted by $u_i(p)$, is defined similarly to (2.1). We define

$$U = \{(u_1(p),\ u_2(p)) \mid p \in P\}.$$

U is the set of all feasible payoff pairs in G. A payoff pair $u = (u_1, u_2) \in U$ is called an outcome of G.

Definition 2.2. Let $u = (u_1, u_2) \in U$.
(1) u is said to be *weakly Pareto efficient* iff there exists no $v = (v_1, v_2) \in U$ such that

$$v_i > u_i, \quad i = 1,2.$$

(2) u is said to be *Pareto efficient* iff there exists no $v = (v_1, v_2) \in U$ such that

$$v_i \geq u_i, \quad i = 1,2$$

with at least one strict inequality.

We will assume the following about the game G.

Assumption 2.1. (1) There exists a unique Nash equilibrium point $q^* = (q_1^*, q_2^*)$ of G. We normalize the equilibrium payoffs for players as $u_i(q^*) = 0$ for all $i = 1,2$.
(2) The set of weakly Pareto efficient outcomes is equal to that of Pareto efficient outcomes.

Finally, we define the Nash bargaining solution for the two-person cooperative game G.

Definition 2.3. A payoff pair $u^* = (u_1^*, u_2^*) \in U$ is called *the Nash bargaining solution* of G with the disagreement point $v = (0,0)$ iff

$$u_1^* u_2^* = \max_{\substack{(u_1,u_2)\in U \\ u_i \geq 0,\ i=1,2}} u_1 u_2 . \tag{2.2}$$

Remark that the Nash bargaining solution is a unique solution for (2.2) under Assumption 2.1.(2).

3. A Noncooperative Bargaining Model

We present a noncooperative bargaining model in which two players are allowed to negotiate for a long-term contract about their correlated strategies for G over infinitely many periods. If they are not bound by any contract, the players can have negotiations governed by the following rule at every period $t = 1,2,\ldots$.

<u>Rule</u>:

(1) A random move selects the first proposer i (= 1,2) with equal probability.

(2) The first proposer i proposes to player j ($\neq i$) a correlated strategy $p^i \in P$ and $T^i \in N^*$, where N^* is the set of positive integers including the infinite number ∞. Here, T^i represents a term of contract, i.e., a number of periods for which the correlated strategy $p^i \in P$ should be taken successively.

(3) Player j either accepts or rejects player i's proposal (p^i,T^i). If he accepts it, the two players reach a binding agreement that they will take p^i from period t to period $t+T^i-1$. After they have followed the agreement, they will negotiate again for a new contract at period $t+T^i$ by the same rule. If player j rejects the proposal, he can make a counterproposal.

(4) Player j counterproposes a pair $(p^j,T^j) \in P \times N^*$.

(5) Player i either accepts or rejects player j's proposal (p^j,T^j). If he accepts it, the same rule as in (3) is applied. If he rejects it, negotiations at period t break down.

(6) After negotiations have broken down, players select their pure strategies s_1,s_2 for G independently, and then period t ends. Each player is informed of all moves at this period and they will negotiate again at the next period $t+1$ by the same rule as at period t.

Formally, our repeated bargaining game Γ described above can be defined as an extensive game of infinite length and with perfect information except that players select their pure strategies for the game G simultaneously at every period if negotiations break down. Let Γ^t $(t = 1,2,...)$ denote the bargaining game at period t described by the rule (1)-(6) above. In the extensive form of Γ, a path from the origin of Γ to a random move at the beginning of Γ^t is called the history of Γ^t, denoted by h^t. To clarify the dependency on the history, we write $\Gamma^t(h^t)$ instead of Γ^t.

A strategy σ_i for player i (= 1,2) in Γ is a function which assigns his behavior to his every move in Γ. Let $\sigma_i(h^t)$ be the strategy for player i in $\Gamma^t(h^t)$ induced by σ_i. The strategy $\sigma_i(h^t)$ is divided into two component-strategies $\sigma_i^1(h^t)$ and $\sigma_i^2(h^t)$, which are employed when player i is selected as the first and second (last) proposers, respectively, by the random move.

Each $\sigma_i^k(h^t)$ (k = 1,2) prescribes the following decisions to player i:

(1) a proposal in $P \times N^*$,

(2) a response rule to player j's proposal, which is a function from $P \times N^*$ to {accept, reject},

(3) a mixed strategy for G which he selects if negotiations break down.

Player i is free to make these decisions dependent on the history of the game within period t as well as before period t. Let Σ_i be the set of strategies σ_i for player i in Γ. Let $\Sigma = \Sigma_1 \times \Sigma_2$. A strategy pair $\sigma = (\sigma_1, \sigma_2) \in \Sigma$ uniquely determines a sequence of probability distributions on $S = S_1 \times S_2$ for all periods $t = 1,2,\ldots$, denoted by $a(\sigma) = \{p_\sigma^t\}_{t=1}^\infty$. Therefore, a sequence of expected payoff pairs

$$u(\sigma) = \{u^t(\sigma)\}_{t=1}^\infty, \quad u^t(\sigma) = (u_1(p_\sigma^t),\ u_2(p_\sigma^t)), \quad t \geq 1$$

is also determined by σ. In this paper, we assume that players 1 and 2 evaluate the sequence $u(\sigma)$ of expected payoff pairs according to the discounted sum of their own expected payoffs at all periods and that they have a common discount factor δ, $0 \leq \delta < 1$. That is, the payoff function for player i (= 1,2) for Γ is defined by

$$F_i(\sigma) = \sum_{t=1}^{\infty} \delta^{t-1} u_i(p^t) \qquad (3.1)$$

where $a(\sigma) = \{p^t\}_{t=1}^\infty$. We call $F_i(\sigma)$ the discounted payoff for player i in Γ when σ is employed. For our analysis, it will be convenient to introduce the normalized discounted payoff $(1-\delta)F_i(\sigma)$ for player i since we can show that $((1-\delta)F_1(\sigma),\ (1-\delta)F_2(\sigma)) \in U$ for all $\sigma \in \Sigma$.

A Nash equilibrium point of Γ can be defined in the same way as in Definition 2.1. We also define a subgame perfect equilibrium point of Γ in a usual manner.

Definition 3.1. A strategy pair $\sigma^* = (\sigma_1^*, \sigma_2^*) \in \Sigma$ is said to be ***a subgame perfect equilibrium point*** of Γ iff it induces a Nash equilibrium point on every subgame of Γ as well as Γ itself.

The noncooperative solution concept which we will employ in this paper is a stationary subgame perfect equilibrium point.

Definition 3.2. A subgame perfect equilibrium point $\sigma^* = (\sigma_1^*, \sigma_2^*)$ of Γ is said to be *stationary* iff for all $i = 1,2$ and all $t = 1,2,\ldots$ the strategy $\sigma_i^{*t}(h^t)$ induced on $\Gamma^t(h^t)$ by σ_i^* depends neither on history h^t nor on t.

In a stationary equilibrium point, each player's behavior is the same for the bargaining games at all periods. The purpose of our analysis is to characterize a stationary subgame perfect equilibrium point of Γ for all discount factor δ, $0 \le \delta < 1$. In what follows, we simply call a stationary subgame perfect equilibrium point of Γ an equilibrium point.

For an equilibrium point σ of Γ, we can define in a similar way to (3.1) the discounted payoff which player i $(= 1,2)$ obtains in the remainder of Γ when player j $(\neq i)$ becomes the last proposer in the bargaining game at period t. We call this discounted payoff, denoted by $F_i^t(\sigma|j)$, the conditional discounted payoff for player i at the equilibrium point σ when player j becomes the last proposer at period t. It is clear from Definition 3.2 that $F_i^t(\sigma|j)$ is independent of history h^t of the game before period t and also of t. In the next section, it will be shown that $F_i^t(\sigma|j)$ is independent of history of the game within period t. Therefore, we will write $F_i(\sigma|j)$ instead of $F_i^t(\sigma|j)$.

Finally, we remark that the rule of our repeated bargaining game Γ can be applied to an alternative formulation of the Nash bargaining problem simply given by a pair of the feasible set U and the disagreement point $c = (c_1, c_2) \in U$. In this framework, players propose utility pairs $u = (u_1, u_2) \in U$ to each other at every period instead of correlated strategies and obtain the disagreement payoffs if negotiations break down. Our analysis does not change at all in this framework if the feasible set U is compact and convex and satisfies Assumption 2.1.(2).

4. The Equilibrium Point of Γ

In this section, we will prove that for any discount factor δ, $0 \le \delta < 1$, an equilibrium point of our repeated bargaining game Γ exists and the equilibrium discounted payoff pair for players is uniquely determined. We will also show that in equilibrium the normalized discounted payoff pair and the payoff pairs which the players demand when they become the last proposers, called the equilibrium demand pairs, all converge to the Nash bargaining solution of G as the discount factor δ goes to 1. The disagreement point of the Nash bargaining solution is given by the unique Nash equilibrium payoff pair $(0,0)$ of G. The "backward induction" argument in the theory of an extensive game plays an important role in our proof of the result.

Lemma 4.1. Let $\sigma^* = (\sigma_1^*, \sigma_2^*)$ be an equilibrium point of Γ. Then, σ^* induces the (unique) Nash equilibrium point $q^* = (q_1^*, q_2^*)$ of G at every period t (= 1,2,...) when negotiations break down.

Proof. Suppose that negotiations break down at period t. Since σ^* is stationary, players 1 and 2 conform to σ^* from period t+1, indpendent of the outcome at period t. Therefore, σ^* must induce a Nash equilibrium point of G at period t after negotiations have broken down. Otherwise, it contradicts that σ^* is a subgame perfect equilibrium point of Γ. Q.E.D.

Lemma 4.2. Let $\sigma^* = (\sigma_1^*. \sigma_2^*)$ be an equilibrium point of Γ. Then, we have

$$(1-\delta)F_i(\sigma^*) \geq 0, \qquad i = 1,2. \tag{4.1}$$

Proof. Suppose that $(1-\delta)F_i(\sigma^*) < 0$ for some $i = 1,2$. Let us consider the following strategy σ_i of player i. At every period t = 1,2,..., he rejects any proposal of the opponent, and proposes (q^*, ∞), and selects q_i^* whenever negotiations break down. From Lemma 4.1, we can show that

$$F_i(\sigma_i, \sigma_j^*) = (\frac{1}{1-\delta})u_i(q^*) = 0 > F_i(\sigma^*).$$

A contradiction. Q.E.D.

Next, we will investigate what correlated strategy each player proposes when he becomes the last proposer at every period. We will use the following notations for $i = 1,2$ and $p \in P$: $F(\sigma^*) = (F_1(\sigma^*), F_2(\sigma^*))$, $F(\sigma^*|i) = (F_1(\sigma^*|i), F_2(\sigma^*|i))$, $u(p) = (u_1(p), u_2(p))$.

The following lemma is crucial to our analysis.

Lemma 4.3. Let $\sigma^* = (\sigma_1^*, \sigma_2^*)$ be an equilibrium point of Γ. Then, we have for $i = 1,2$

$$(1-\delta)F(\sigma^*|i) = u(p^i) \tag{4.2}$$

where p^i is a solution of the maximization problem:

$$(P_i) \qquad \max_{p \in P} u_i(p) \quad \text{subject to}$$

$$u_j(p) \geq \delta(1-\delta)F_j(\sigma^*), \quad j \neq i. \tag{4.3}$$

Proof. We examine the optimal proposal of player i when he becomes the last proposer. Suppose that player i proposes $(p,T) \in P \times N^*$ to player j $(\neq i)$. If player j accepts it, he will obtain the discounted payoff

$$u_j(p) + \ldots + \delta^{T-1}u_j(p) + \delta^T F_j(\sigma^*) = \frac{1-\delta^T}{1-\delta} u_j(p) + \delta^T F_j(\sigma^*).$$

After their contract has ended at period $t+T-1$, players i and j conform to σ^* from period $t+T$ since σ^* is stationary. On the other hand, if player j rejects (p,T), his discounted payoff will be

$$0 + \delta F_j(\sigma^*).$$

The players select the Nash equilibrium point q^* of G after negotiations have broken down and from the next period they will conform to σ^*. Then, the best response of player j to player i's proposal (p,T) is given by

$$\begin{array}{ll} \text{accept} & \text{if } u_j(p) > K \\ \text{accept or reject} & \text{if } u_j(p) = K \\ \text{reject} & \text{if } u_j(p) < K \end{array} \tag{4.4}$$

where $K = \frac{\delta-\delta^T}{1-\delta^T}(1-\delta)F_j(\sigma^*)$. Consider the following maximization problem:

$$(GP_i) \quad \begin{array}{l} \max\limits_{(p,T)\in P\times N^*} \; (1-\delta^T)\, u_i(p) + \delta^T(1-\delta)F_i(\sigma^*) \\ \text{subject to } \; u_j(p) \geq \frac{\delta-\delta^T}{1-\delta^T}(1-\delta)F_j(\sigma^*), \quad j \neq i \end{array} \tag{4.5}$$

where $\delta^\infty = 0$. Let m_i and M_i $(i = 1,2)$ denote the maximum values of the problems (P_i) and (GP_i), respectively.

Claim 1: $m_i = M_i$.

By putting $T = \infty$ in (GP_i), we can prove $m_i \leq M_i$. For any $(p,T) \in P\times N^*$ satisfying (4.5), select $p' \in P$ such that

$$u(p') = (1-\delta^T)u(p) + \delta^T(1-\delta)F(\sigma^*) \in U. \tag{4.6}$$

Then, we have

$$u_j(p') = (1-\delta^T)u_j(p) + \delta^T(1-\delta)F_j(\sigma^*)$$

$$\geq \delta(1-\delta)F_j(\sigma^*)$$

$$u_i(p') = (1-\delta^T)u_i(p) + \delta^T(1-\delta)F_i(\sigma^*).$$

Therefore, $m_i \geq M_i$. This completes the proof of the claim.

Claim 1 implies: if $p^* \in P$ is a solution of (P_i), then (p^*,∞) is a solution of (GP_i) and conversely if (p,T) is a solution of (GP_i), then the $p' \in P$ defined by (4.6) is a solution of (P_i). Therefore, it suffices us to prove that

$$(1-\delta)F(\sigma^*|i) = (1-\delta^T)u(p) + \delta^T(1-\delta)F(\sigma^*) \tag{4.7}$$

for a solution (p,T) of (GP_i).

<u>Claim 2</u>: $0 \leq (1-\delta)F_i(\sigma^*|i) \leq m_i = M_i$.

The inequality of the left-hand side can be proved in the same way as in Lemma 4.2. If player i proposes $(p,T) \in P \times N^*$ such that

$$\frac{1-\delta^T}{1-\delta} u_i(p) + \frac{\delta^T}{1-\delta} (1-\delta)F_i(\sigma^*) > \frac{M_i}{1-\delta}$$

then we have

$$u_j(p) < \frac{\delta-\delta^T}{1-\delta^T} (1-\delta)F_j(\sigma^*),$$

since M_i is the maximum value of (GP_i). Then player j rejects (p,T) from (4.4) and player i obtains the discounted payoff $\delta F_i(\sigma^*)$. Since there exists $p \in P$ such that $u(p) = (1-\delta)F(\sigma^*)$ and it satisfies (4.5) for any $T \in N^*$, we have

$$\delta F_i(\sigma^*) \leq F_i(\sigma^*) \leq \frac{M_i}{1-\delta}. \tag{4.8}$$

This completes the proof of the claim.

Consider the case that the Nash equilibrium payoff $(0,0)$ of G is Pareto efficient. Lemma 4.2 implies $(1-\delta)F_i(\sigma^*) = 0$ for all $i = 1,2$ since $(1-\delta)F(\sigma^*) \in U$. We have the two possibilities with respect to the play of σ^* when player i becomes the last proposer: his proposal $(p,T) \in P \times N^*$ is either (i) accepted or (ii) rejected. Since the Nash equilibrium payoff (0,0) is Pareto efficient, we must have $u(p) = (0,0)$ in case (i). Therefore, $F_j(\sigma^*|i) = 0$ for all $j = 1,2$. In case (ii), we also have $F_j(\sigma^*|i) = \delta F_j(\sigma^*) = 0$ for all $j = 1,2$. Since it is clear from Assumption 2.1.(2) that $u(p) = 0$ for any solution p of (GP_i), we have (4.7).

Next consider the case that the Nash equilibrium payoff (0,0) of G is not Pareto efficient. Then, from Assumption 2.1.(2) there exists some $r \in P$ such that $u_i(r) > 0$ for all $i = 1,2$. Let $p^* \in P$ such that $(1-\delta)F(\sigma^*) = u(p^*)$.

<u>Claim 3</u>: $\delta F_i(\sigma^*) < \frac{M_i}{1-\delta}$.

From (4.8), it suffices to prove $m_i = M_i > 0$. When $F_j(\sigma^*) = 0$, r satisfies (4.3) and we have $m_i \geq u_i(r) > 0$. Assume $F_j(\sigma^*) > 0$. For any t, $0 \leq t < 1$, define

$$p^t = (1-t)r + tp^* \in P.$$

For sufficiently large $t < 1$, we have

$$u_j(p^t) = (1-t)u_j(r) + t\, u_j(p^*) > \delta(1-\delta)F_j(\sigma^*).$$

and

$$u_i(p^t) = (1-t)u_i(r) + t\, u_i(p^*) > 0.$$

Therefore, we have $m_i > 0$ if we put $p = p^t$ in (P_i). This completes the proof of the claim.

<u>Claim 4</u>: $\frac{m_i}{1-\delta} \leq F_i(\sigma^*|i)$.

Let $p^i \in P$ be a solution of (P_i). Assume that $u_j(p^i) > \delta(1-\delta)F_j(\sigma^*)$. If player i proposes (p^i,∞), then player j accepts it from (4.4) and player i obtains the discounted payoff $\frac{u_i(p^i)}{1-\delta} = \frac{m_i}{1-\delta}$. This implies $F_i(\sigma^*|i) \geq \frac{m_i}{1-\delta}$. Assume that $u_j(p^i) = \delta(1-\delta)F_j(\sigma^*)$. Define $p' \in P$ by

$$p' = \begin{cases} r & \text{if } F_j(\sigma^*) = 0 \\ p^* & \text{if } F_j(\sigma^*) > 0. \end{cases}$$

Then, $u_j(p') > \delta(1-\delta)F_j(\sigma^*)$. Define

$$p'^t = (1-t)\ p' + tp^i, \quad 0 \leq t < 1.$$

For any $\varepsilon > 0$, there exists a sufficiently large $t < 1$ such that

$$u_j(p'^t) > \delta(1-\delta)F_j(\sigma^*),$$

$$u_i(p'^t) = (1-t)u_i(p') + t\ u_i(p^i) > \frac{m_i}{1-\delta} - \varepsilon.$$

If player i proposes (p'^t,∞), player j accepts it from (4.4). Therefore, we have

$$F_i(\sigma^*|i) \geq u_i(p'^t) > \frac{m_i}{1-\delta} - \varepsilon.$$

Since ε is any positive number, this completes the proof of the claim.

Claims 2 and 4 imply $F_i(\sigma^*|i) = m_i/(1-\delta)$. This together with Claim 3 shows that on the play of σ^* player i's last proposal (p,T) is accepted and thus it is a solution of (GP_i). Therefore, we have (4.7). Q.E.D.

From Lemma 4.3, we can see that the optimal proposal of each player when he becomes the last proposer can be found in the class of infinitely long-term contracts and that it can be characterized as a solution of a simple maximization problem (P_i). Lemma 4.3 implies that in an equilibrium point of Γ the players reach an agreement at period 1.

Now we can proceed one step further in the "backward induction" procedure. Let $\sigma^* = (\sigma_1^*,\sigma_2^*)$ be an equilibrium point of Γ. Similarly to $F_i(\sigma^*|j)$, $i,j = 1,2$, we can define the conditional discounted payoff $G_i(\sigma^*|j)$ for player i at σ^* when player j becomes the first proposer. $G_i(\sigma^*|j)$ does not depend on t since σ^* is stationary.

<u>Lemma 4.4</u>. Let $\sigma^* = (\sigma_1^*,\sigma_2^*)$ be an equilibrium point of Γ. Then, for $i,j = 1,2$, $i \neq j$, we have

$$G(\sigma^*|j) = F(\sigma^*|i). \tag{4.9}$$

<u>Proof</u>. If player j's proposal is rejected by player i on the play of σ^*, then (4.9) is obvious. Therefore, assume that player j's proposal (p,T) is accepted on the play of σ^*. Then, we have

$$(1-\delta)G(\sigma^*|j) = (1-\delta^T)u(p) + \delta^T(1-\delta)F(\sigma^*) \in U.$$

Let p^i be a solution of (P_i) in Lemma 4.3. From Lemma 4.3, we can see that player j can obtain the discounted payoff $1/(1-\delta)u_j(p^i) = F_j(\sigma^*|i)$ by proposing (p^i,∞), regardless of player i's response. Therefore, we have

$$(1-\delta)G_j(\sigma^*|j) \geq (1-\delta)F_j(\sigma^*|i). \tag{4.10}$$

On the other hand, player i can obtain $F_i(\sigma^*|i)$ by rejecting player j's proposal. Therefore, we have

$$(1-\delta)G_i(\sigma^*|j) \geq (1-\delta)F_i(\sigma^*|i). \tag{4.11}$$

From Lemma 4.3 and Assumption 2.1.(2), $(1-\delta)F(\sigma^*|i) \in U$ is Pareto efficient. Therefore, from (4.10) and (4.11) we have (4.9). Q.E.D.

Theorem 4.5. Let $\sigma^* = (\sigma_1^*,\sigma_2^*)$ be an equilibrium point of Γ. Then,

$$F(\sigma^*) = 1/2\ (F(\sigma^*|1) + F(\sigma^*|2))$$

$$(1-\delta)F(\sigma^*|i) = u(p^i), \quad i = 1,2 \tag{4.12}$$

where p^i is a solution of

$$(P_i) \qquad \max_{p\in P}\ u_i(p) \quad \text{subject to} \quad u_j(p) \geq \delta(1-\delta)F_j(\sigma^*), \quad j \neq i.$$

Proof. By definition,

$$F(\sigma^*) = 1/2\ (G(\sigma^*|1) + G(\sigma^*|2)).$$

Then, the theorem follows from Lemmas 4.3 and 4.4. Q.E.D.

Theorem 4.5 shows that the discounted payoffs for players in an equilibrium point of Γ are determined by their conditional discounted payoffs in the equilibrium point when they become the last proposers in a period. As we can see from the proof of Lemma 4.4, this fact does not depend on the number of successive proposals which are allowed within one period by the breakdown of negotiations. Therefore, even if we change our rule of the game Γ such that any fixed number of proposals can be made by two players in turn within one period, an equilibrium point of the game is determined by (4.12). The essential feature of the rule of our bargaining game Γ is that every player has an equal opportunity to make the last proposal at every period.

In what follows, we will solve (4.12) for any discount factor δ, $0 \leq \delta < 1$. We will introduce some notations. Let U be the feasible set in G. Define $U^+ = \{(u_1,u_2) \in U | u_i \geq 0, \ i = 1,2\}$ and let U_i^+ be the projection of U^+ onto the i-th coordinate axis for $i = 1,2$. We will define the two functions h and g on U_1^+, U_2^+, respectively:

$$h(x) = \max_y \ y \quad \text{subject to} \quad (x',y) \in U^+, \quad x' \geq x,$$

$$g(y) = \max_x \ x \quad \text{subject to} \quad (x,y') \in U^+, \quad y' \geq y.$$

We can show that h and g are continuous decreasing functions on U_1^+ and U_2^+, respectively. For $i = 1,2$, let $u_i \in U^+$ be the feasible payoff pair in which player i enjoys his maximum payoff in U^+. Assumption 2.1.(2) guarantees that such an u_i exists uniquely. We call u_i player i's maximum point in U^+. We have

$$u_1 = (g(a_2),a_2) \quad \text{for some} \quad a_2 \in U_2^+$$

$$u_2 = (a_1,h(a_1)) \quad \text{for some} \quad a_1 \in U_1^+.$$

<u>Lemma 4.6</u>. Let $\sigma^* = (\sigma_1^*,\sigma_2^*)$ be an equilibrium point of Γ. If the Nash equilibrium payoff $(0,0)$ is Pareto efficient, then $F(\sigma^*) = F(\sigma^*|i) = (0,0)$ for all $i = 1,2$.

<u>Proof</u>. From Lemma 4.2 and Assumption 2.1.(2), we have $F_i(\sigma^*) = 0$ for all $i = 1,2$. This implies that the Nash equilibrium point $q^* = (q_1^*,q_2^*)$ is a solution of (P_i) for $i = 1,2$. Then, Theorem 4.5 implies $F(\sigma^*|i) = (0,0)$ for all $i = 1,2$. Q.E.D.

If the unique Nash equilibrium payoff $(0,0)$ of G is Pareto efficient, there exists no room for negotiations between two players. They just select the Nash equilibrium point of G at every period in Γ. In what follows, we will assume the following without loss of generality:

<u>Assumption 4.7</u>. (1) The Nash equilibrium payoff $(0,0)$ is not Pareto efficient.

(2) The Nash product of the maximum point for player 2 is greater than or equal to

that of the maximum point for player 1, i.e.,

$$a_1 h(a_1) \geq g(a_2)a_2.$$

Let $(u^*,h(u^*))$ be the Nash bargaining solution with the disagreement point $(0,0)$, that is, u^* is a unique solution of

$$\max_{a_1 \leq u \leq g(a_2)} u\, h(u).$$

Then, for any x with $a_1 \leq x \leq u^*$, there exists a unique $t \in U_1^+$ such that

$$x\, h(x) = t\, h(t), \quad t \geq x.$$

We can show that $t = t(x)$ is a continuous decreasing function on $a_1 \leq x \leq u^*$. Let $k \in U_2^+$ be a unique solution of $a_1 h(a_1) = g(k)\, k$. To explain the notations introduced here, we draw the typical picture of the feasible set of G in Figure 4.1.

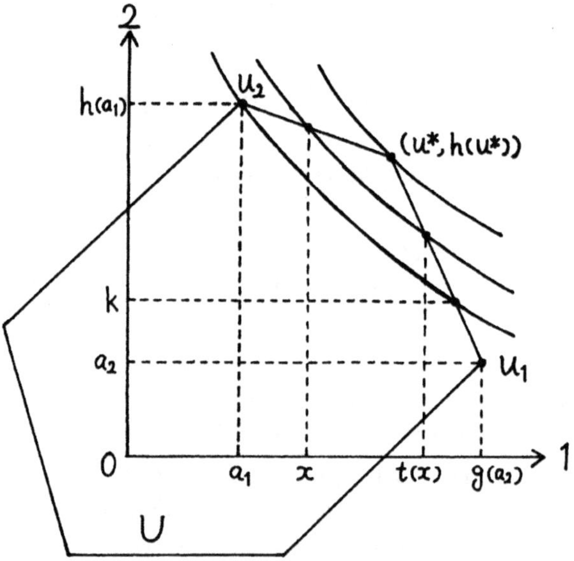

Figure 4.1: The feasible set of the game G.

Lemma 4.8. Equation (4.12) has a unique solution $(F(\sigma^*|1), F(\sigma^*|2))$ such that

(1) $(1-\delta)F(\sigma^*|1) = (g(a_2),a_2)$, $(1-\delta)F(\sigma^*|2) = (a_1,h(a_1))$ for $0 \le \delta \le \frac{2a_2}{h(a_1)+a_2}$

(2) $(1-\delta)F(\sigma^*|1) = \left[g(\frac{\delta}{2-\delta}h(a_1)), \frac{\delta}{2-\delta}h(a_1)\right]$,

$(1-\delta)F(\sigma^*|2) = (a_1,h(a_1))$ for $\frac{2a_2}{h(a_1)+a_2} < \delta \le \frac{2k}{h(a_1)+k}$

(3) $(1-\delta)F(\sigma^*|1) = \left[g(\frac{\delta}{2-\delta}h(x)), \frac{\delta}{2-\delta}h(x)\right]$,

$(1-\delta)F(\sigma^*|2) = (x,h(x))$ for $\frac{2k}{h(a_1)+k} < \delta < 1$

where x is a unique solution of

$$\delta = \frac{2x}{x+t(x)} .$$

Let $x = x(\delta)$. Then

$$\lim_{\delta\to 1} (x(\delta), h(x(\delta))) = (u^*,h(u^*))$$

where $(u^*,h(u^*))$ is the Nash bargaining solution of G with the disagreement point (0.0).

Proof. In the maximization problem (P_i), $i = 1,2$, there are the following four possible cases depending on whether or not player i's payoff at the solution $p^i \in U$ is smaller than his maximum payoff in U.

$$(1-\delta)F_1(\sigma^*|1) = g(a_2), \quad (1-\delta)F_2(\sigma^*|2) = h(a_1) \tag{4.13}$$

$$(1-\delta)F_1(\sigma^*|1) < g(a_2), \quad (1-\delta)F_2(\sigma^*|2) = h(a_1) \tag{4.14}$$

$$(1-\delta)F_1(\sigma^*|1) = g(a_2), \quad (1-\delta)F_2(\sigma^*|2) < h(a_1) \tag{4.15}$$

$$(1-\delta)F_1(\sigma^*|1) < g(a_2), \quad (1-\delta)F_2(\sigma^*|2) < h(a_1). \tag{4.16}$$

Since $F_i(\sigma^*|i)$, $i = 1,2$, is the maximum value of the problem (P_i), (4.13), (4.14), (4.15), and (4.16) imply

$$\delta(1-\delta)F_1(\sigma^*) \le a_1, \quad \delta(1-\delta)F_2(\sigma^*) \le a_2 \tag{4.17}$$

$$\delta(1-\delta)F_1(\sigma^*) \le a_1, \quad \delta(1-\delta)F_2(\sigma^*) > a_2 \tag{4.18}$$

$$\delta(1-\delta)F_1(\sigma^*) > a_1, \quad \delta(1-\delta)F_2(\sigma^*) \le a_2 \tag{4.19}$$

$$\delta(1-\delta)F_1(\sigma^*) > a_1, \quad \delta(1-\delta)F_2(\sigma^*) > a_2 , \tag{4.20}$$

respectively. First assume (4.13). Then, from (4.12) we have

$$\begin{aligned} (1-\delta)F_1(\sigma^*) &= 1/2(a_1 + g(a_2)), \\ (1-\delta)F_2(\sigma^*) &= 1/2(h(a_1) + a_2). \end{aligned} \tag{4.21}$$

From (4.17) and (4.21),

$$0 \le \delta \le \min\left[\frac{2a_1}{a_1+g(a_2)}, \frac{2a_2}{h(a_1)+a_2}\right] = \frac{2a_2}{h(a_1)+a_2} \quad \text{(from Assumption 4.7.(2))}.$$

Next assume (4.14). Then, from (4.12), we have

$$(1-\delta)F_1(\sigma^*) = 1/2(a_1 + g(\delta(1-\delta)F_2(\sigma^*)))$$

$$(1-\delta)F_2(\sigma^*) = 1/2(h(a_1) + \delta(1-\delta)F_2(\sigma^*)).$$

Therefore,

$$(1-\delta)F_1(\sigma^*) = 1/2(a_1 + g(\tfrac{\delta}{2-\delta}\, h(a_1))) \tag{4.22}$$

$$(1-\delta)F_2(\sigma^*) = \frac{h(a_1)}{2-\delta} . \tag{4.23}$$

From (4.18) and (4.23), we have

$$\delta > \frac{2a_2}{h(a_1)+a_2} . \tag{4.24}$$

Also, from (4.18) and (4.22), we have

$$\frac{2-\delta}{\delta}\, a_1 \ge g\left[\frac{\delta}{2-\delta}\, h(a_1)\right]$$

and thus

$$a_1\ h(a_1) \geq \frac{\delta}{2-\delta}\ h(a_1)\ g\left[\frac{\delta}{2-\delta}\ h(a_1)\right].$$

This implies

$$\frac{\delta}{2-\delta}\ h(a_1) \leq k \tag{4.25}$$

and

$$a_1\ h(a_1) > g(a_2)a_2 \tag{4.26}$$

since $\frac{\delta}{2-\delta} h(a_1) > a_2$. (4.24) and (4.25) imply

$$\frac{2a_2}{h(a_1)+a_2} < \delta \leq \frac{2k}{h(a_1)+k} .$$

By the same argument as above, we can show that (4.15) implies $a_1 h(a_1) < g(a_2)a_2$. Therefore, Assumption 4.7.(2) implies that (4.15) is not possible. Finally assume (4.16). Then, form (4.12) we have

$$\begin{aligned}(1-\delta)F_1(\sigma^*) &= 1/2\ \{\delta(1-\delta)F_1(\sigma^*) + g(\delta(1-\delta)F_2(\sigma^*))\}\\ (1-\delta)F_2(\sigma^*) &= 1/2\ \{h(\delta(1-\delta)F_1(\sigma^*)) + \delta(1-\delta)F_2(\sigma^*)\}\ .\end{aligned} \tag{4.27}$$

Let $x_i^\delta = (1-\delta)F_i(\sigma^*)$, $i = 1,2$. Then, (4.27) is equivalent to

$$\begin{aligned}(2-\delta)x_1^\delta &= g(\delta x_2^\delta)\\ (2-\delta)x_2^\delta &= h(\delta x_1^\delta).\end{aligned} \tag{4.28}$$

From (4.28), we have

$$(2-\delta)x_1^\delta\ h((2-\delta)x_1^\delta) = \delta x_1^\delta\ h(\delta x_1^\delta). \tag{4.29}$$

Since $(2-\delta)x_1^\delta > \delta x_1^\delta$, we must have

$$(2-\delta)x_1^\delta = t(\delta x_1^\delta). \tag{4.30}$$

From (4.20) and (4.30),

$$\frac{\delta}{2-\delta} = \frac{\delta x_1^{\delta}}{t(\delta x_1^{\delta})} > \frac{a_1}{t(a_1)} = \frac{k}{h(a_1)}$$

and therefore,

$$\frac{2k}{k+h(a_1)} < \delta < 1.$$

We will show that

$$\frac{\delta}{2-\delta} = \frac{x}{t(x)}, \tag{4.31}$$

has a unique solution x for any δ with $\frac{2k}{k+h(a_1)} < \delta < 1$.
Then, (4.12) has a unique solution such that

$$(1-\delta)F(\sigma^*|1) = \left[g(\tfrac{\delta}{2-\delta}\, h(x)),\ \tfrac{\delta}{2-\delta}\, h(x)\right]$$

$$(1-\delta)F(\sigma^*|2) = (x,h(x)).$$

For this purpose, we rewrite (4.31) as

$$\delta = \frac{2x}{x+t(x)} . \tag{4.32}$$

We can show that the right-hand side of (4.32) is a monotonically decreasing continuous function of x with $a_1 \leq x \leq u^*$. Therefore, for any δ with $2a_1/(a_1+t(a_1)) = 2k/(k+h(a_1)) < \delta < 1$, there exists a unique solution $x(\delta)$ of (4.32) and we can show that

$$\lim_{\delta \to 1} x(\delta) = u^*.$$

Q.E.D.

In an equilibrium point of Γ, the utility pair which every player demands when he becomes the last proposer at a period is uniquely determined by the conditions (1), (2) and (3) of Lemma 4.8. We call it the equilibrium demand pair of the player. We can observe the following properties of the equilibrium demand pairs for players. When the discount factor δ is very small, two players demand their respective maximum points in the set of feasible individually rational payoff pairs. When the discount factor goes beyond some critical level, the player whose maximum point has the smaller Nash product begins to concede, that is, to decrease his demand. When the discount factor becomes greater, the other player also begins to concede and the two players demand the payoff pairs with the identical Nash products. The greater the discount factor is, the closer their demand pairs are. Finally, their demand pairs converge to the Nash bargaining solution as the discount factor goes to 1.

Up to now, we have not proved the existence of an equilibrium point of Γ. In the next theorem, we will construct an equilibrium point of Γ.

Theorem 4.9. For any discount factor δ, $0 \leq \delta < 1$, an equilibrium point of Γ exists and the equilibrium discounted payoff pair for players is uniquely determined by (4.12) in Theorem 4.5 and (1), (2), (3) in Lemma 4.8.

Proof. From Theorem 4.5 and Lemma 4.8, it suffices us to prove the existence of an equilibrium point of Γ satisfying (1), (2), (3) of Lemma 4.8. For simplicity, we construct an equilibrium point $\sigma^* = (\sigma_1^*, \sigma_2^*)$ of Γ for δ with $2k/(h(a_1)+k) < \delta < 1$ only. The equilibrium strategy for each player is given as follows.

Player 1:

(1) propose (p^1,∞) where $u(p^1) = \left[g(\frac{\delta}{2-\delta}h(x)),\ \frac{\delta}{2-\delta}h(x)\right]$,

(2) for the first proposal (s,T) of player 2,

accept it if $u_1(s) \geq \dfrac{E_{11}}{1-\delta^T} - \dfrac{\delta^T}{1-\delta^T}E_1$

reject it otherwise

and for the last proposal (s,T) of player 2,

accept it if $u_1(s) \geq \dfrac{\delta-\delta^T}{1-\delta^T}E_1$

reject it otherwise,

where $E_{11} = \frac{2-\delta}{\delta}x$ and $E_1 = \frac{1}{\delta}x$ and x is a unique solution of (4.32), and

(3) select the Nash equilibrium strategy in the component game G when negotiations break down.

Player 2:

(1) propose (p^2,∞) where $u(p^2) = (x,h(x))$,

(2) for the first proposal (s,T) of player 1,

accept it if $u_2(s) \geq \frac{E_{22}}{1-\delta^T} - \frac{\delta^T}{1-\delta^T} E_2$

reject it otherwise

and for the last proposal (s,T) of player 1,

accept it if $u_2(s) \geq \frac{\delta-\delta^T}{1-\delta^T} E_2$

reject it otherwise

where $E_{22} = h(x)$ and $E_2 = \frac{1}{2-\delta} h(x)$, and

(3) select the Nash equilibrium strategy in the component game G when negotiations break down.

It can be shown without much difficulty that the strategy combination defined above is an equilibrium point of Γ satisfying (3) of Lemma 4.8. Q.E.D.

We are now in a position to state our main theorem.

Theorem 4.10. There exists a unique (stationary and subgame perfect) equilibrium point of the bargaining game Γ for any discount factor δ, $0 \leq \delta < 1$. The equilibrium demand pairs and the (normalized) discounted equilibrium payoff pair for players all converge to the Nash bargaining solution of the component game G when the discount factor δ goes to 1. The disagreement point of the Nash bargaining solution is given by the unique Nash equilibrium payoff pair of the component game G.

Proof. Follows from Theorems 4.8 and 4.9. Q.E.D.

To conclude this section, we examine an example of the two-person bargaining problem where players negotiate about how to divide a cake of size 1 between them. The feasible set of this bargaining problem is drawn in Figure 4.2 and the disagreement point is (0,0). From Lemma 4.8.(3), we can show that for any discount factor $0 \leq \delta < 1$, the equilibrium demand pairs for both players and the (normalized) discounted equilibrium payoff pair are characterized as follows:

Player 1's demand pair:	$(1-\delta/2, \delta/2)$
Player 2's demand pair:	$(\delta/2, 1-\delta/2)$
The (normalized) discounted payoff pair for players:	$(1/2, 1/2)$.

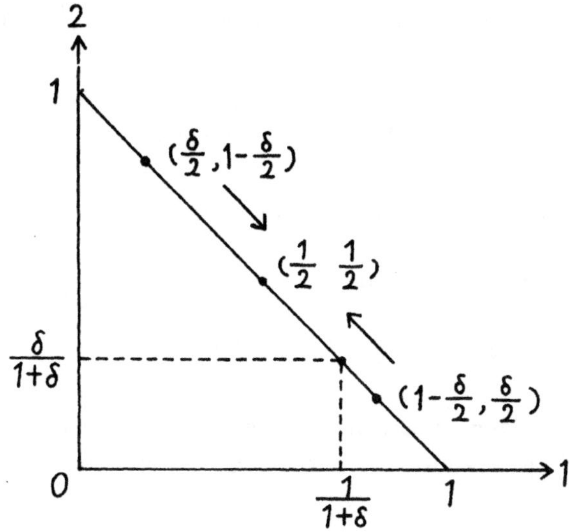

Figure 4.2: The bargaining problem of dividing a cake of size 1.

We remark that in Rubinstein's (1982) model the equilibrium partition of the cake is given by $(1/(1+\delta),\ \delta/(1+\delta))$. The first proposer (player 1) has an advantage over the other. The difference between the negotiation outcomes in both models is caused by the difference between their negotiation processes. In Rubinstein's model, one player (player 1) makes a proposal at period 1 and, if no agreement is reached, then the other player (player 2) makes a proposal at period 2. On the other hand, in our model every player has the equal opportunity to make a proposal at every period. As the discount factor goes to 1, the negotiation outcomes in both models converge to the equal partition (1/2, 1/2).

5. Short-Term Contract vs. Long-Term Contract

In this section, we relax our institutional assumption in the game Γ that any long-term contract is possible between two players, and consider the situation that only short-term contracts, i.e., contracts of term 1, are possible in the game Γ. We investigate how this change of the institutional assumption influences the equilibrium outcome of Γ. We denote by Γ^1 the bargaining game with the same rule as Γ except that only short-term contracts are possible. A stationary subgame perfect equilibrium point of Γ^1 can be defined in the same way as in Γ. We employ the same assumptions and notations as in the previous section.

Lemma 5.1. Let $\sigma^* = (\sigma_1^*, \sigma_2^*)$ be an equilibrium point of Γ^1. Then, we have for $i = 1,2$

$$F(\sigma^*|i) = u(p^i) + \delta\, F(\sigma^*) \tag{5.1}$$

where p^i is a solution of the maximization problem

$$\max_{p \in P} \quad u_i(p)$$
$$\text{subject to} \quad u_j(p) \geq 0, \quad j \neq i, \quad j = 1,2. \tag{5.2}$$

Proof. Suppose that player i proposes $p \in P$ to player j $(\neq i)$ when he becomes the last proposer in a period. If player j accepts it, he will obtain the discounted payoff

$$u_j(p) + \delta\, F_j(\sigma^*).$$

On the other hand, if player j rejects it, player j's discounted payoff will be

$$0 + \delta\, F_j(\sigma^*).$$

Therefore, the best response of player j is:

$$\begin{array}{lll} \text{accept} & \text{if} & u_j(p) > 0 \\ \text{accept or reject} & \text{if} & u_j(p) = 0 \\ \text{reject} & \text{if} & u_j(p) < 0. \end{array} \tag{5.3}$$

Since the Nash equilibrium payoff pair $(0,0)$ is not Pareto efficient, there exists some $p \in P$ such that $u_i(p) > 0$ for all $i = 1,2$. If player i proposes p, then player j accepts it from (5.3) and player i's discount payoff is

$$u_i(p) + \delta\, F_i(\sigma^*) > \delta\, F_i(\sigma^*).$$

Therefore, since σ^* is a subgame perfect equilibrium point of Γ^1, σ^* requires that player i proposes a solution p^i of (5.2) and player j accepts it. Therefore, we have (5.1). Q.E.D.

Theorem 5.2. For any δ, $0 \leq \delta < 1$, the game Γ^1 has a unique equilibrium discounted payoff given by

$$(1-\delta)F(\sigma^*) = (u(p^1) + u(p^2))/2. \tag{5.4}$$

where $p^i (i=1,2)$ is a solution of the maximization problem (5.2) in Lemma 5.1.

Proof. Similarly to Lemma 4.4, we can prove

$$G(\sigma^*|j) = F(\sigma^*|i), \quad i,j = 1,2, \quad i \neq j.$$

Therefore, we have

$$F(\sigma^*) = (F(\sigma^*|1) + F(\sigma^*|2))/2. \tag{5.5}$$

(5.1) and (5.5) imply (5.4). Q.E.D.

Theorem 5.2 shows that, when only short-term contracts are possible, both players demand their maximum points regardless of the discount factor. As a result, the (normalized) equilibrium discounted payoff pair is neither Pareto efficient nor the Nash bargaining solution of G.

6. Concluding Remarks

We have presented a noncooperative bargaining model for a two-person game in normal form in which the Nash bargaining solution can be implemented as a unique stationary subgame perfect equilibrium point in the limiting case that the common discount factor of payoffs goes to 1. We have constructed our bargaining model based on the supergame model by allowing players to negotiate for a long-term contract on actions at every period (if they are not bound by any existing contract) before they select their actions noncooperatively. In our earlier work (Okada, 1988a), we obtained the noncooperative implementation of a Pareto efficient and individually rational outcome in this model under the criterion of average payoffs. Furthermore, in this paper, we have obtained the implementation of the Nash bargaining solution under the criterion of discounted payoffs by introducing a chance move at the beginning of every period to select the first proposer with equal probability. We could say from these results that our bargaining model provides us with a unified view of noncooperative implementations of individually rational outcomes (which corresponds to the Folk Theorem of a supergame), and Pareto efficient and individually rational outcomes, and the Nash bargaining solution in a two-person game in normal form.

To conclude the paper, we have some remarks. We have assumed that the underlying two-person game has a unique Nash equilibrium point. If there are many Nash equilibrium points in the game, the uniqueness of a stationary subgame perfect equilibrium point does not hold in our bargaining game. If we assume a stronger condition for an equilibrium point than Definition 3.2 to guarantee that the same Nash equilibrium point is selected whenever negotiations break down, then we have as many stationary subgame perfect equilibrium points in the bargaining game as Nash equilibrium points in the component game. They implement Nash bargaining solutions with different disagreement points.

Let us consider the "battle of sexes" game of which payoff matrix is given by Figure 6.1. This game has two pure strategy equilibrium points (α_1,β_1), (α_2,β_2) and one mixed strategy equilibrium point ((3/5, 2/5), (2/5, 3/5)).

	β_1	β_2
α_1	2 , 1	-1 , -1
α_2	-1 , -1	1 , 2

Figure 6.1: The "battle of sexes".

The Nash bargaining solutions with these equilibrium points as the disagreement points are pure strategy combinations (α_1,β_1), (α_2,β_2), and a correlated strategy which selects (α_1,β_1) and (α_2,β_2) with equal probability, respectively. Therefore, we need an equilibrium selection theory for the underlying game in order to obtain a unique outcome of negotiations. For example, we may require that in a symmetric game two rational players select a symmetric equilibrium point in case of the breakdown of negotiations. Then, we can obtain a correlated strategy selecting (α_1,β_1) and (α_2,β_2) with equal probability as a unique outcome of negotiations. A general theory of the equilibrium selection is developed by Harsanyi and Selten (1988).

Finally, in this paper, we have considered a two-person bargaining model. If there are more than two players in the game, the situation will be much more complicated since the players are confronted not only with the problem of payoff distributions but also with the problem of coalition formation. In a subsequent paper, we will try to extend our bargaining model to n-person games and to investigate a noncooperative implementation of a cooperative solution in an n-person game in characteristic function form (Okada, 1988b).

Acknowledgments

I would like to thank the members of the Research Project "Game Theory in the Behavioral Sciences" at ZiF, University of Bielefeld, for their helpful comments. Specially, I am grateful to K. Binmore, J. Friedman, W. Güth, R. Selten, and A. Shaked for their encouragement and valuable suggestions. Of course, errors remain my own. Financial support and hospitality from ZiF at University of Bielefeld is gratefully acknowledged.

References

Binmore, K. (1980). Nash bargaining theory, parts I, II and III, Discussion papers. London School of Economics. Also, in: Binmore, K. and P. Dasgupta (eds.), The Economics of Bargaining, Basil Blackwell 1987.

Harsanyi, J.C. and R. Selten (1988). A general theory of equilibrium selection in games. Cambridge: The MIT-Press.

Moulin, H. (1982). Bargaining and noncooperative implementation, Discussion paper. Ecole Polytechnique.

Nash, J.F. (1950). The bargaining problem. Econometrica **18**: 155-162.

Nash, J.F. (1953). Two-person cooperative games. Econometrica **21**: 128-140.

Okada, A. (1988a). A two-person repeated bargaining game with long-term contracts. This volume.

Okada, A. (1988b). A noncooperative bargaining model for the core in n-person characteristic function games. unpublished.

Rubinstein, A. (1982). Perfect equilibrium in a bargaining model, Econometrica **50**: 97-109.

A TWO-PERSON REPEATED BARGAINING GAME WITH LONG-TERM CONTRACTS

by

Akira Okada

Abstract: Does a noncooperative equilibrium point necessarily lead to a Pareto efficient outcome in a supergame if binding agreements on actions are possible among players? We present a two-person repeated bargaining game in which players can negotiate for a long-term contract on their actions in the supergame model. We show that a subgame perfect equilibrium point of our game necessarily leads to a Pareto efficient outcome if the equilibrium strategies for both players have *zero-memory*. We also point out that the question above is answered negatively if the equilibrium strategies for players have *complete memory*.

1. Introduction

This paper is concerned with under what conditions cooperation can be realized in noncooperative situations where players seek to maximize their own payoffs. The *Folk Theorem* in the theory of a supergame, where an identical component game is repeated infinitely many times, provides us with an answer to the question. The Folk Theorem states that the average payoffs of Nash equilibrium points in a supergame are the feasible individually rational payoffs in the component game (Aumann, 1981). It shows that a cooperative outcome, i.e., a Pareto efficient and individually rational outcome in a game can be attained by players' noncooperative behavior if the game is repeated infinitely many times. The mechanism for realizing cooperation is the punishing behavior which players incorporate into their equilibrium strategies for the supergame. When one player deviates from cooperation, all other players punish him by selecting the minimax strategies against him in the remainder of the game. This punishment motivates the player to realize cooperation.

On the other hand, the answer provided by the Folk Theorem is weak in a sense that many Pareto inefficient outcomes are also attained by equilibrium points in a supergame. The Folk Theorem contains an embarrassing result that the set of all Nash equilibrium payoffs is verly large in the supergame (Kaneko, 1982). Rubinstein (1979) proved that the Folk Theorem remains true even if we adopt a subgame perfect equilibrium point which has been commonly recognized as a more appropriate equilibrium concept for a repeated game since the pioneering work of Selten (1973, 1975).

In this paper, we ask a further question about noncooperative implementation of cooperation in the framework of a supergame: Does a noncooperative equilibrium point necessarily lead to a Pareto efficient outcome in a supergame if they are allowed to negotiate for their actions at every period and to reach a binding agreement? For this purpose, we will present a two-person repeated bargaining game based on the supergame model in which players can negotiate for actions at every period before they choose their own actions independently. In negotiations, the players are allowed to reach an agreement not only about which action pair they should choose, but also about how long they should take it from now on. That is, they can make a long-term contract on their actions. Once they reach an agreement, it is assumed to be enforced. In this paper, instead of considering the problem of enforcement, we simply assume the existence of some outside mechanism such as police and court, in practice, to enforce an agreement.

Although our game situation with the institutional assumption above is conventionally classified into the cooperative game, the analysis will be done in the framework of a noncooperative game by modeling a negotiation process in the extensive form. Nash (1951) and recently Harsanyi (1978) suggested that a cooperative game should be analyzed as a noncooperative game by constructing a bargaining game in extensive form and by studying its equilibrium points. This approach enables us to explain cooperation among players by their noncooperative utility maximization under various institutional assumptions.

The rule of our bargaining game is as follows. If they are not bound by any contract, two players 1 and 2 can negotiate for actions at every period of the game. They in turn propose action pairs and terms of contract. Player 1 first makes a proposal to player 2. Then, player 2 decides whether he accepts it or not. If player 2 accepts it, they will take the agreed action pair for as many periods as they agreed. If player 2 rejects it, he can make a counterproposal to player 1. Player 1 also decides whether he accepts it or not. If he rejects it, their negotiations at the present period break down and they must select their own actions independently. Afterwards, the game proceeds to the next period and the same process will be repeated infinitely many times. We employ the limit of average payoffs as the players' preference in our repeated game.

Even in our repeated bargaining game where agreements on actions are enforceable, punishing behavior of players may play an important role in cooperation. In order to investigate the role of punishments in our repeated bargaining game, we will introduce a concept of memory for a strategy. The memory of a strategy restricts its dependency upon the history of the game and thus it restricts the punishments by each player on the other. We will characterize the stationary outcomes of subgame perfect equilibria under the assumption that players' equilibrium strategies have complete or incomplete memory.

We will show that the set of all stationary outcomes of subgame perfect equilibria in our repeated bargaining game coincides with the set of all Pareto efficient and individually rational outcomes in the component game if the memory of both players' equilibrium strategies is incomplete in the following way: there exists at least one period t^* (≥ 2) such that any history of the game from period 1 to period t^*-1 does not affect players' behavior from period t^*. We will also point out that the set of all stationary outcomes of subgame perfect equilibria contains many Pareto inefficient outcomes if players equilibrium strategies have complete memory.

2. The Repeated Bargaining Game

Let $G = (S_1, S_2; f_1, f_2)$ be a two-person game in normal form, where S_i is the set of player i's actions and f_i his payoff function. We assume that for $i = 1,2$ S_i is a compact set and f_i is a real-valued continuous function on the Cartesian product $S = S_1 \times S_2$. An action pair $s = (s_1, s_2) \in S$ is called an outcome in G.

The *minimax payoff* for player i in G is defined to be

$$v_i = \min_{s_j} \max_{s_i} f_i(s_i, s_j), \qquad j \neq i.$$

An action for player j attaining v_i is called the *minimax action for player j against player i*, denoted by m_j^i. An outcome $s \in S$ is said to be *individually rational* if $f_i(s) \geq v_i$ for all $i = 1,2$, and to be (weakly) *Pareto efficient* if there exists no $\bar{s} \in S$ such that $f_i(\bar{s}) > f_i(s)$ for all $i = 1,2$. Let E be the set of all outcomes $s \in S$ which are both individually rational and Pareto efficient. Then, we define

$$\underline{w}_i = \min_{s \in E} f_i(s), \qquad i = 1,2,$$

which is the least payoff for player i when all individually rational and Pareto efficient outcomes are selected. For $i = 1,2$, we also define

$$\bar{w}_i = \max_{s \in E} f_i(s) \qquad \text{subject to } f_j(s) = \underline{w}_j,\ j \neq i.$$

An outcome attaining $\bar{w}_i$ is denoted by $p^i \in E$. It can easily be proved that $f_i(s) \leq \bar{w}_i$ for all individually rational outcomes $s \in S$ and all $i = 1,2$.

Now, we present a repeated bargaining game in which the game G is played infinitely many times. Unlike the model of a supergame of G, if they are not bound by any contract, two players can negotiate for a long-term contract on actions at the beginning of every period. The rule of our bargaining game at every period t $(= 1,2,\ldots)$ is as follows.

(1) Player 1 first proposes to player 2 an outcome $s^1 \in S$ and a number T^1 of periods for which s^1 should be taken successively. We allow the case that $T^1 = \infty$.

(2) Then, player 2 decides whether he accepts player 1's proposal (s^1,T^1) or rejects it. If player 2 accepts it, the two players reach a binding agreement that they will take s^1 from period t to period $t + T^1-1$. After they take the agreed outcome s^1 for the agreed T^1 periods, they will negotiate again at period $t + T^1$ by the same rule. If player 2 rejects the proposal, the game proceeds to the next move.

(3) Player 2 proposes an outcome $s^2 \in S$ and a term T^2 of contract to player 1.

(4) Player 1 decides whether he accepts player 2's proposal (s^2,T^2) or not. If he accepts it, the same rule as in (2) is applied. If he rejects it, negotiations at period t break down.

(5) After negotiations have broken down, players 1 and 2 choose their own actions s_1 and s_2 independently, and then period t ends. Each player is informed of all moves at this period and they will negotiate again at the next period $t+1$ by the same rule as above.

Formally, our repeated bargaining game Γ can be described as an extensive game of infinite length and with perfect information except that players select their actions simultaneously at every period if negotiations break down. The path from the origin to a move in the extensive form of Γ is called the *history* of the move. Let h^t denote a history of the player 1's first move at period t when no contract is enforced, and let $\Gamma^t(h^t)$ be the bargaining game at period t with the history h^t which is described by (1) - (5) above.

A strategy for player i $(= 1,2)$ in Γ is a function which assigns his behavior to his every move in Γ. Following the conventional notation in the theory of a supergame, we represent a strategy for player i in Γ as $\sigma_i = \{\sigma_i^t\}_{t=1}^{\infty}$ where each σ_i^t prescribes player i's behavior of the following kinds at period t, contingent to the history of the game: (a) his proposal, i.e., an outcome and a term of contract, (b) his response to player j's proposal, and (c) his action which he should take when negotiations break down. The set of all strategies σ_i $(i = 1,2)$ for player i in Γ is denoted by Σ_i. We put $\Sigma = \Sigma_1 \times \Sigma_2$. A strategy pair $\sigma = (\sigma_1,\sigma_2) \in \Sigma$ uniquely determines a sequence of outcomes in G, denoted by $a(\sigma) = \{s^t\}_{t=1}^{\infty}$ where each $s^t \varepsilon S$ $(t = 1,2,...)$ is taken at period t. In this paper, we assume that players evaluate the sequence $a(\sigma)$ according to the inferior limit of the means of the payoffs at all periods. That is, the payoff function for player i in Γ is defined by

$$F_i(\sigma) = \underline{\lim}_{T\to\infty} (1/T) \sum_{t=1}^{T} f_i(s^t)$$

where $a(\sigma) = \{s^t\}_{t=1}^{\infty}$.

Definition 2.1. A strategy pair $\sigma^* = (\sigma_1^*,\sigma_2^*) \varepsilon \Sigma$ is said to be a (Nash) *equilibrium point* of Γ *if* $F_i(\sigma_i^*,\sigma_j^*) \geq F_i(\sigma_i,\sigma_j^*)$ for all $\sigma_i \varepsilon \Sigma_i$ and all $i = 1,2$. An equilibrium point σ^* of Γ is said to be *stationary* if $a(\sigma^*) = (s^*,s^*,...)$ for some $s^* \varepsilon S$.

Regarded as an extensive game, the bargaining game Γ has the following five types of subgames at every period t, which correspond to the respective stages within period t described in the rule of Γ: (1) Γ_1^t ; the subgame starting with player 1's proposal, (2) Γ_2^t ; the subgame starting with player 2's response, (3) Γ_3^t ; the subgame starting with player 2's counterproposal, (4) Γ_4^t ; the subgame starting with player 1's response, (5) Γ_5^t ; the subgame after negotiations have broken down.

We define a subgame perfect equilibrium point of Γ in a usual manner (Selten, 1973 and 1975).

Definition 2.2. A strategy pair $\sigma^* = (\sigma_1^*,\sigma_2^*) \varepsilon \Sigma$ is said to be a *subgame perfect equilibrium point* of Γ if it induces an equilibrium point on every subgame of Γ as well as Γ itself.

Finally, we introduce the concept of memory of an equilibrium point. Let $\sigma = (\sigma_1,\sigma_2) \varepsilon \Sigma$, $\sigma_i = \{\sigma_i^t\}_{t=1}^{\infty}$, $i = 1,2$, be a strategy pair of the players, and let $\Gamma^t(h^t)$ be the bargaining game at period t with a history h^t. We denote by $\sigma^t(h^t)$ the strategy pair in $\Gamma^t(h^t)$ induced by σ.

Definition 2.3. Let $M = \{M^t\}_{t=2}^{\infty}$ be a sequence of sets of integers such that $M^t \subseteq \{1,...,t-1\}$ for every $t = 2,3,...$. An equilibrium point $\sigma^* = (\sigma_1^*,\sigma_2^*)$ of Γ is said to have ***memory*** M if the following condition holds for every $t \geq 2$: for any two histories h^t and $\tilde{h}^t$ of the first move at period t, $\sigma^{*t}(h^t) = \sigma^{*t}(\tilde{h}^t)$ if the histories h^t and $\tilde{h}^t$ have the same projections on M^t, i.e., the plays of the game at all periods in M^t are the same on both h^t and $\tilde{h}^t$.

<u>Definition 2.4</u>. Let $\sigma^* = (\sigma_1^*, \sigma_2^*) \in \Sigma$, $\sigma_i^* = \{\sigma_i^{*t}\}_{t=1}^{\infty}$, $i = 1,2$, be an equilibrium point of Γ with memory $M = \{M^t\}_{t=2}^{\infty}$.

(1) σ^* is said to have *zero-memory* if $\sigma^{*t}(h^t)$ is independent of h^t (i.e., $M^t = \phi$ for all t) and also independent of t.

(2) σ^* is said to have *complete memory* if

$$M^t = \{1,\ldots,t-1\} \quad \text{for every} \quad t = 2,3,\ldots$$

and to have *incomplete memory* otherwise.

In order to see how the concept of memory for an equilibrium point can restrict players' punishments, we will consider what kind of punishments each player may inflict on the other in Γ. Suppose that negotiations break down at period t. Then, he may immediately punish the other at the final stage of the period, say, by selecting the minimax action against the other. Furthermore, he may continue to punish the other after period t, say, by proposing an undesirable outcome for the other. We call the former an *intra-period* punishment and the latter an *inter-period* punishment. The memory for an equilibrium point defined above does not restrict any intra-period punishment, but may restrict inter-period punishments. In an equilibrium point $\sigma = (\sigma_1, \sigma_2)$ with memory $M = \{M^t\}_{t=2}^{\infty}$, the players may punish each other at every period t only if they deviate from the equilibrium path at some period in M^t.

In the next section, we will examine how the stationary outcome of an equilibrium point of Γ is affected by the memory of it. Here we remark that we do not impose any limitation on the memory of a permissible strategy of a player in Γ. As we can see in Definition 2.3, an equilibrium point with memory M is defined in the same way as in Definition 2.1, i.e., under the condition that all permissible strategies for players have complete memory. Therefore, our concept of memory of an equilibrium point does not imply *bounded rationality* of a player or any limitation of his (or her) memorizing ability. Our aim for introducing the memory of an equilibrium point is to investigate the role of punishments in a noncooperative equilibrium point of our repeated bargaining game Γ, and more specifically to examine whether or not punishments are useful for players to realize cooperation in Γ.

3. Characterizations of Equilibria with Memory M

Theorem 3.1. Let $M = \{M^t\}_{t=2}^{\infty}$ be an arbitrary memory of an equilibrium point of Γ. There exists a Nash equilibrium point $\sigma^* = (\sigma_1^*, \sigma_2^*)$ with memory M such that $a(\sigma^*) = (s^*, s^*, \ldots)$ if and only if $s^* \in S$ is individually rational.

Proof. We define a strategy σ_i^* for player i in Γ as follows: at every period $t = 1, 2, \ldots$, (i) propose (s^*, ∞) to player j ($\neq$i), (ii) when player j proposes (s,T), accept it if $s = s^*$ and reject it otherwise, and (iii) when negotiations break down, select the minimax action m_i^j against player j. It can be shown that the strategy pair $\sigma^* = (\sigma_1^*, \sigma_2^*)$ is a Nash equilibrium point with zero-memory and thus with any memory such that $a(\sigma^*) = (s^*, s^*, \ldots)$. Since the proof of the only-if part is not so difficult, we omit it. Q.E.D.

Theorem 3.1 shows that the set of all stationary outcomes of Nash equilibrium points in the bargaining game Γ are identical to that of the supergame of G in which no explicit negotiations are allowed between two players. This means that, as long as we adopt the Nash equilibrium point as a noncooperative solution concept for Γ, the possibility of negotiations and of binding agreements is not sufficient for selecting Pareto efficient outcomes. This disappointing result comes from the fact that a Nash equilibrium point may include an incredible threat by each player to reject any proposal of the other unless it is equal to his proposal. Theorem 3.1 also shows that the limitation on memory of a Nash equilibrium point of Γ has no influence on the stationary outcome of it.

Next we will characterize the stationary outcomes of subgame perfect equilibria of Γ. Let us start with the simplest case of zero-memory.

Theorem 3.2. There exists a subgame perfect equilibrium point σ^* of Γ with zero-memory such that $a(\sigma^*) = (s^*, s^*, \ldots)$ if and only if $s^* \in S$ is both Pareto efficient and individually rational.

Proof. if-part: We define a strategy σ_i^* for player i in Γ as follows: at every period t, (i) propose (s^*, ∞) to player j ($\neq$ i), (ii) when player j proposes (s,T), accept it if $s = s^*$ or $f_i(s) > f_i(s^*)$, and reject it otherwise, and (iii) when negotiations break down, select the minimax action m_i^j against player j. Clearly, $\sigma^* = (\sigma_1^*, \sigma_2^*)$ has zero-memory and $a(\sigma^*) = (s^*, s^*, \ldots)$. We will show by the backward induction that σ^* induces equilibrium points on every type of subgames in Γ.

(1) Γ_1^{t+1}: σ^* induces the sequence $(s^*, s^*, \ldots)$ of action pairs on the subgame Γ_1^{t+1}. Assume that player i (i = 1,2) alone deviates from σ^*. At every period $t' \geq t+1$, any proposal (s,T) of player i with $f_i(s) > f_i(s^*)$ is rejected by player j since $f_j(s^*) \geq f_j(s)$ from the Pareto efficiency of s^*. Moreover, when

negotiations break down, player i's payoff will be at most v_i because player j selects the minimax action m_j^i against player i. Since $f_i(s^*) \geq v_i$, player i cannot increase his payoff at any period t'. Since player j's strategy σ_j^* is independent of the history of the game, player i cannot increase his payoff at any succeeding periods by the same reason as above. Therefore, σ^* induces an equilibrium point on the subgame Γ_1^{t+1}.

(2) Γ_5^t: After negotiations have broken down at period t, the equilibrium point will be employed in the subgame Γ_1^{t+1} independent of the history at period t since σ^* has zero-memory. Because the payoffs at period t do not affect the average payoffs for players, σ^* induces an equilibrium point on Γ_5^t.

(3) Γ_4^t: It suffices to show that σ_1^* prescribes the optimal response of player 1 to player 2's counterproposal (s^2,T^2), provided that σ^* is employed after he makes any response. For simplicity, let us consider only the case that $s^2 = s^*$ or $f_1(s^2) > f_1(s^*)$. If player 1 accepts (s^2,T^2), his average payoff will be $f_1(s^2)$ when $T^2 = \infty$, and $f_1(s^*)$ when $T^2 < \infty$. If he rejects (s^2,T^2), his average payoff will be at most $f_1(s^*)$. Therefore, in this case, it is the optimal response for player 1 to accept (s^2,T^2).

(4) Γ_3^t and (5) Γ_2^t: Similarly to (1) and (3), we can prove that σ^* induces equilibrium points on Γ_3^t and Γ_2^t, respectively.

<u>Only-if part</u>. Let $\sigma^* = (\sigma_1^*,\sigma_2^*)$ be a subgame perfect equilibrium point of Γ with zero-memory such that $a(\sigma^*) = (s^*,s^*,\ldots)$. From Theorem 3.1, s^* must be individually rational. Assume that s^* is not Pareto efficient. Then, there exists some $\tilde{s} \in S$ such that $f_i(\tilde{s}) > f_i(s^*)$ for all $i = 1,2$. Suppose that player 2 rejects player 1's proposal at period 1 and counterproposes $(\tilde{s},\infty)$ to player 1. If player 1 accepts the proposal, his average payoff will be $f_1(\tilde{s})$ ($> f_1(s^*)$). On the other hand, if he rejects it, the sequence $(s^*,s^*,\ldots)$ of action pairs will happen from the next period by σ^* since σ^* has zero-memory. Since σ^* is a subgame perfect equilibrium point, σ_1^* requires player 1 to accept player 2's proposal $(\tilde{s},\infty)$. Then, player 2's average payoff increases from $f_2(s^*)$ to $f_2(\tilde{s})$, which contradicts that σ^* is a subgame perfect equilibrium point of Γ.

Q.E.D.

By introducing subgame perfectness and zero-memory, we can eliminate Pareto inefficient outcomes from the set of noncooperative equilibria of Γ. The condition of zero-memory means that each player does not inflict any inter-period punishment on the other even if he deviates from the equilibrium point. We will show that this condition is crucial to Theorem 3.2. Assume that two players agree to take a Pareto inefficient outcome s' in the remainder of the game in a subgame perfect equilibrium point σ^*. Since s' is not Pareto efficient, there exists an outcome $\tilde{s}$ more profitable to both players than s'. It would be natural for each player to propose $\tilde{s}$ with the expectation that the other would accept it. Suppose that player 2 makes a new proposal to take $\tilde{s}$ at every period from now on. What is the optimal response of player 1 to this proposal? If player 1 accepts it, player 1's average payoff will be $f_1(\tilde{s})$. On the other hand, if player 1 rejects the proposal, their negotiations break down and they will negotiate again at the next period. Therefore, the optimal response of player 1 to the proposal crucially depends on what will happen at future periods after he has rejected it. The original agreement may not be reached in re-negotiations if the history of the game affects players' behavior in the equilibrium point σ^*. But, if σ^* has zero-memory, σ^* leads both players to reach the original agreement again at the next period after negotiations have broken down. In this case, it is optimal for player 1 to accept player 2's proposal, and thus player 2 can increase his average payoff from $f_2(s')$ to $f_2(\tilde{s})$. This shows that only Pareto efficient outcomes can be attained in subgame perfect equilibrium points of Γ with zero-memory.

We can weaken the condition of zero-memory in Theorem 3.2 as follows.

<u>Theorem 3.3</u>. Let memory $M = \{M^t\}_{t=2}^{\infty}$ satisfy the following condition: there exists at least one period t^* (≥ 2) such that

$$M^{t^*} = \phi \quad \text{and} \quad M^t \subseteq \{t^*,\dots,t-1\} \quad \text{for all} \quad t > t^*. \tag{A}$$

Then there exists a subgame perfect equilibrium point σ^* of Γ with memory M such that $a(\sigma^*) = (s^*,s^*,\dots)$ if and only if $s^* \in S$ is both Pareto efficient and individually rational.

<u>Proof</u>. Since a subgame perfect equilibrium point with zero-memory has memory M satisfying (A), the if-part is obvious from Theorem 3.2. Let $\sigma^* = (\sigma_1^*,\sigma_2^*)$ be a subgame perfect equilibrium point with memory M satisfying (A). Assume that $a(\sigma^*) = (s^*,s^*,\dots)$ for some Pareto inefficient outcome $s^* \in S$. Then, there exists some $\tilde{s} \in S$ such that $f_i(\tilde{s}) > f_i(s^*)$ for $i = 1,2$. Let $F_1^{t^*}(\sigma^*)$ be player 1's average payoff for σ^* in the subgame $\Gamma_1^{t^*}$ starting with player 1's proposal at

period t^*. From (A), $F_1^{t^*}(\sigma^*)$ is independent of the history of the game before period t^*. The two cases are possible: (i) $f_1(\bar{s}) > F_1^{t^*}(\sigma^*)$ and (ii) $f_1(\bar{s}) \leq F_1^{t^*}(\sigma^*)$. In case (i), let us consider player 2's deviation from σ^* such that he rejects any proposal of player 1 and counterproposes $(\bar{s},\infty)$ from period 1 to period t^*-1. If player 1 accepts player 2's proposal at some period $t(< t^*-1)$, then player 2 can increase his average payoff from $f_2(s^*)$ to $f_2(\bar{s})$. This contradicts the fact that σ^* is a subgame perfect equilibrium point of Γ. Therefore, it suffices to consider the case that negotiations break down at all periods before period t^*-1. At period t^*-1, it is optimal for player 1 to accept player 2's counterproposal $(\bar{s},\infty)$ since $f_1(\bar{s}) > F_1^{t^*}(\sigma^*)$. Therefore, player 2 can increase his average payoff from $f_2(s^*)$ to $f_2(\bar{s})$. Again, we have a contradiction. In case (ii), let us consider the same deviation of player 1 from σ^* as that of player 2 in case (i). By the same reason as in case (i), it is sufficient for us to assume that negotiations break down at all periods before period t^*-1. Then, player 1's average payoff is $f_1(\bar{s})$ or $F_1^{t^*}(\sigma^*)$, depending on whether or not player 2 accepts his proposal $(\bar{s},\infty)$ at period t^*-1. Whichever happens, player 1 will obtain his average payoff greater than $f_1(s^*)$. A contradiction. Q.E.D.

The condition (A) in Theorem 3.3 means that any history of the game from period 1 to period t^*-1 does not affect players' behavior from period t^*. Therefore, in a subgame perfect equilibrium point with memory M satisfying (A), they do not inflict inter-period punishments on each other for their past behavior before period t^*.

The next theorem examines what outcome happens in a subgame perfect equilibrium point of Γ if it has complete memory.

Theorem 3.4. There exists a subgame perfect equilibrium point σ^* of Γ with complete memory such that $a(\sigma^*) = (s^*,s^*,\ldots)$ if $f_i(s^*) \geq \underline{w}_i$ for all $i = 1,2$.

Proof. We construct the subgame perfect equilibrium point $\sigma^* = (\sigma_1^*,\sigma_2^*)$ of Γ required in the theorem. Given any Pareto efficient and individually rational outcome $s \in E$, let $\sigma(s)$ be a subgame perfect equilibrium point of Γ with zero-memory such that $a(\sigma(s)) = (s,s,\ldots)$. We have already proved the existence of such an equilibrium point in Theorem 3.2. Let $\Gamma^t(h^t)$ be the bargaining game at period t with a history h^t. The strategy σ_i^* for player $i(= 1,2)$ prescribes the following behavior on $\Gamma^t(h^t)$, depending on whether or not negotiations break down on the history h^t.

Case (1): No breakdown of negotiations happens on the history h^t (the case $t = 1$ is included).

For player 1: (i) propose (s^*,∞) to player 2, (ii) respond to player 2's proposal (s^2,T^2) in the way that (iia) when player 1 himself proposed s^* at period t, accept it if $f_1(s^2) > \bar{w}_1$ and $T^2 = \infty$, and reject it otherwise, and (iib) when player 1 did not propose s^*, reject it if $f_1(s^2) < \underline{w}_1$ and $T^2 = \infty$, and accept it otherwise, and (iii) when negotiations break down, select the minimax action m_1^2 against player 2.

For player 2: (i) propose (s^*,∞) if player 1 proposed s^* at period t and (p^2,∞) otherwise, (ii) when player 1 proposes (s^1,T^1), accept it if $s^1 = s^*$ or, $f_2(s^1) > \bar{w}_2$ and $T^1 = \infty$, and reject it otherwise, and (iii) when negotiations break down, select the minimax action m_2^1 against player 1.

Case (2): Negotiations break down on the history h^t.

Let $\tilde{t}$ $(\leq t-1)$ be the first period at which negotiations break down on h^t. Then, at period t, players employ $\sigma(p^1)$ if player 1 proposed s^* at period $\tilde{t}$, and $\sigma(p^2)$ otherwise.

Clearly, we have $a(\sigma^*) = (s^*,s^*,\ldots)$. Without much difficulty we can prove that $\sigma^* = (\sigma_1^*,\sigma_2^*)$ is a subgame perfect equilibrium point of Γ if $f_i(s^*) \geq \underline{w}_i$, $i = 1,2$. Q.E.D.

Theorem 3.4 shows that the set of all stationary outcomes of subgame perfect equilibria of Γ with complete memory generally contains many Pareto inefficient outcomes of the component game G. In a subgame perfect equilibrium point with complete memory, each player is free to punish the other for deviation from the equilibrium point at previous periods. In order to be "credible" to the other, such a punishment itself needs to be a part of a subgame perfect equilibrium point of the subgame after deviation has happened. The essence of Theorem 3.4 is in that the players can employ as the punishments subgame perfect equilibrium points with zero-memory constructed in Theorem 3.2 of which stationary outcomes are both Pareto efficient and individually rational. The punishment level on player i $(= 1,2)$ is his least payoff $\underline{w}_i$ in all Pareto efficient and individually rational payoffs. This punishment level $\underline{w}_i$ on player i can be attained by the action pair $p^j \in E$ $(j \neq i)$, and then player j enjoys his highest payoff $\bar{w}_j$ in all feasible individually rational outcomes.

Let us examine why a Pareto inefficient outcome s^* is realized by a subgame perfect equilibrium point with complete memory. Suppose that player 2 rejects player 1's proposal s^* at some period t and counterproposes a new outcome $\tilde{s}$ which is more efficient (in the sense of Pareto) than s^*. In case of complete memory, unlike Theorem 3.2, player 1 may reject this proposal "strategically" with the expectation that at all future periods he will be able to obtain the higher payoff $\bar{w}_1$ than that at $\tilde{s}$ by committing himself to obey the subgame perfect equilibrium point $\sigma(p^1)$ with the stationary payoff vector $(\bar{w}_1,\underline{w}_2)$. If $\sigma(p^1)$ is employed after period t, player 2's average payoff will be $\underline{w}_2$. Therefore, if $f_2(s^*) \geq \underline{w}_2$, player 2 may be dissuaded from proposing the more efficient outcome $\tilde{s}$ than s^*. The similar argument holds for player 1. In this way, the players may agree to take a Pareto inefficient outcome.

To conclude this section, we provide a condition on an equilibrium point which guarantees the only-if part of Theorem 3.4.

Definition 3.5. An equilibrium point σ^* of Γ is said to be *subgame stationary* if for any subgame of Γ there exists an outcome $s \in S$ such that σ^* induces the outcome s at almost all periods (i.e. except finitely many periods) in the subgame.

Theorem 3.6. There exists a subgame perfect and subgame stationary equilibrium point σ^* of Γ with complete memory such that $a(\sigma^*) = (s^*,s^*,\ldots)$ if and only if $f_i(s^*) \geq \underline{w}_i$ for all $i = 1.2$.

Proof. Since the subgame perfect equilibrium point σ^* constructed in the proof of Theorem 3.4 is subgame stationary, we can prove the if-part from Theorem 3.4. Let σ^* be a subgame perfect and subgame stationary equilibrium point of Γ such that $a(\sigma^*) = (s^*,s^*,\ldots)$. Assume that $f_i(s^*) < \underline{w}_i$ for some $i = 1,2$. Without loss of generality, we can put $i = 1$. Suppose that player 1 deviates from σ^* at period 1 by proposing (p^2,∞). Let Γ_3^1 be the subgame of Γ which starts after player 2 rejects (p^2,∞). Since σ^* is subgame stationary, there exists some $s \in S$ such that for every $i = 1,2$ player i's average payoff for σ^* in Γ_3^1 is equal to $f_i(s)$. Since σ^* is a subgame perfect equilibrium point of Γ, s must be individually rational. Then, we have $f_2(p^2) \geq f_2(s)$. If $f_2(p^2) > f_2(s)$, then it is optimal for player 2 to accept player 1's proposal (p^2,∞) and thus player 1's average payoff increases from $f_1(s^*)$ to $\underline{w}_1 = f_1(p^2)$. A contradiction. We assume $f_2(p^2) = f_2(s)$. In this case, we can show that s is Pareto efficient. From the definition of $\underline{w}_i$, we have $f_1(s) \geq \underline{w}_1$. Therefore, player 1 can increase his average payoff by proposing (p^2,∞), regardless of player 2's response to it. A contradiction.

Q.E.D.

4. Conclusion

Our results show that a noncooperative equilibrium point does not necessarily lead to a Pareto efficient outcome in a supergame even if we introduce the strong institutional assumption that any agreements on actions are enforceable (Theorems 3.4 and 3.6). In other words, the assumption of binding agreements is not sufficient for selecting Pareto efficient equilibrium outcomes in the supergame model. This negative result is caused by the punishment which each player persistently inflicts on the other whenever the other deviates from an equilibrium point. In fact, if the two players do not incorporate such inter-period punishments into their equilibrium strategies, they necessarily lead to a Pareto efficient outcome in a subgame perfect equilibrium point of our repeated bargaining game (Theorems 3.2 and 3.3). The role of punishments in our bargaining game is very contrastive to that in the supergame model where an equilibrium point with zero-memory (i.e. without any punishments) results in the play that the players select a Nash equilibrium point of the component game at every period.

Finally, we discuss a generalization of our results. In this paper, we have paid attention only to pure strategy outcomes in the component game. However, if we allow a more general form of a long-term contract on actions in our bargaining game, we can extend our results to jointly-mixed strategy outcomes. Assume that players can agree to take any sequence of actions for the current and future periods rather than a sequence of identical actions in our bargaining game Γ. We denote such a modified game by $\tilde{\Gamma}$. Let U be the convex hull of the set $\{(f_1(s),f_2(s)) | s \in S\}$. U is the set of players' expected payoff vectors attained by their jointly-mixed strategies. Let E_U be the set of all Pareto efficient outcomes $u = (u_1,u_2)$ in U such that $u_i \geq v_i$ for all $i = 1,2$ where v_i is the player i's minimax payoff in G with respect to pure strategies. We define the punishment level on player i $(= 1,2)$ by $\underline{w}_i = \min\{u_i | u = (u_i,u_j) \in E_U\}$. Then we can prove the following theorems.

<u>Theorem 4.1</u>. There exists a subgame perfect equilibrium point σ^* of the modified game $\tilde{\Gamma}$ with zero-memory such that $F(\sigma^*) = u^*$ if and only if $u^* \in E_U$.

<u>Theorem 4.2</u>. There exists a subgame perfect equilibrium point σ^* of the modified game $\tilde{\Gamma}$ with complete memory such that $F(\sigma^*) = u^*$ if and only if $u^* \in U$ and $u_i^* \geq \underline{w}_i$ for all $i = 1,2$.

Since the proofs of these theorems are similar to those of Theorems 3.2, 3.4 and 3.6, we omit them.

Acknowledgments

I would like to thank the members of the Research Project "Game theory in the Behavioral Sciences" at ZiF, University of Bielefeld, for their helpful comments, and specially I am grateful to James Friedman and Shmuel Zamir for reading an earlier version of the paper very carefully and for providing many useful suggestions. Of course, errors remain my own. Financial support and hospitality from ZiF at University of Bielefeld is gratefully acknowledged.

References

Aumann, R.J. (1981). Surveys of repeated games. In: Essays in Game Theory and Mathematical Economics in Honor of Oskar Morgenstern, Mannheim/Wien/Zürich: Bibliographisches Institut.

Harsanyi, J.C. (1978). A solution theory for noncooperative games and its implications for cooperative games. In: P.C. Ordeshook (ed.), Game Theory and Political Science. New York: New York Univ. Press.

Kaneko, M. (1982). Some remarks on the folk theorem in game theory. Math. Soc. Sci. **3**: 281-290.

Nash, J.F. (1951). Noncooperative games. Ann. Math. **54**: 286-295.

Rubinstein, A. (1979). Equilibrium in supergames with the overtaking criterion. J. Econ. Th. **21**: 1-9.

Selten, R. (1973). A simple model of imperfect competition, where 4 are few and 6 are many. Int. J. Game Th. **2**: 141-201.

Selten, R. (1975). Reexamination of the perfectness concept for equilibrium points in extensive games. Int. J. Game Th. **4**: 25-55.

THREE APPROACHES TO BARGAINING IN NTU GAMES[1]

ELAINE BENNETT

This paper presents a noncooperative model of bargaining in characteristic function games and relates its outcomes to those of a cooperative model and a bargaining theory model. Despite the differences in the approach of these three models and the resulting differences in the nature of their solutions, all three models make similar predictions of bargaining outcomes.

1. INTRODUCTION

In many situations in Economics, there are gains from cooperation and conflict over how these gains should be shared. Examples include trade in an exchange economy, firm formation and profit sharing in a production conomy, and formation of jurisdictions and production of public goods in a local public good economy. For each of these situations we would like to predict the coalitions that are likely to form and the rewards that agents are likely to receive for their participation.

These and many other situations can be modeled as characteristic function games with sidepayments (TU games) or without sidepayments (NTU games). An NTU game specifies for each potential coalition the set of attainable utility vectors available to it. Since each coalition can select only one utility vector from this set, there is conflict over which utility vector will be chosen. In the situations we consider, each agent can participate in at most one coalition "at a time", so participation in a particular coalition entails a cost: the foregone opportunity to partici-

1. I would like to thank the other members of the research group on Game Theory at the Center for Interdisciplinary Studies at the University of Bielefeld, especially James Friedman and Reinhard Selten, for their comments. I would also like to thank William Zame for his extensive comments on a earlier draft. I am wrateful to the Center for Interdisciplinary Studies at the University of Bielefeld for its support and hospitality and to Franz Weissing for his care and attention. This reasearch was supported, in part, by the National Science Foundation grant SES-8706631.

pate in another coalition. Hence, players in each coalition must bargain over the distribution of gains within the coalition and one important element in the bargaining will be the players' opportunities in other coalitions.

This paper presents a noncooperative model and an overview of two other models of bargaining in NTU games, a cooperative model and a "bargaining theory" model. These models represent three very different ideas about how to model bargaining in NTU games. Despite the differ nc s in their approach and the resulting differences in the very nature of what is considered a solution, all three models make similar predictions of the bargaining outcomes.

The *aspiration model* takes a cooperative game theory approach. The focus here is on the problem each player faces in setting an appropriate reservation price for his coalitional participation. This approach assumes that each player sets a price for his coalitional participation and that the bargaining that takes place within coalitions is bargaining over the "reasonableness" of these prices. A solution in the aspiration model is a price vector that satisfies certain bargaining criteria.

The *multilateral bargaining model* takes a bargaining theory approach. The focus here is on the problem each potential coalition faces in deciding on the distribution of payoff within the coalition. This approach assumes that each coalition, *given* a disagreement vector, bargains systematically to an agreement utility vector; and that *given* agreements in all other coalitions, each coalition systematically determines its disagreement vector. A solution in the multilateral bargaining problem is a sp cification of an agreement payoff vector for *each* coalition which is consistent with the bargaining in *every* coalition.

The *proposal-making model* takes a noncooperative approach. The focus here is to make explicit each step in the bargaining process. The bargaining process begins when nature randomly selects the first player to have the initiative. A player with the initiative can propose a coalition, a payoff vector, and a member of the coalition to respond to the proposal (or else pass the initiative to another player). A player responding to a proposal can either accept the proposal and pass it on for consideration by another member of the coalition, or else reject the proposal and take the initiative to make an alternative proposal and select a play r to respond.... A solution in the proposal-making model is a strategy for each player which is "rational" given the strategies of others and given the rules of the bargaining procedure. The rationality notion we adopt is that of stationary subgame perfection.

The solutions of each of these models make predictions about the likely coalitions and the payoff distributions that result from each characteristic function gam . In the aspiration model the solution is a price v ctor. Giv n such a pric vector, the coalitions that ar likely to form are those that can afford their players' prices; within each coalition, payoff is distributed according to their players' prices. In the multilateral bargaining model, a solution is a consistent set of agreements, only some of which are feasible. Given such agreements, the coalitions that are likely to form are the coalitions whose agreements are feasible; within each coalition, payoff is distributed according to these agreements. In the proposal-making model a solution is a strategy profile. Given such a strategy profile, the likely coalitions and payoff distributions are those corresponding to proposals that are made and accepted with positive probability.

We show that, in all three models, underlying each solution is a price vector, i.e., a specification of a reservation price (in utility terms) for each player. In the aspiration model the price vector *is* the solution. The price vectors underlying solutions of the multilateral bargaining model and the proposal-making model are less obvious. In multilateral bargaining models, we call p the *price vector* of a multilateral bargaining solution if, in every "possible" coalition, every player's agreement payoff is his price. In the proposal-making model we call p the *price vector* of a solution if, in every proposal that has positive probability of occurring, each player is paid his price.

The common feature of these models is that, despite the differences in approach and the nature of solutions, all three agree on the range of possible price vectors and possible coalitions which are the outcomes of bargaining.

The paper is organized in the following way. Following the Introduction, Section 2 provides basic notation and definitions. Sections 3 and 4 provide overviews of the cooperative and the bargaining theory approaches: Section 3 describes the aspiration model and Section 4 the multilateral bargaining model. Section 5 presents a noncooperative model of bargaining called the proposal-making model, while Section 6 characterizes its stationary subgame perfect outcomes. A discussion of related literature in Section 7 completes the paper.

2. UNIQUE OPPORTUNITIES AND NTU GAMES

Unique Opportunities

In this paper we focus attention on situations in which there is a unique opportunity in the sense that, once a set of players commits themselves to forming a coalition (for a stated distribution of payoff), no additional gains from any further coalition formation remain. For example:

- after elections in parliamentary systems where no party obtains a majority of the seats, various coalitions of parties can form a government, but only one government can form,
- various consortiums of companies can launch a communications satellite, but currently there is only demand for one geostationary satellite.

In the first example, the characteristic function will reflect the fact that no two disjoint coalitions can both form a (majority) government. In the second example, however, disjoint coalitions could be profitabl even though only the first coalition to form will actually reap the rewards. The limitations implied by having a "unique opportunity" are restrictions *not* on the class NTU games to be considered but rather on the class of environments to which the theory applies. The unique opportunity assumption is crucial to the description of the proposal-making model, but not for the description of the aspiration or multilateral bargaining models. However, the interpretation of their solutions would be different in the absence of the unique opportunity assumption.

NTU Games

A game in characteristic function form with nontransferable utility (an ***NTU game*** or ***nonsidepayment game***) is a pair $\langle N,V \rangle$ where $N = \{1,\ldots,n\}$ is a nonempty set of ***players*** and V, the ***characteristic function***, assigns to each nonempty subset S of N (a ***coalition***) a convex, compact subset $V(S)$ of R^S_+ which contains the origin and is ***strongly comprehensive*** (i.e., if $x \in V(S)$, $y \in R^S_+$, $y_i \leq x_i$ for each $i \in S$ and $y \neq x$, then $y \in \mathring{V}(S)$, the interior of $V(S)$ with respect to R^S_+. Let $\mathcal{C}$ denote the set of coalitions. It is convenient to 0-normalize the game by requiring $V([i]) = \{ 0 \}$ for each player $i \in N$. We do not require that the game be superadditive.

If x, y are in R^S we write $x \leq y$ if $x_i \leq y_i$ for each $i \in S$; we write $x < y$ if $x \leq y$ and $x \neq y$, and $x << y$ if $x_i < y_i$ for each i. If T is a subset of S, then by x^T we mean the restriction of x to T (thinking of vectors in R^S as functions from S to R).

We interpret a vector in $V(S)$ as a vector of utilities which the coalition S can achieve without the cooperation of players not in S. Strong comprehensiveness allows for free disposal of utility and means that for each feasible utility vector for a coalition, any player receiving a nonzero utility can improve the utility of all other players by sacrificing some of his own utility. That is, if $x \in V(S)$ and there is a v ctor $y \in V(S)$ such that $y > x$ then there exists a vector $z \in V(S)$ such that $z >> x$. When $V(S)$ is strongly comprehensive the weak and strong Par to efficient boundaries of $V(S)$ coincide.

3. THE ASPIRATION MODEL

In situations where the agreement to form a coalition is a binding commitment, members of a coalition are unlikely to make such a commitment without a parallel commitment to the choice of a utility vector (from the coalition's attainable utility vectors) which each member finds "acceptable" in light of his other opportunities in other coalitions. As Shapley and Shubik (1972, p. 116) put it,

> "A prudent 'economic' man playing this game would be loath to enter a partnership for a stated share of the proceeds until he had satisfied himself that more favorable terms could not be obtained elsewhere. We can imagine that each player would set a price on his participation, and that no contracts would be signed until the prices on both sides of each partnership formed are in harmony..."

The aspiration approach to bargaining assumes that each player sets a reservation utility level (his "price") for his coalitional participation, and that the bargaining that takes place is bargaining over the "reasonableness" of these prices.

Each player sets his price with some knowledge or expectation of the prices that other players demand. We expect players to lower their prices whenever they must and raise their prices whenever they can. The aspiration approach assumes that a player lowers his prices when *no* coalition can afford to pay him and his partners their prices; and that a player in a coalition will raise his price if the coalition can afford to pay its players their prices and still have payoff leftover.

Formally, we regard vectors $p \in R_+^N$ as vectors of prices and p_i as player i's price. We say that player i's price is ***realizable*** at p if there exists a coalition S containing player i such that $p^S \in V(S)$. If every player's price is realizable at p we call p ***realizable.*** We say player i's price is ***maximal*** if for every coalition S containing i, and every vector q^S in V(S) it is not the case that $q^S > p^S$. If every player's price is maximal at p we say that p is ***maximal.*** (A price vector is maximal if no player can raise his payoff demand in any coalition without making his price "unrealizable".) If a vector p is both realizable and maximal we call p an ***aspiration.*** Bennett and Zame [1988] prove that the set of aspirations is nonempty for every NTU game.

Realizability and maximality are minimal desiderata for "reasonable" reservation prices. Several authors have imposed additional desiderata in order to make more precise predictions of bargaining outcomes. See, for example, Albers [1974, 1980] for TU games or Bennett and Zame [1988] for NTU games.

Outcomes

For each price vector $p \in R^N_+$, define $\mathcal{C}(p)$ to be the set of coalitions that can afford to pay their players their prices, i.e.,

$$\mathcal{C}(p) = \{ S \in \mathcal{C} \mid p^S \in V(S) \}.$$

For each player $i \in N$, we define $\mathcal{C}_i(p)$ to be the set of coalitions in $\mathcal{C}(p)$ which contain i. For some values of p, $\mathcal{C}(p)$ or $\mathcal{C}_i(p)$ may be empty; however if p is realizable then every $\mathcal{C}_i(p)$ (and therefore $\mathcal{C}(p)$) is nonempty. If the price vector p is maximal then each player i finds the coalitions in his $\mathcal{C}_i(p)$ equally desirable because he can obtain his price from any coalition in $\mathcal{C}_i(p)$ and, given the prices of other players, he can't obtain more. Hence, whenever p is an aspiration, all players can agree that all the "desirable" coalitions to form are those in $\mathcal{C}(p)$.

The selection of a particular aspiration p is the prediction that one of the coalitions in $\mathcal{C}(p)$ is likely to form and that each player in the coalition is paid his price.

4. THE MULTILATERAL BARGAINING MODEL

The multilateral bargaining approach views an NTU game as a set of interrelated bargaining problems. This approach decomposes the factors that lead to the choice of a particular payoff vector in each coalition into internal considerations (such as the coalition's standards of fair division) and external considerations (such as the opportunities available to each of its members in other coalitions). This approach assumes that the effects of these factors can be summarized by two functions; the first takes players' outside opportunities and produces a utility vector referred to as the "outside option vector"; the second takes outside option vector, and produces an agreement utility vector. (That these effects are summarized by functions simply means that the same data always lead to the same resolution.) These functions are viewed as part of the description of the problem just as the initial endowments and utility functions are part of the description of an exchange economy.

Given an outside option vector, a coalition's bargaining function reflects the coalition's standards of fair division, the institutional rules for bargaining within the coalition, and the bargaining skills of its members. Let $f^S : R^S \to R^S$, denote the ***bargaining function for S.***

We assume that the bargaining function for each coalition $S \in \mathcal{C}$ satisfies the following conditions:

1. For each S and each outside option vector $d^S \in V(S)$, the agreement vector for S, $x^S = f^S(d^S)$ satisfies:

a. **Individual Rationality:** $x^S \geqslant d^S$, and

b. **Pareto Optimality:** x^S is on the Pareto optimal frontier[2] of $V(S)$.

2. **Continuity:** For each S, the bargaining function, f^S, is a continuous function of its outside option vector.

3. **Agreeing to Disagree:** If for S the outside option vector $d^S \notin V(S)$, then the agreement vector for S is the outside option vector, i.e., $x^S = f^S(d^S) = d^S$.

Given d^S, the pair $(V(S),d^S)$ is a "bargaining problem" in the sense of Nash [1950]. For $d^S \in V(S)$ the conditions given above are minimal requirements for solutions to the bargaining problem and $f^S(d^S)$ has the usual interpretation as the agreement utility vector that results from bargaining within the coalition. In condition 3, we have extended each bargaining solution to allow for outside option vectors outside the attainable set because we are going to determine an outside option vector d^S for each coalition S, to reflect its members' opportunities in other coalitions. The outside option vector determined in this way will lie outside $V(S)$ if, for members of S, the opportunities available in other coalitions are more attractive than those available in S. In this case, the members of S "agree to disagree", i.e., they agree to settle for their outside opportunities. We view the agreement to disagree as the result of bargaining; the players of S negotiate but their "final offers" aren't compatible given the resources S. In setting the agreement vector equal to the outside option vector we are assuming that, as a result of bargaining, each player would be willing to form the coalition for his component of the outside option vector if some other player would take the necessary loss to make the utility vector feasible for the coalition.[3] (For example, if the coalition [1,2] can divide 3 in any way it chooses and the outside option vector is (2,2), agreeing to disagree here means that each player would be willing to form the coalition if he could obtain 2 (i.e., if the other player would accept a payoff of only 1.)

---

2. Strong comprehensiveness implies that the weak and strong Pareto frontiers of each $V(S)$ coincide.

3. The assumption that players will settle for their ouside options when their outside option vector is infeasible is in keeping with the usual assumptions of solution concepts for the simple bargaining problem. Virtually every solution concept for the simple bargaining problem (e.g., the Nash bargaining solution) requires players to settle for their outside option utility levels whenever the outside option vector is on the Pareto boundary of $V(S)$.

We now turn to the question of how outside options are determined. In multilateral bargaining, if the players in one coalition fail to reach an agreement may enter into other coalitions. Thus, for each coalition S and for each player $i \in S$, we want to use as i's component of the outside option vector, d^S, the utility he would receive if he broke off negotiations in S and took the initiative to form his best alternative coalition. Of course, i's alternatives depend on the agreements that will be reached in other coalitions. We assume that the players in S make accurate (and therefore identical) conjectures about these agreements. To see what this implies, fix a (conjectured) agreement vector x^T for each coalition $T \neq S$; given these conjectured agreement vectors what are the utilities of player i's alternatives? If $i \in T$ and the agreement vector x^T is attainable for the coalition T, then player i can certainly obtain x_i^T in the coalition T. However, if the agreement vector x^T is not attainable for T, player i cannot obtain as much as x_i^T. In view of our previous discussion about the meaning of "agreeing to disagree", the most that player i can obtain in T is the largest utility which allows all of the other members of T to obtain their agreement utilities. That is, the utility to player i of the unattainable agreement x^T in the coalition T is $\max \{ t_i \mid x^T/t_i \in V(T)\}$. (We use x^T/t_i to denote the vector obtained from x^T by replacing the i-th component by t_i). If there is no value of t_i for which $x^T/t_i \in V(T)$, then the unattainable agreement x^T has no utility for player i; by convention we agree to take 0 to be the maximum in this case.

Formally, given agreements $\{x^T\}_{T \neq S}$ in all other coalitions, we define the ***outside option vector*** $d^S(\{x^T\}_{T \neq S})$ for the coalition S in the following way. For each $i \in S$ and each coalition $T \neq S$ with $i \in T$, set:

$$u_i^T(x^T) = \begin{cases} x_i^T, & \text{if } x^T \in V(T) \\ \max \{ 0,\ t_i \mid x^T/t_i \in V(T) \}, & \text{otherwise.} \end{cases}$$

and

$$d_i^S(x) = \max \{ u_i^T(x^T) \mid i \in T \text{ and } T \neq S\}.$$

By definition, d^S is a function from collections $\{x^T\}_{T \neq S}$ to R_+^S; however it is convenient to view d^S as a function defined over collections of agreement vectors for *all* coalitions (although d^S won't depend on x^S). We refer to the function d^S as the ***outside option function*** for the coalition S.

Although we allow different coalitions to have different bargaining functions, we assume that all coalitions use the same outside option function. Hence, a complete description of a multilateral bargaining problem specifies a set of coalitions $\mathcal{C}$, and for each coalition $S \in \mathcal{C}$, an attainable utility set $V(S)$ and a bargaining function f^S. We refer to the triple $\langle N,V,f \rangle$ as a ***multilateral bargaining problem.***

A solution for the multilateral bargaining problem is a agreement payoff vector for each coalition which is consistent with the description of bargaining in every coalition. Formally, the vector $x = \{x^S\}_{S \in \mathcal{C}}$, is a ***consistent conjecture*** (or simply a solution) for bargaining problem $\langle N,V,f \rangle$ if for every coalition $S \in \mathcal{C}$, $x^S = f^S(\ d^S(x)\ )$.

A consistent conjecture is a stable set of beliefs about the outcomes of the bargaining because, given the nature of bargaining in each coalition, no player can improve his payoff in any coalition by renegotiating his utility allocation.

Prices

Let x be a consistent conjecture for the bargaining problem $\langle N,V,f \rangle$. We say that player i's ***price*** is p_i if he obtains exactly p_i as his agreement utility from every one of his coalitions. We say that the price vector p ***generates*** x, if for every player i and for every coalition S containing player i, we have $x_i^S = p_i$.

Bennett [1985] shows that for every bargaining problem $\langle N,V,f \rangle$:

1. There exists a consistent conjecture for $\langle N,V,f \rangle$;
2. Every consistent conjecture is price generated;
3. Every price vector that generates a consistent conjecture is an aspiration.

Moreover, for every NTU game $\langle N,V \rangle$ and every aspiration p for $\langle N,V \rangle$, there exists a bargaining rule f and a vector $x = \{x^S\}_{S \in C}$ such that x is a consistent conjecture for $\langle N,V,f \rangle$ and p generates x.

Outcomes

Given a multilateral solution $x = \{x^S\}_{S \in C}$, the "possible" coalitions to form are those with feasible agreement vectors (i.e., coalitions that *don't* agree to disagree). If the multilateral solution x has the price vector p, the coalitions that are likely to form are precisely the coalitions in $C(p)$.

The selection of a particular multilateral solution x, with price vector p is the prediction that one of the coalitions in $C(p)$ will form and that within the coalition that forms, each player is paid his price.

5. THE PROPOSAL-MAKING MODEL

The proposal-making model takes a noncooperative approach to modeling bargaining in NTU games. The proposal-making model specifies a particular bargaining procedure as an extensive form game with perfect information.

In this model players bargain by making, accepting and rejecting proposals. A ***proposal*** (S,q^S) made by player i specifies a proposed coalition, S (any coalition in C containing i) and a proposed payoff distribution, q^S, (any payoff vector $q^S \in V(S)$).

The game is played according to the following rules:

0. Nature has the first move. Nature selects each player $i \in N$ with probability $a_0(i) > 0$. The selected player has the initiative at the next move and follows rule 1.

1. **The Initiator Role.** The player with the initiative, call him i, can either pass the initiative or make a proposal and name a responder. If i passes the initiative to j, player j will have the initiative at the next move and follow rule 1. If player i makes the proposal (S,q^S), he also designates a player in S to respond to the proposal. The next move is made by the designated responder who follows rule 2.

2. **The Responder Role.** The designated responder, call her j, either accepts or rejects the proposal. If j accepts the proposal (S,q^S) and every other player in S has already accepted then the game ends (see 3.), otherwise j designates any player in S who has not yet accepted as the next responder (this player then follows rule 2). If j rejects the proposal she takes the initiative and follows rule 1.

3. **Game's End.** If a proposal, call it (S,q^S), is accepted by every player in S, the game ends. The players in S obtain their components of q^S and players not in S obtain nothing.

4. Infinite plays result in a payoff of 0 for each player.

A sketch of the game tree is provided as Figure 1.

Nodes and Actions

After the initial move by nature, at each nonterminal node there is a player whose turn it is to move. Let η^0 denote the node at which it is nature's move (the initial node). For each player i let $\mathcal{H}_i$ denote the set of nodes where it is player i's turn to move. At $\eta \in \mathcal{H}_i$ either i has the initiative or is responding to a proposal made by another player. Let $\mathcal{H}_i^I$ be the set of nodes at which i has the initiative and let $\mathcal{H}_i^R$

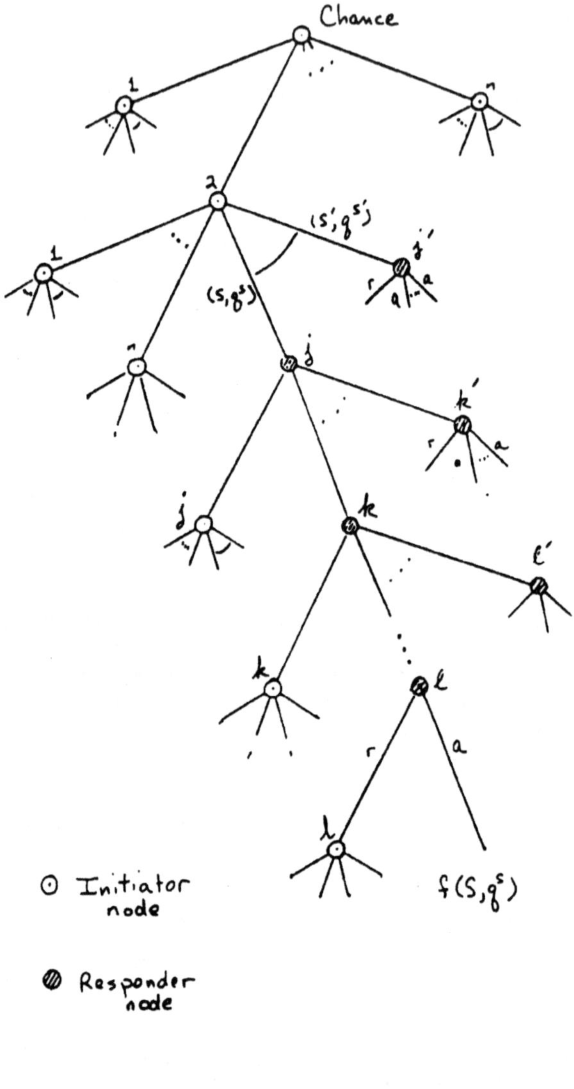

FIGURE 1: The Game Tree of the Proposal-Making Model

be the set of nodes at which i responds to a proposal. Let Z denote the set of terminal nodes.

For each player i, and each node $\eta \in \mathcal{H}_i$, let $A_i(\eta)$ denote the set of actions available to i at η. At $\eta \in \mathcal{H}_i^I$ player i has the initiative and $A_i(\eta)$ contains two type of actions: player i can designate which player $j \in N$ will have the initiative (and the next move), or else player i can make a proposal, call it (S,q^S), and designate a player $j \in S$, to respond to the proposal (this responder will have the next move). At $\eta \in \mathcal{H}_i^R$ player i is responding to a proposal, call it (S,q^S), and $A_i(\eta)$ again contains two types of actions: player i can reject the proposal (this action leads to a node where player i has the initiative) or else player i can accept the proposal and designate a player $k \in S$ who has not yet accepted to be the next responder. (If, at η, every other player in S has already accepted the proposal, acceptance of the proposal leads to a terminal node ζ where the game ends with the formation of the coalition S.)

Strategies and Stationarity

A behavioral strategy for player i, σ_i, specifies for each node η in $\mathcal{H}_i$ a probability distribution $\sigma_i(\eta)$ with finite carrier over the actions in $A_i(\eta)$ i.e., at each node in $\mathcal{H}_i$, $\sigma_i(\eta)$ assigns a positive probability to at most a finite number of actions in $A_i(\eta)$. Let $\sigma = (\sigma_1,\ldots,\sigma_n)$ denote a strategy profile, i.e., a strategy for each player, let $\sigma_{)i(} = (\sigma_1,\ldots,\sigma_{i-1},\sigma_{i+1},\ldots,\sigma_n)$ denote a strategy for everyone except i. Let Σ_i denote the set of strategies for player i, and let Σ denote the set of all strategy profiles.

Informally, the strategy σ_i is a stationary strategy for player i if it is independent of the history of the game. By this we mean that σ_i makes the same prescription at every node at which player i has the initiative and that σ_i makes the same prescription at every node at which player i is responding to the same proposal by the same proposer with the same set of players having already accepted the proposal.

To formalize this notion let $\mathcal{H}_i^R(S,q^S,j,T)$ denote the set of nodes at which player i is to respond to the proposal (S,q^S) made by player j and already accepted by the players in T. We say σ_i is a ***stationary strategy for i*** if:

1. For $\eta, \eta' \in \mathcal{H}_i^I$, $\sigma_i(\eta) = \sigma_i(\eta')$.
2. For $\eta, \eta' \in \mathcal{H}_i^R(S,q^S,j,T)$, $\sigma_i(\eta) = \sigma_i(\eta')$.

The profile σ is a ***stationary strategy profile*** if for each player i, σ_i is a stationary strategy for i.

Expected Payoffs and Equilibria

Given a strategy profile σ, a nonterminal node η and a terminal node ζ, we let $Pr_\sigma(\zeta \mid \eta)$ be the probability that ζ is reached starting from η given that players follow the strategy profile σ. Since we have required strategies to have finite carrier at each node, it is easily seen that $Pr_\sigma(\zeta \mid \eta)$ is well defined. (Notice, that in particular, that $Pr_\sigma(\zeta \mid \eta) = 0$ if ζ does not follow η.) For the probability that, starting from η, the game does not terminate[4], given that players follow σ, we write $Pr_\sigma(\infty \mid \eta) = 1 - \Sigma_\zeta Pr_\sigma(\zeta \mid \eta)$.

The payoff function h specifies the payoff each player receives at each terminal node. If acceptance of the proposal (S,q^S) leads to the terminal node ζ, h specifies that each player in S receives his component of q^S and each player who is left out receives 0. Formally, we define the ***payoff function*** $h : Z \to R^N$ by $h_i(\zeta) = q_i^S$ for $i \in S$ and $h_i(\zeta) = 0$ for $i \in S$ where (S,q^S) is the proposal whose acceptance led to ζ.

Since $h_i(\zeta)$ is the payoff to player i at the terminal node ζ and 0 is the payoff to player i if the game does not terminate, player i's expected payoff at η given the strategy profile σ, $E_i(\eta \mid \sigma)$ is given by $E_i(\eta \mid \sigma) = \Sigma_\zeta h_i(\zeta) Pr_\sigma(\zeta \mid \eta)$.

We say that the terminal node ζ has a *positive probability of occurring* given σ if $Pr_\sigma(\zeta \mid \eta^0) > 0$. We say the proposal (S,q^S) has a *positive probability of acceptance* given σ if acceptance of (S,q^S) leads to a terminal node ζ that has a positive probability of occurring given σ. (Recall that η^0 is the initial node of the game.)

The strategy σ_i is a ***best reply*** to $\sigma_{)i(}$ at η if for every strategy $\sigma'_i \in \Sigma_i$, $E_i(\eta \mid \sigma) \geq E_i(\eta \mid \sigma/\sigma'_i)$. The strategy profile σ is a ***Nash equilibrium*** if for every player $i \in N$, σ_i is a best reply to σ at the initial node, η^0.

The strategy profile σ is a ***subgame perfect equilibrium*** if for every player i, and every nonterminal node η, the strategy σ_i is a best reply to $\sigma_{)i(}$ starting at η. (Since every nonterminal node in the game is the initial node of a subgame, this definition agrees with the usual notion of subgame perfection.)

---

4. Although the sum $\Sigma_\zeta Pr_\sigma(\zeta \mid \eta)$ extends over an uncountable set of terminal nodes, at most countably many terms are non-zero; this point is made clear in the Appendix.

6. STATIONARY SUBGAME PERFECT EQUILIBRIUM OUTCOMES

According to the rules of the game, any player who is responding to a proposal can take the initiative by simply rejecting the current proposal. For a proposal to be "acceptable" every player in the proposed coalition must be offered a utility at least as high as his expected payoff as an initiator; we therefore refer to player's expected payoff as initiator as his "reservation price". Formally, for σ_i a stationary strategy for player i, and $\eta_i \in H_i$ be a node at which player i has the initiative, we define player i's ***reservation price*** $p_i(\sigma)$ to be i's expected payoff as an initiator, $p_i(\sigma) = E_i(\eta_i \mid \sigma)$. (Since σ is a stationary strategy p_i is independent of the choice of initiator node.) We refer to $p(\sigma) = (p_1(\sigma),\ldots,p_n(\sigma))$ as the ***reservation price vector*** of σ.

In the following we use $(p/t_i)^S$ to denote the vector containing one component for each player in S whose components are the components of p except for player i whose component is t_i.

Theorem 6.1: *If $p = (p_1,\ldots,p_n)$ is the reservation price vector of a stationary subgame perfect strategy profile, then for each player i, $p_i = \max\{ u_i^S(p_{-i}) \mid S \in C_i \}$ where $u_j^S(p_{-j})$ is given by $u_i^S(p_{-i}) = \max\{ t_i \mid (p/t_i)^S \in V(S) \}$.*

Proof: We first show that player i's reservation price is at least the value p_i given above. Player i can ensure the cooperation of each partner $j \in S$ by proposing a payoff q_j^S which is strictly greater than j's expected payoff if his refuses, i.e., j's reservation price (otherwise backward induction would show that the strategy of one of the players in S was not subgame perfect). Since player i has a strategy, which for any $\epsilon > 0$ guarantees player i a payoff of $\max\{ t_i \mid (p/t_i)^S \in V(S) \} - \epsilon$, his utility for forming the coalition S, $u_i^S(p_{-i})$, must be at least $\max\{ t_i \mid (p/t_i)^S \in V(S)\}$. (If the set $\{ t_i \mid (p/t_i)^S \in V(S) \}$ is empty player i will not propose the coalition S since i can obtain 0 without the cooperation of other players; in this case for simplicity we set $u_i^S(p_{-i}) = 0$.) Since player i can propose any coalition $S \in C_i$ his reservation price must be at least $u_i(p_{-i}) = \max\{ u_i^S(p_{-i}) \mid S \in C_i \}$. Since any proposal that assigns a player $j \in S$ *less* than his reservation price will be rejected, $u_i(p_{-i})$ is also the largest payoff player i can obtain by making an acceptable proposal.

To show that $u_i(p_{-i})$ is i's reservation price we must show that i's expected utility would not be higher if he either passed the initiative or made an unacceptable proposal. Consider any subgame that follows player i passing the initiative or having his proposal rejected and consider any terminal node that is reached with positive probability (given

SSP strategies in the subgame) in which player i participates in the successful coalition. Let j be the player who made the successful proposal. If $i = j$ the previous argument shows that his payoff is no more than $u_i(p_{-i})$. For $j \neq i$ to be willing to make the proposal, say (S,q^S), his payoff must be at least his reservation price, i.e., $q_j^S \geq p_j$. For the other players $k \neq i$ to be willing to accept the proposal, $q_k^S \geq p_k$. Hence player i's reservation price $p_i = u_i(p_{-i})$.∎

Theorem 6.2: *The reservation price vector of every stationary subgame perfect SSP strategy profile is an aspiration.*

Proof: We show that if p is a reservation price vector of a SSP strategy profile then p is realizable and maximal for each player i. By Theorem 6.1, we have that player i's price p_i satisfies $p_i = \max \{ u_i^S \mid S \in C_i \}$ where $u_i^S = \max \{ t_i \mid (p/t_i)^S \in V(S) \}$. If $p_i > 0$, p is realizable for player i in any coalition S^* in which the maximum u_i^S is obtained; if $p_i = 0$ then p is realizable the coalition $[i]$. The vector p is maximal for player i because for each coalition $S \in C_i$ t_i is the largest feasible residual in the coalition S and u_i is the largest of these values. Since p is realizable for each player i, p is an aspiration.∎

Theorem 6.3: *Every aspiration p is the reservation price vector of an stationary subgame perfect strategy profile.*

Proof: For each vector $p \in R^N$ we describe a class of stationary strategy profiles such that each strategy profile in the class has a reservation price vector equal to p and show that if p is an aspiration then these strategy profiles are subgame perfect.

Let $p \in R_+^N$. We say that σ_i^p is a *price strategy* (for the price vector p) for player i if:

1. When i has the initiative he assigns 0 probability to passing the initiative and assigns positive probability to making a proposal (S,q^S) only when $q^S = p^S$.
2. When player i responds to the proposal (S,q^S) he accepts if $q_j^S \geq p_j$ for every player j of S (including i) who has not yet accepted the proposal and rejects the proposal otherwise.

We say that $\sigma^p \in \Sigma$ is a *price strategy profile* if for every player i, σ_i^p is a price strategy. For each $p \in R^N$, let Σ^p denote the set of all price strategy profiles corresponding to the vector p. Notice that each strategy supports only one price vector p but many strategies support that same p; these strategies differ by the probabilities that players with the initiative assign to making various proposals (i.e., for player i the probabilities he assigns to making each of the proposals in the set $\{ (S,p^S) \mid S \in C_i(p) \}$).

It is easy to see that for any price strategy profile σ^p, p is its reservation price vector of σ^p since only proposals made with positive probability given σ^p are of the form (S,p^S) and every player accepts such a proposal.

Now fix an aspiration p, and fix any $\sigma^p \in \Sigma^p$. To show that σ^p is subgame perfect we show that for every player i, σ^p_i is a best response to $\sigma^p_{)i(}$, starting at any decision node for i. This requires three lemmas (below). Lemma 6.4 shows that if i makes an "acceptable" proposal at any of his initiator nodes, then his proposed payoff is not greater than p_i. Lemma 6.5 shows that σ^p_i is a best response to $\sigma^p_{)i(}$ at any of his initiator nodes. Lemma 6.6 completes the proof by showing that σ^p_i is a best response to $\sigma^p_{)i(}$ at any of his responder nodes.

Lemma 6.4: *If player i, makes a proposal (S,q^S) that has a positive probability of being accepted given $\sigma^p{}_{)i(}$, then $q^S_i \leq p_i$.*

Proof of Lemma 6.4: Suppose player i makes the proposal (S,q^S). (The vector q^S need not be compatible with σ^p_i.) Since (S,q^S) is a proposal, $q^S \in V(S)$. Suppose first that $S = [i]$. By maximality of aspirations $q^S_i \leq p_i$. Suppose instead $S \neq [i]$. If (S,q^S) has a positive probability of acceptance by the first responder ($\neq i$) (who by assumption is following a price strategy for p) then $q^{S-i} \geq p^{S-i}$. By maximality it must be the case that $q^S_i \leq p_i$.∎

Lemma 6.5: *For every player i, σ^p_i is a best response to $\sigma^p{}_{)i(}$, at any initiator node $\eta_i \in \mathcal{H}^I_i$.*

Proof of Lemma 6.5: For any $\sigma' \in \Sigma_i$ and let $\sigma' = \sigma^p/\sigma'$. Consider any terminal node ζ that is reached from η_i with positive probability given σ'. Let (S,q^S) be the proposal made by player k at node η whose acceptance leads to ζ. If $k = i$, then Lemma 6.4 implies that $h_i(\zeta) = q^S_i \leq p$. If $k =/i$, then (since k is following σ^p), $q^S = p^S$. In this case $h_i(\zeta) = p_i$ if $i \in S$ and $h_i(\zeta) = 0 \leq p_i$ if $i \notin S$. Since the choice of ζ was arbitrary, we conclude that $E_i(\eta_i \mid \sigma') \leq p_i = E_i(\eta_i \mid \sigma^p)$. Hence σ^p_i is a best response to $\sigma^p_{)i(}$ at every $\eta_i \in \mathcal{H}^I_i$.∎

Lemma 6.6: *For every player i, σ^p_i is a best response to $\sigma^p_{)i(}$ at any responder node $\eta_i \in \mathcal{H}^R_i$.*

Proof of Lemma 6.6: We first show that if i accepts the proposal (S,q^S) which is rejected with probability 1 by some future respondent (call her j), then i's expected payoff (given $\sigma^P_{)i(}$) is at most p_i. To see why this is true notice that player j, having rejected the proposal will follow σ^P_j, and make a proposal $(S",q^{S"})$ with $q^{S"} = p^{S"}$ and this proposal will accepted for certain if $i \notin S"$. This would result in a payoff to i of $0 \leq p_i$. If $i \in S"$, and i accepts, so will everyone else, so i's payoff is p_i for certain. If i rejects $(S",q^{S"})$, Lemma 6.5 implies that his expected payoff is at most p_i.

To complete the proof, let $\eta_i \in \mathcal{H}^R_i$ and suppose at η_i player i is responding to a proposal (S,q^S) and the players in S' (a subset of S) have not yet accepted it. We consider three cases depending on whether or not $q^{S'}$ is at least as large as $p^{S'}$.

Case 1: $q^{S'} \geq p^{S'}$. In this case $E_i(\eta_i \mid \sigma^P) = q^S_i$. If player i accepts the proposal she obtains q^S_i, if she rejects the proposal, her expected payoff is (by Lemma 6.5) at most $p_i \leq q^{S'}_i$. Hence σ^P_i is a best response to $\sigma^P_{)i(}$ at η_i.

Case 2: $q^{S'}_k < p_k$ for some $k \in S'-i$. In this case $E_i(\eta_i \mid \sigma^P) = p_i$. If player i rejects the proposal, she takes the initiative and Lemma 6.5 guarantees her expected payoff to be at most p_i; if she accepts the proposal, player k will certainly reject the proposal and, as we earlier argued, i's expected payoff is at most $p_i = E_i(\eta_i \mid \sigma^P)$, so σ^P_i is a best response.

Case 3: $q^{S'-i} \geq p^{S'-i}$ and $q^{S'}_i < p_i$. In this case $E_i(\eta_i \mid \sigma^P) = p^S_i$. If player i accepts the proposal she obtains $q^S_i < p_i$; if she rejects the proposal, her expected payoff is (by Lemma 6.5) at most $p_i \leq q^{S'}_i$. Hence σ^P_i is a best response to $\sigma^P_{)i(}$ at η_i.

Since in each case σ^P_i is a best reply to $\sigma^P_{)i(}$, σ^P_i is a best reply at every responder node -- this completes the proof of Lemma 6.6. Lemmas 6.5 and 6.6, together prove that σ^P_i is a best response at every node for player i, --this completes the proof of Lemma 6.6, and with it the proof of Theorem 6.3.∎

Prices

Associated with each strategy profile σ is a set of proposals, call it $P(\sigma)$, corresponding to the terminal nodes that have positive probability of being reached given σ. The proposals in $P(\sigma)$ are the proposals that have a positive probability of being made and accepted given σ. We call the proposals in $P(\sigma)$ the ***acceptable proposals*** given σ.

We say that the vector p *generates* the set of acceptable proposals $P(\sigma)$ if in every acceptable proposal every player is paid according to p.

Formally, the vector $p \in R^N$ *gen rat s* $P(\sigma)$ if for every $(S,q^S) \in P(\sigma)$, $q^S = p^S$.

Theorem 6.7: ***The reservation price vector of each stationary subgam perfect strategy profile generates its set of acceptable proposals.***

Proof: To see that the reservation price vector $p = p(\sigma)$ of each SSP σ generates $P(\sigma)$ we first note that each player's expected payoff is at least as high when he has the initiative as when any other player has the initiative (otherwise he would pass the initiative to this other player); i.e., for $\eta \in \mathcal{H}_i^I$, $\eta_j \in \mathcal{H}_j^I$, $E_i(\eta_i \mid \sigma) \geq E_i(\eta_j \mid \sigma)$.

Let ζ be any terminal node that occurs with positive probability given σ. (There is a terminal node with positive probability since a strategy profile that leads only to infinite plays cannot be subgame perfect.) Let i be the proposer and let (S,q^S) be the proposal whose acceptance led to ζ and let η be the node at which i makes this proposal. Consider the path from η to ζ. Since (S,q^S) is accepted with positive probability $q_j^S \geq p_j$ for every $j \neq i$ in S. We claim that $q_i^S \geq p_i$, also. To see this, note that, since i made the proposal (S,q^S) (with positive probability) at η, his expected payoff given this action must be at least p_i. However, his expected payoff given this action is a weighted average of his payoff given that his proposal is accepted (which is q_i^S) and his expected payoff given that his proposal is rejected. If the proposal were rejected by responder j, player i would expect to obtain $E_i(\eta_j|\sigma)$ for $\eta_j \in \mathcal{H}_j^I$ which, as we have noted above is at most p_i. So i's expected payoff is a weighted average of q_i^S and a number no larger than p_i. Hence for i's expectation to be p_i, $q_i^S \geq p_i$. Hence $q^S \geq p^S$. Since p is maximal, $q^S \in V(S)$, and V(S) is strongly comprehensive, $p^S = q^S$. Hence p is the price vector of σ .∎

We have proved that the set of all reservation price vectors of stationary subgame perfect equilibria is the set of aspirations; it follows that the set of all price vectors that generate SSP possible outcomes is also the set of aspirations.

Outcomes

For each SSP strategy profile σ, let $C(\sigma)$ be the set of coalitions that have a positive probability of being formed given σ, i.e.,

$$C(\sigma) = \{ S \mid (S,q^S) \in P(\sigma) \text{ for } q^S \in R^S \}.$$

One might expect that if p generates $P(\sigma)$ then $C(\sigma) = C(p)$. However this is not the case. While $C(\sigma)$ is always a subset of C(p), when there are "many" coalitions in C(p) players can discriminate against a particular coalition by never proposing it. If players do not discriminate (i.e., each player as initiator assigns positive probability to proposing every coalition in $C_1(p)$), then $C(\sigma) = C(p)$ and the three models agree on the entire set of predicted coalitions.

7. RELATED LITERATURE

1. Selten [1981] presented the proposal-making model for TU games (modeled as a recursive rather than an extensive form game) and proved the analog of the Section 6 results. After showing that equilibrium points are aspirations (Selten calls them semistable demand vectors), he uses equilibrium selection arguments to isolate a particular set of aspirations, which he refers to as the set of stable demand vectors. Selten's equilibrium selection arguments could be carried over directly to the NTU case; the analog of Selten's stable demand vectors are the bargaining aspirations of Bennett and Zame [1988].

2. One way to avoid the multiplicity of equilibria in noncooperative bargaining models is to introduce friction into the bargaining procedure. For instance, following Rubinstein [1981], one might assume that drawing up a proposal takes time and that time matters (i.e., players discount the utility of future agreements). This is the approach taken by Binmore [1985] and by Chatterjee et al [1987].

Binmore [1985] presents a noncooperative model and a bargaining theory model for a limited class of 3-player NTU games (games in which the coalition on the whole earns nothing). Binmore discusses several alternative noncooperative models for this class of games including one he calls the "telephone" bargaining model. Apart from details, the telephone bargaining model is the proposal-making model with discounting. Binmore "tests" this model on a particular 3-player TU example and finds its (unique) outcome "unintuitive". Binmore then dismisses the model and goes on to the "market demand" model -- a model unrelated to the model presented here. (Binmore [1985] also presents a bargaining theory model which is quite similar in spirit to the multilateral bargaining model presented here.)

Chatterjee et al [1987] presents a noncooperative model for the class of TU games, which, apart from details, is again the proposal-making model with discounting. Although differences in "detail" can result in substantive differences in the equilibria of the model, we believe their results to be suggestive of the type of results that would obtained (for the class of TU games) if this friction were introduced into the proposal-making model. In the model of Chatterjee et al, the bargaining outcomes for each TU game are generated by a unique price vector; and this price vector is an aspiration. Unfortunately the unique aspiration chosen is often not the most intuitively plausible price vector for the game.

Although the nonuniqueness of the equilibrium of the proposal-making model is unsatisfying, the results of Chatterjee et al and Binmore suggest that time pressure is not the "right" friction in this context. Other frictions and other ways of selecting among price vectors in the proposal-making model should be investigated.

3. Bennett [1990] shows that, in the absence of stationarity "anything" can happen in the proposal-making model-- i.e., any individually rational payoff distribution for any coalition is a possible outcome of subgame perfect equilibrium strategies.

4. Various solution concepts on the space of aspirations have been proposed, independently and in various guises, by a number of authors. The *aspiration core* for TU games was first identified by Cross [1967]; it was later proposed independently by Albers [1974], by Turbay [1977], and by Bennett and Wooders [1979]. Bennett [1980] recognized it as the extension of the core solution concept to the space of aspirations. The *aspiration bargaining set* for TU games was first proposed by Albers [1974], and independently by Bennett [1980, 1983], who recognized it as the extension of the bargaining set to the space of aspirations. The extension of the aspiration bargaining set to the class of NTU games is in Bennett and Zame [1988]. Other aspiration solutions (kernel and nucleolus variants, in particular) are described in Albers [1980] and Bennett [1980].

REFERENCES

The Aspiration Approach

Albers, W., "Zwei Losungskonzepte fÜr Kooperative Mehrpersonspiele, die auf Anspruchnisiveaus der Spieler Basieren", *OR-Verfarhen* 21, 1-13, 1974.

Albers, W., "Core- and Kernel-variants Based on Imputations and Demand Profiles", in: *Game Theory and Related Topics*, O. Moeschlin and D Palaschke (eds.), North-Holland, Amsterdam, 1980.

Bennett, E., *Coalition Formation and Payoff Distribution in Coop rative Games*, Ph.D. dissertation, Northwestern University, 1980.

Bennett, E., "The Aspiration Approach to Predicting Coalition Formation and Payoff Distribution in Sidepayment Games", *International Journal of Game Theory*, 12, 1-28, 1983.

Bennett, E., "A New Approach to Predicting Coalition Formation and Payoff Distribution in Characteristic Function Games", in *Coalitions and Collective Actions*, Manfred J. Holler, (ed.), Springer Verlag, Wurzburg, 1984.

Bennett, E. and M. Wooders, "Income Distribution and Firm Formation", *Journal of Comparative Economics*, 3, 304-317, 1979.

Bennett, E. and W. Zame, "Bargaining in Cooperative Games", *International Journal of Game Theory* 17, 279-300, 1988.

Cross, J., "Some Theoretical Characteristics of Economic and Political Coalitions" *Journal of Conflict Resolution*, 11, 184-195, 1967.

Turbay, G., *On Value Theories for N-person Cooperative Games*, Ph.D. Dissertation, Rice University, 1977.

The Multilateral Bargaining Approach

B nnett, E., "Multilateral Bargaining Problems", Economics Department, Working Paper Series, University of Kansas, Lawrence, Kansas, 1985.

Bennett, E., "Multilateral Nash Bargaining Solutions", Game Theory in the Behavioral Sciences Working Paper 3, Center for Interdisciplinary Studies, University of Bielefeld, West Germany, 1987.

Bennett, E., "Consistent Bargaining Conjectures in Marriage and Matching", *Journal of Economic Theory* 45, 392-407, 1988b.

Binmore, K., "Bargaining and Coalitions" in *Game Theoretic Models of Bargaining*, A. E. Roth (ed.), Cambridge University Press, Cambridge, 1985.

The Noncooperative Approach

Bennett, E., "A Folk Theorem for the Proposal-Making Model", in this volume, 1990.

Binmore, K., "Bargaining and Coalitions" in *Game Theoretic Models of Bargaining*, A. E. Roth (ed.), Cambridge University Press, Cambridge, 1985.

Chatterjee, K., B. Dutta, D. Ray, and D. Sengupta, "A Non-cooperative Theory of Coalitional Bargaining", mimeo, 1987.

Selten, R., "A Non-cooperative Model of Characteristic Function Bargaining", in *Essays in Game Theory and Mathematical Economics in Honor of Oskar Morgenstern*, V. Boehm and H. Nachtkamp (eds.), Wissenschaftsverlag Bibliographisches Institut Mannheim, Wien - Zurich, 131-151, 1981.

Other References

Shapley L., and M. Shubik, "The Assignment Game 1: The Core", *International Journal of Game Theory* 1, 111-130, 1972.

Rubinstein A., "Perfect Equilibrium in a Bargaining Model", *Econometrica* 50, 97-110, 1981.

APPENDIX

For each node η, let $T(\eta)$ be the set of nodes that follow η. Write $T_1(\eta)$ for the subset of $T(\eta)$ consisting of immediate successors of η, $T_2(\eta)$ for the immediate successors of elements of $T_1(\eta)$, etc. Note that $T(\eta)$ is the union over all $T_i(\eta)$.

Fix a strategy profile σ; recall that we require σ to have a finite support at each node. For $\eta' \in T_1(\eta)$, write $Pr_\sigma(\eta' \mid \eta)$ for the conditional probability that η' is reached from η given that players follow the strategy profile σ. This defines a probability measure on $T_1(\eta)$ with finite support. Hence, we may define for $\eta'' \in T_2(\eta)$ the conditional probability $Pr_\sigma(\eta'' \mid \eta)$ that η'' is reached from η given that players follow the strategy profile σ, obtaining a probability measure on $T_2(\eta)$ with finite support. Continuing in this way, we obtain, for each n, a probability measure on $T_n(\eta)$ with finite support.

Let Z be the set of terminal nodes. Write $Z(\eta)$ for the set of terminal nodes in $T(\eta)$ and $Z_n(\eta)$ for the set of terminal nodes in $T_n(\eta)$. Note that

$$\sum_{\xi \in Z_1(\eta)} Pr_\sigma(\xi \mid \eta)$$

is the probability that the game terminates in one step, beginning from η, that

$$\sum_{\xi \in Z_2(\eta)} Pr_\sigma(\xi \mid \eta)$$

is the probability that the game terminates in two steps, beginning from η, etc. (Note that each of these sums is in fact finite, since each of the probability measures we have constructed has finite support.)

For ξ a terminal node not following η, set $Pr_\sigma(\xi \mid \eta) = 0$. The probability that the game terminates, starting from η, is

$$\begin{aligned} \sum Pr_\sigma(\xi \mid \eta) &= \sum_{\xi \in Z(\eta)} Pr_\sigma(\xi \mid \eta) \\ &= \sum_n \sum_{\xi \in Z_n(\eta)} Pr_\sigma(\xi \mid \eta) \end{aligned}$$

The last double sum is countable, since each of the measures $Pr_\sigma(\cdot \mid \eta)$ has finite support. Thus $Pr_\sigma(\infty \mid \eta) = 1 - \Sigma Pr_\sigma(\xi \mid \eta)$, the probability that the game does not terminate (i.e., that the strategy σ leads to an infinite play), beginning from η, is clearly well-defined.

FOLK THEOREMS FOR THE PROPOSAL-MAKING MODEL[1]

Elaine Bennett

In the tradition of folk theorems, this paper shows that nearly anything can be a bargaining outcome of the proposal-making model.

1. INTRODUCTION

Selten [1981] presents the following bargaining procedure (the "proposal-making" procedure) for characteristic function games. Nature selects a player to be given the initiative. A player with the initiative can pass the initiative to another player or make a proposal (consisting of a proposed coalition and a payoff vector for the coalition). Players in the proposed coalition sequentially consider the proposal; if all of them accept, the game ends with the formation of the coalition; any player who rejects the proposal takes the initiative (and can pass the initiative or make a proposal...). If the games ends with the acceptance of the proposal to form the coalition S with the payoff vector q^S then the outcome of the game is that the coalition S forms; players in S obtain their components of q^S while other players obtain nothing. Selten considers the class of sidepayment games (TU games) and formulates this bargaining procedure as a recursive game. Bennett [1990] considers the broader class of nonsidepayment games (NTU games) and formulates the bargaining procedure as an extensive form game. Both Bennett and Selten characterize the stationary subgame perfect equilibrium outcomes of this bargaining procedure.

One might expect the restriction to stationary strategies to induce a certain uniformity across the outcomes of each subgame perfect strategy

---

1. I would like to thank the other members of the research group on Game Theory at the Center for Interdisciplinary Studies at the University of Bielefeld, especially James Friedman and Reinhard Selten, for their comments. I would also like to thank William Zame for his extensive comments. I am grateful to the Center for Interdisciplinary Studies at the University of Bielefeld for its support and hospitality and to Franz Weissing for his care and attention. This research was also supported, in part, by the National Science Foundation grant SES-8706631.

profile; this is indeed true. If, in a stationary subgame perfect equilibrium, the same coalition is proposed (and accepted) with positive probability (at different terminal nodes), then the proposed payoffs must be equal. Moreover, if the same player appears in two *different* coalitions that are accepted with positive probability then the player has the same proposed payoff. Sets of outcomes that have this uniformity property we call *price generated* since each player would appear to have a reservation payoff level, i.e., a reservation price.

The outcomes of stationary subgame perfect strategy profiles proposals are also coalitionally rational; i.e., if the proposal to form S with payoff vector q^S is such an outcome then no subcoalition of S can unilaterally improve upon its members payoffs.

In the absence of stationarity what can be said about the subgame perfect outcomes? The present paper shows that neither coalitional rationality nor uniformity of payoffs is maintained. Indeed, we show that nearly anything can happen. Such results are commonly referred to as folk theorems. Our first folk theorem asserts that any individually rational and realizable vector can be the price vector of a subgame perfect strategy profile. In particular, given a coalition S and an individually rational payoff vector q^S which is feasible for S, there is a strategy profile σ such that the proposal (S,q^S) has a positive probability of occuring, given that players follow σ. Our second folk theorem asserts that given a coalition S and an individually rational payoff vector q^S which is feasible for S, there is a strategy profile σ such that (S,q^S) is the *unique* outcome with a positive probability of occuring, given σ if and only if 0 is individually rational for every player not in S. In a concluding example we show that the uniformity of payoffs is also lost; the outcomes of subgame perfect strategy profiles need not be price-generated.

2. THE FOLK THEOREMS

We assume that the reader is familiar with Bennett [1990] and use the same notation and definitions with a few exceptions. In order to highlight the role of the individual rationality assumption, we do not assume that the game is 0-normalized. We use $\bar{v}_i$ to denote the maximum payoff player i can obtain on his own, i.e., $\bar{v}_i = \max \{ t \mid t \in V(i)\}$. We say that the vector $p^S \in R^S$ is ***individually rational*** if for every player $i \in S$, $p_i \geq \bar{v}_i$. In Bennett [1990], we referred to the proposals in $P(\sigma)$ (the set of proposals that are made and accepted with positive probability given that players follow the strategy profile σ) as the *possible outcomes* of σ; in this paper we refer to the proposals in $P(\sigma)$ as the ***bargaining outcomes*** of σ.

The basic building block for the strategies we will construct are "price strategies". Let $p \in R^N_+$. We say that σ^p_i is a *pric strategy* (with parameter p) for player i if:

1. At each of i's initiator nodes, σ^p_i assigns 0 probability to passing the initiative to another player.

2. At each of i's initiator nodes, if σ^p_i assigns a nonzero probability to making the proposal (S,q^S) then $q^S = p^S$. (We do not insist that σ^p_i assigns positive probability to all proposals (S,q^S) with $q^S = p^S$.)

3. At each of his responder nodes, player i responds to the proposal (S,q^S) by accepting whenever $q^S_j \geqslant p_j$ for every player j of S (including i) who has not yet accepted the proposal and rejecting otherwise.

We say that $\sigma^p \in \Sigma$ is a ***price strategy profile*** (with parameter p) if for every player i, σ^p_i is a price strategy with parameter p. For each $p \in R^N$, let Σ^p denote the set of all price strategy profiles corresponding to the vector p. Notice that many price strategy profiles correspond to a single parameter; they differ by the probabilities assigned to proposing the various coalitions.

The following is our first folk theorem.

Theorem 1: ***Every individually rational and realizable payoff vector is the price vector of a subgame perfect equilibrium strategy profile.***

Proof: Outline of the proof. Let r be individually rational and realizable. We construct a strategy profile σ^* with price vector r, (i.e., $r = p(\sigma^*)$) and prove that it is subgame perfect. The construction proceeds in two steps. In the first step, for every proposer and every proposal, we find a "punishment" aspiration with the property that any price strategy profile with that parameter "punishes" the player for making the proposal. In the second step we select a price strategy for r and for each punishment aspiration and label every nonterminal node with either r or the appropriate punishment aspiration. For each node η, the desired strategy profile σ^* is given by $\sigma^*(\eta) = \sigma^{p*}(\eta)$ where $p*$ is the label of the node η, i.e., if it is i's turn to move at η, σ^* prescribes the action specified by the price strategy for player i with parameter $p*$ at the node η. Three lemmas are then used to show that this σ^* is subgame perfect.

St p 1: For each proposer $\mathfrak{f}$ and each proposal (S,q^S) we construct a punishment aspiration $\beta(S,q^S,\mathfrak{f})$ such that: (1) $\beta_{\mathfrak{f}}(S,q^S,\mathfrak{f}) = \bar{v}_{\mathfrak{f}}$; (2) there is a player $i*$ in S with $\beta_{i*}(S,q^S,\mathfrak{f}) \geqslant q^S_{i*}$ (such a player would not lose by rejecting (S,q^S) and enforcing the punishment aspiration $\beta(S,q^S,\hat{\imath})$; and (3) if $S \neq [\hat{\imath}]$ and q^S is individually rational for player $\hat{\imath}$ then $i^* \neq \hat{\imath}$.

For each coalition S and each player $k \in S$ we define a function $\gamma^S{}_k : R_+^N \to R_+$ by:

$$\gamma_k^S(x) = 0, \text{ if } (x_1,\ldots,x_{k-1}, 0, x_{k+1},\ldots,x_n)^S \notin V(S);$$

$$\gamma_k^S(x) = \sup\{ t \mid (x_1,\ldots,x_{k-1}, t, x_{k+1},\ldots,x_n)^S \in V(S)\} \text{ otherwise.}$$

We define $\gamma_k : R_+^N \to R_+$ by:

$$\gamma_k(x) = \sup\{ \gamma_k^S(x) \mid k \in S, \ S \text{ in } N\}.$$

$\gamma_k(x)$ is the largest value for t consistent with the vector $(x_1,\ldots,x_{k-1}, t, x_{k+1},\ldots,x_n)$ being realizable for player k. (Keep in mind that $0 \in V(k)$ so that $(x_1,\ldots,x_{k-1}, 0, x_{k+1},\ldots,x_n)$ is always realizable for player k.)

For each player k let $\phi_k(x)$ be the vector obtained from x by replacing x_k by $\gamma_k(x)$; i.e., $\phi_k(x) = (x_1,\ldots,x_{k-1}, \gamma_k(x), x_{k+1},\ldots x_n)$. Thus each ϕ_k is a map from R_+^N onto itself.

Each map ϕ_k enjoys the following properties:

(i) $\phi_k(x)$ is realizable and maximal for player k.

(ii) if x is realizable for player k, then $\phi_k(x) \geq x$;

(iii) if x is maximal for player k, then $\phi_k(x) \leq x$;

(iv) if x is maximal for player ℓ, then $\phi_k(x)$ is maximal for player ℓ (whether $\ell = k$ or $\ell \neq k$).

The first three properties are easily checked. To see that the fourth is true, write $y = \phi_k(x)$ and suppose that y is not maximal for ℓ. Then there is a coalition S containing player ℓ with $y^S \in \text{INT } V(S)$. If $k \in S$, this implies that y is not maximal for player k, which contradicts property (i). On the other hand, since applying ϕ_k to x alters at most the k-th component, if $k \notin S$ then $y^S = x^S$ which means that $x^S \in V(S)$ and this contradicts the assumption that x is maximal for player ℓ.

For each $k = 1, 2,\ldots,n$ define $\Phi^k: R_+^N \to R_+^N$ as follows:

$$\Phi^k = \phi_k \circ \phi_{k-1} \circ\ldots\phi_2 \circ \phi_1$$

Let $\bar{m} \in R_+$ be a number sufficiently large that each V(S) fits in a cube with a diagonal of length $\bar{m}$. For each coalition S let m^S ($\in R^S$) denote the vector which assigns $\bar{m}$ to each player in S .

Let (S,q^S) be a proposal of player $\hat{\imath}$ and set $s = |S|$. We renumber the players (if necessary) so that $\hat{\imath} = s$, $S = \{1,2,\ldots,s\}$, and $N\text{-}S = \{s+1,\ldots,n\}$.

Define $x \in R^N$ by $x_i = q_i^S$ for $i = 1,\ldots,s-1$, $x_s = \bar{v}_i$ and $x_i = \bar{m}$ for $i = s+1,\ldots,n$. Set $z = \Phi^s(x)$. By repeated application of properties (i) and (iv) (above) we see that z is maximal for all $i \in S$ and realizable for s.

We first prove that player $s = \hat{\imath}$ is limited to his individually rational payoff, i.e., $z_s = \bar{v}_s$. Suppose not; then, since z is maximal for s, $z_s > x_s = \bar{v}_s$. By (i), $z = \Phi^s(x)$ is realizable for s; let $S"$ be a coalition containing s with $z^{S"} \in V(S")$. Clearly $S"$ cannot contain a player from $N\text{-}S$ (since each player in $N\text{-}S$ demands $\bar{m}$). Thus $S"$ is a subset of S. Since $z_s > \bar{v}_s$, $S" \neq [s]$. Let $i \neq s$ be a player of $S"$. Repeated applications of (i) and (iv) show that $\Phi^{s-1}(x)$ is maximal for player i hence $((\Phi^{s-1}(x))^{S"-s}, \bar{v}_s)$ -- an element of $R^{S"}$ -- is not in the interior of $V(S")$. Since $z_s > \bar{v}_s$, the vector $z^{S"} = ((\Phi^{s-1})^{S"-s}, z_s)$ cannot be in $V(S")$, a contradiction. This contradiction establishes that $z_s = \bar{v}_s$.

We now select a player to fill the rôle of $i*$. There are two cases to consider.

Case 1: $q_{\hat{\imath}}^S \geq \bar{v}_{\hat{\imath}}$ and $S \neq [\hat{\imath}]$. Then $x^S \in V(S)$ and by property (ii), there is at least one player $i \neq \hat{\imath}$ in S such that $z_i \geq x_i$. Let $i*$ be the last such player. By property (i), $\Phi^{i*}(x)$ is realizable for player $i*$. Let $S*$ be a coalition for which $\Phi^{i*}(x)$ is realizable. Since player $i*$ was the last player with $z_i \geq x_i$, and $z_s = \bar{v}_s$ and since each ϕ_j only changes player j's component, $z^{S*} \in V(S*)$.

Cas 2: Either $q_{\hat{\imath}}^S < \bar{v}_{\hat{\imath}}$ or $(S,q^S) = ([\hat{\imath}],\bar{v}_i)$. For either possibility $z_s \geq q_i^S$. Let $i* = \hat{\imath}$ and let $S* = [s]$.

Renumbering the players, if necessary, we assume that $i* = 1$. We next show that $\Phi^n(z)$ is an aspiration. Repeated applications of properties (i) and (iv) proves that $z = \Phi^s(x)$ is maximal for every player in S. By construction of $\bar{m}$, z is maximal for every player in N-S. Since z is realizable and maximal for $i* = 1$, properties (ii) and (iii) imply $\phi^1(z) = z$ so $\Phi^1(z)$ is realizable for player 1. By property (iv), $\Phi^1(z)$ is maximal for player 1. By (iii), $\Phi^2(z) \leq \Phi^1(z) = \phi^1(z)$ so $\Phi^2(z)$ is also realizable for player 1. Continuing in this way, we see that $\Phi^k(z)$ is realizable and maximal for all players with indices k and below. Hence $\Phi^n(z)$ is realizable and maximal for all players and therefore an aspiration. Set $\beta(S,q^S,\hat{\imath}) = \beta = \Phi^n(z)$. By construction $\beta_{i*} \geq x_{i*}$ and therefore $\beta_{i*} \geq q_{i*}^S$. An argument similar to the one above shows that if $\hat{\imath} \neq i*$, $\beta_{\hat{\imath}} = z_{\hat{\imath}} = \bar{v}_{\hat{\imath}}$. By construction β satisfies (1)

$\beta_{\hat{\imath}} = \bar{v}_{\hat{\imath}}$, (2) there exists an $i^* \in S$ with $\beta_{i^*} \geq q^S_{i^*}$, and (3) if q^S is individually rational for i and $S \neq [\hat{\imath}]$ then $i^* \neq \hat{\imath}$. Hence β is the desired aspiration. We will say that $\beta = \beta(S,q^S,\hat{\imath})$ is the ***punishment aspiration*** for the proposal (S,q^S) by player i. This completes step 1.

Step 2: Loosely speaking we can describe the strategy profile that implements the price vector r in the following way. Every player makes proposals and accepts proposals consistent with r until a player, call him i, deviates by making a proposal (S,q^S) which is not consistent with r (all other deviations from the prescribed strategy are ignored). Then all players make proposals and accept proposals consistent with the punishm nt aspiration $\beta(S,q^S,i)$ until a player, call him i' deviates by making a proposal $(S',q^{S'})$ which is not consistent with $\beta(S,q^S,i)$. Then all players make proposals and accept proposals consistent with the punishment aspiration $\beta(S',q^{S'},i')$. The play continues in this way with the last deviator (if any) being punished until the end (if it ends) of the game.

In order to make this precise we next label every node η with a vector $p^*(\eta) \in R^N$ which we refer to as the "required" payoff vector at that node. Fix a node $\bar{\eta}$ in the tree. We label $\bar{\eta}$ by labeling every node in the path from the initial node to $\bar{\eta}$. The initial node is labeled r and every subsequent node on the path is labeled r until we come to $\bar{\eta}$ or to a node η_1 where some player, call him i_1 makes a proposal, call it (S,q^S), such that $q^S \neq r^S$. The node η_1 is labeled r and the following node is labeled $\beta(S,q^S,i_1)$. Every subsequent node is labeled $\beta(S,q^S,i_1)$ until we come a node η_2 at which a player, call her i_2, makes a proposal, call it $(S',q^{S'})$ with $q^{S'} \neq p_1^{S'}$. In this case the node η_2 is labeled $\beta(S,q^S,i_1)$ and the following node is labeled $\beta(S',q^{S'},i_2)$. Continuing in this way, we eventually reach and label the node $\bar{\eta}$. In this way we label each node η with the required payoff vector, $p^*(\eta)$.

For each $p \in R^N$, fix a price strategy profile $\sigma^p \in \Sigma^p$. Define σ^* by $\sigma^*_i(\eta) = \sigma^p_i(\eta)$ for $p = p^*(\eta)$ for every player i and every node for player i. We assert that σ^* is a subgame perfect strategy profile with a price vector of r.

To see that σ^* implements r notice that each player's first initiator node is labeled r, so each player is to play according to σ^r and therefore make proposals of the form (S,r^S). Given these actions, the succeeding responder nodes are all labeled r so each responder is to play according to σ^r and therefore accepts the proposal (S,r^S). Continuing in this way, it is clear that all nodes along the equilibrium path are labeled r so every proposal made with positive probability is form (S,r^S). Hence r is the price vector of σ^r.

To show that σ^* is subgame perfect, we show that for each player i, and each decision node of player i, σ^*_i is a best response to $\sigma^*_{)i(}$. The proof proceeds in three Lemmas. The first asserts that if i makes an acceptable proposal then his payoff is no more than his component of the node-label. The second asserts that σ_i^* is a best response to $\sigma^*)i($ at i's initiator nodes and the third asserts that σ_i^* is a best response at i's responder nodes.

Lemma 1: *If i makes a proposal (S,q^S) that has a positive probability of being accepted given $\sigma^*)i($, then $q_i^S \leq p_i^*(\eta)$.*

Proof of Lemma 1: Suppose first that $S = [i]$. Since every label is individually rational (all aspirations are individually rational and r is individually rational) $q_i^S \leq p_i^*(\eta)$ as desired. (Recall that all proposed payoff vectors must be feasible for their coalitions.) Suppose next that player i makes the proposal (S,q^S) at the (initiator) node η with the label $p^*(\eta)$ and designates player $j \neq i$ as the responder (either as part of the proposal or else after responding himself) and that this action leads to the node η_j, a responder node for player j. If $q^S = p^*(\eta)^S$ then clearly $q_i \leq p_i^*(\eta)$. If $q^S \neq p^*(\eta)^S$, then $p^*(\eta_j) = \beta(S,q^S,i)$ is a punishment aspiration. Since player j follows σ^*_j, j accepts only if $q^{S-i} \geq p^*(\eta_j)^{S-i}$. Since $p^*(\eta_j)$ is an aspiration and therefore maximal, if j accepts the proposal with positive probability, it must be that $q_i^S \leq p_i^*(\eta_j)$. Since $p^*(\eta_j)$ is a punishment aspiration for player i, $p_i^*(\eta_j) = \bar{v}_i$. Since every label, and in particular $p^*(\eta_j)$, is individually rational, $q_i^S \leq p_i^*(\eta_j)$ so $q_i^S \leq p_i^*(\eta)$, as desired.■

Lemma 2: *The strategy σ_i^* is a best response to $\sigma^*)i($ at every initiator node for player i.*

Proof of Lemma 2: Consider any $\sigma' \in \Sigma_i$ and consider any terminal node ζ that is reached from η_i with positive probability given σ^*/σ'. Let (S,q^S) be the proposal made by player k at node η whose acceptance leads to ζ. As we have shown in the previous lemma, if $k = i$ then $q_i^S \leq p^*(\eta)$. If $k \neq i$, he follows σ^* so $q^S = p^*(\eta)$. Hence $h_i(\zeta) = p_i^*(\eta)$ if $i \in S$ and $h_i(\zeta) = 0$ otherwise.

However since every player $j \neq i$ is following σ^* if the node η has positive probability of being reached from η_i given σ^*/σ', then the label of η is either $p^*(\eta_i)$ or $\beta(S',q^{S'},i)$ for some $S' \in C_i$, and some $q^{S'} \in V(S')$. In either case $p_i^*(\eta) \leq p^*(\eta_i)$. Hence $E_i(\eta_i|\sigma^*/\sigma') \leq p_i^*(\eta_i) = E_i(\eta_i|\sigma^*)$, so σ_i^* is a best response to $\sigma^*)i($ at the initiator node η_i. ■

Lemma 3: *The strategy σ_i^* is a best response to $\sigma^*)i($ at every responder node for player i.*

Proof of Lemma 3: At the node η, i is to respond to a proposal (S,q^S); let S' be the set of players in S (including player i) who have not yet accepted the proposal. In order to see that σ_i^* is a best response to $\sigma^*_{)i(}$ at η, it is convenient to consider three cases.

Case 1: $q_j^S \geq p_j^*(\eta)$ for every $j \in S'$. In this case, σ^* calls for every player in S' to accept (S,q^S) so $E_i(\eta \mid \sigma^*) = q_i^S \geq p^*{}_i(\eta)$. On the other hand, if player i rejects (S,q^S), then he becomes the initiator at a node η' that is an immediate successor to η. By Lemma 2, σ_i^* is a best response to $\sigma^*_{)i(}$ at the node η', so player i's expected payoff at η' is at most $p^*(\eta')$. However, given our labeling, $p^*(\eta') = p^*(\eta)$, so i's expected payoff at η' is at most $p_i^*(\eta)$. We conclude that if player i deviates from σ_i^* and rejects (S,q^S), then his expected payoff is at most $p_i^*(\eta)$. Therefore $E_i(\eta \mid \sigma_i^*) \leq p_i^*(\eta) \leq q_i^S = E_i(\eta \mid \sigma^*)$; i.e., σ_i^* is a best response to $\sigma^*_{)i(}$ at η.

Case 2: $q_i^S < p^*{}_i(\eta)$ and $q_j^S \geq p_j^*(\eta)$ for every $j \in S'-[i]$. In this case, σ_i^* calls for player i to reject (S,q^S) and make some (counter) proposal consistent with $p^*(\eta)$. Since this counter proposal will certainly be accepted (given that all other players follow σ^*), $E_i(\eta \mid \sigma^*) = p_i^*(\eta)$. On the other hand, if player i deviates from $\sigma^*{}_i$ and accepts (S,q^S), so will all players in $S'-[i]$, so player i's expected payoff will be q_i^S. Since $q_i^S < p_i^*(\eta) = E_i(\eta \mid \sigma^*)$, it is again the case that σ_i^* is a best response to $\sigma^*_{)i(}$ at η.

Case 3: $q_j^S < p_j^*$ for some $j \in S'-[i]$. In this case, σ_i^* calls for player i to reject (S,q^S) so $E_i(\eta \mid \sigma^*) = p_i^*(\eta)$. If player i deviates from σ^* be accepting the proposal (S,q^S), it will certainly be rejected by some player $k \in S'-[i]$. (Indeed it will be rejected by whomever i designates as the next respondent.) Player k will then become initiator and, playing according to σ^* will make a proposal of the form $(T, p^*(\eta)^T)$. If $i \notin T$, then this proposal will certainly be accepted by all players in T, so i's payoff will be $0 \leq p_i^*(\eta)$. If $i \in T$, then this proposal will be accepted by all players in $T-[i]$, so it follows from cases 1 and 2 that i's expected payoff will be at most $p_i^*(\eta) = E_i(\eta \mid \sigma^*)$. Once again, σ_i^* is a best response to $\sigma^*_{)i(}$ at η. This completes the proof of Lemma 3 and with it the proof of Theorem 1.■

We now turn to our second folk theorem.

Theorem 2: *Let (S,r^S) be a proposal. There exist a subgame perfect strategy profile σ^* such that (S,r^S) is the unique bargaining outcome for σ, if and only if $r = (r^S,0^{N-S})$ is individually rational for all players.*

Proof: To show that there is a subgame perfect strategy profile for which

(S,r^S) is the unique bargaining outcome, we alter players' price strategies and then follow the arguments in the proof of Theorem 1.

In Step 2 of the proof of Theorem 1, we associated to each $p \in R^N$, an arbitrary price strategy profile σ^p. We alter the assignment in the following ways: (1) For $p = r$ and for each player i in S, we select the particular price strategy for player i that assigns probability 1 to proposing (S,r^S) whenever i has the initiative. (2) For every parameter p and every player i *not* in S we alter player i's strategy so that whenever player i has the initiative, he passes the initiative to one of the players in S.[2] With these alterations, we may continue as in Theorem 1 to define the composite strategy σ^*. Notice first that (S,r^S) is the unique bargaining outcome of σ^* -- since (S,r^S) is the only proposal made with positive probability along the equilibrium path. To show that σ^* is subgame perfect, follow the proofs of the Lemmas keeping the following points in mind. The altered strategies for players $i \notin S$ are "as good" as the unaltered strategies, since either way (given the strategies of other players) player i's expected payoff is his individually rational level 0. No player (in or not in S) can take advantage of these altered strategies in order to improve his payoff because any proposal made which would give the proposer a payoff higher than his componenet of the current node label will be punished.

To complete the proof, we show why $(r^S,0^{N-S})$ must be individually rational. If (S,r^S) is the unique bargaining outcome of a strategy profile σ, then each player $i \in S$ obtains exactly r_i^S when nature selects him as initiator and each player $i \in N-S$ obtains 0 when nature selects him as initiator. Since player i can always obtain at least $\bar{v}_i$ when his is the initiator (by proposing $([i],\bar{v}_i)$) and nature selects each player to the initiator with positive probability, in order for σ_i to be a best response to $\sigma_{)i(}$ we must have that $r_i^S \geq \bar{v}_i$ for $i \in S$ and $0 \geq \bar{v}_i$ for $i \notin S$.■

The first folk theorem asserts that every individually rational and realizable vector is the price vector of a subgame perfect strategy profile. It is easy to construct subgame perfect strategy profiles that aren't price-generated. For example, for each player i, choose a price vector $s(i) \in R^N$ that is individually rational and realizable. Consider the strategy profile which is identical to the one constructed in Theorem 1 except that the price vector to be supported (i.e., the vector "r") depends on the player initially chosen by nature, i.e., in the subgame in which the i-th player is chosen, all players follow the strategy construct-

2. More formally, player $i \notin S$ assigns 0 probability to making a proposal and positive probability to passing the initiative to player j only if $j \in S$.

ed in Theorem 1 for $r = s(i)$. Since no player controls the choice of nature, this composite strategy profile is also subgame perfect. Since different subgames have different price vectors, no single price vector generates its bargaining outcomes.

REFERENCES

Bennett, E., "Three Approaches to Bargaining in NTU games", to appear in this volume, 1990.

Selten, R., "A Non-cooperative Model of Characteristic Function Bargaining", in Essays in Game Theory and Mathematical Economics in Honor of Oskar Morgenstern, V. Boehm and H. Nachtkamp (eds.), Wissenschaftsverlag Bibliographisches Institut Mannheim, Wien - Zurich, 131-151, 1981.

A NONCOOPERATIVE MODEL OF BARGAINING IN SIMPLE SPATIAL GAMES

James D. Laing

The bargaining problem facing players who must reach a collective decision can be modeled precisely as a noncooperative game. This essay uses a noncooperative game-th oretic approach to analyze bargaining in a class of games *without* sidepayments in which a point from a Euclidean set of decision alternatives is to be selected in accordance with a simple collective decision rule (Laing, Nakabayashi, and Slotznick, 1983), such as any weighted or unweighted majority rule. It builds from foundations established in Selten's (1981) noncooperative analysis of bargaining in zero-normalized, one-stage characteristic function games. It models simple spatial bargaining games, and characterizes noncooperative stationary equilibrium strategies and their relation to demand equilibria (Albers, 1975, 1987).*

The results demonstrate that simple spatial games differ in an important respect from the characteristic function games analyzed by Selten. In these characteristic function games, every member of a coalition that enacts any outcome consistent with the demand equilibrium realizes his quota exactly. If this condition obtains, then any stable demand vector can be implemented noncooperatively by a pure stationary strategy under which any proposal that satisfies all reponders' demands is accepted with certainty. But many simple spatial games violate the condition. Typically, some outcome that is consistent with the demand equilibrium exceeds someone's quota. In this case, any path-independent strategy that implements the demand equilibrium must instruct some responder who is offered his quota to reject the proposal with positive probability. Thus, complications arise at the knife's edge in simple spatial bargaining games.

The essay proceeds as follows. Section 1 represents a simple spatial bargaining problem as a noncooperative game in recursive form. Section 2 summarizes the basic theory of stationary equilibrium strategies in noncooperative games. Section 3 establishes the essay's principal results for simple spatial bargaining games in general. Section 4 characterizes the path independent strategies that implement the stable demands in two specific games. Section 5 concludes the discussion.

---

*I conducted this project as a member of Reinhard Selten's research group on "Game Equilibrium Models," Zentrum für interdisziplinäre Forschung, University of Bielefeld, Federal Republic of Germany, January 15 through June 30, 1988. I am grateful to Selten, Wulf Albers, and the other members of this stimulating group for many fruitful discussions, and to Albers and an anonymous referee for helpful comm nts on an earlier draft. Also, I wish to acknowledge support provided by the National Science Foundation (Grant SES87-09476) and by the Department of Decision Sci nces, the Wharton School, and Research Fund of the University of Pennsylvania.

1 SIMPLE SPATIAL BARGAINING GAMES

Section 1 models a class of simple games without sidepayments in which one element from a spatially represented set of alternatives must be enact d as the decision outcome in accordance with a nondictatorial collective decision rule and explicit rules of procedure that govern the process of negotiations. Subsection 1.1 identifies some formal characteristics of the games at issue in this paper by defining a general type of cooperative decision problem that is represented in effectiveness form. 1.2 defines a recursive mechanism that can be used to govern negotiations in such a decision problem. 1.3 specifies properties of each play r's payoff function over the alternative decision outcomes. 1.4 models the resulting decision problem as a noncooperative game of perfect information in recursive form.

1.1 A Collective Decision Problem in Effectiveness Form

In a simple collective decision problem a set $N=\{1,\ldots,n\}$ of players must choose one element from an arbitrary set X of decision alternatives in accordance with a simple collective decision rule (Laing, Nakabayashi, and Slotznick, 1983). In this essay I assume that that X is a compact and convex subset of finite-dimensional Euclidean space. A simple collective decision rule may be defined by specifying for each set $S \subseteq N$ of players the subset $E_S \subseteq X$ of alternatives that S, acting as a coalition, is empowered to enact as the decision outcome. The rule is *simple* if the possible coalitions of players may be partitioned into the sets of (W) *winning*, (L) *losing*, and (B) *blocking* coalitions such that $S \epsilon W$ iff $E_S = X$, $S \epsilon L$ iff $E_S = \emptyset$, and $S \epsilon B$ iff $E_S = z$, where $z \epsilon X$ denotes the default option representing the status quo. The rule is *superadditive* if every superset of a winning coalition is a winning coalition (and every subset of a losing coalition is a losing coalition). Such a rule is *essential* if both $N \epsilon W$ and $X \backslash z \neq \emptyset$, and *proper* if for every partition P of N into coalitions and for every pair of distinct coalitions S and S' in P, $E_S \cap E_{S'} = \emptyset$ or $E_S \cup E_{S'} = z$. The noncooperative bargaining model specified below applies to collective decision problems in which the rule is simple, superadditive, essential, and proper. I assume also that the rule is *nondictatorial*, in that every winning coalition contains mor than one player. For the special case in which the n players' roles under the rules are symmetric, the decision is governed by m/n majority rule, where m is any int ger such that $n/2 < m \leq n$, and S is a winning coalition iff S contains at least m members.

1.2 A Noncooperative Bargaining Mechanism

This subsection characterizes informally, th n sp cifies in detail, a recursive game form for conducting negotiations in such cooperative games. The procedure differs from the one employed by Selt n(1981), but is in the same spirit.

Preview of the mechanism. Let me outline the basic style of the game form specified precisely below. The game begins with a chance move by Nature that determines which player will first have the "initiative": an opportunity to initiate a proposal. At any step in the process only the player who has the floor may send a message. The player who has the floor occupies just one of two positions — initiator or responder. As *initiator*, player i can choose to either

(1) *pass* the initiative to any other player $j \in N\backslash i$ or pass to a chance selection of initiator by Nature, or

(2) make a *proposal* (x,S) in which $i \in S$ both offers an alternative x that S can enact and specifies the order in which players in $S\backslash i$ are queued for responding to the proposal.

As *responder* to a proposal (x,S) from i, player $r \in S\backslash i$ can choose to either

(1) *accept* the proposal, or

(2) *reject* (thus cancelling) the proposal and assume the initiator position.

If in round t all responders agree to the proposal (x,S), then the game ends with this as the final outcome. This procedure incorporates a parameter t^+ that can be set to limit the maximimum number of rounds in the negotiation process, such that if no final decision is enacted by the end of round number t^+, then the default outcome z becomes the final result. A new *round* in the negotiation process begins when any player assumes the initiator position.

The mechanism. In the following procedure, the symbol "M:" identifies a specification of the mechanism, whereas "P:" represents a choice opportunity for the player who has the floor to choose one of the options preceeded by "•".

0 Initialize Parameters (START)

M: Specify any simple collective decision problem in effectiveness form, $(N,X,z,\{E_S\})$. For every subset S containing more than one player, set $E_S := E_S \cup z$. (The justification for this modification of E_S is given below.) Stipulate that any proposal initiated by player i to enact the outcome (x,S) is admissible if $i \in S$ and $x \in E_S \neq \emptyset$. Set the round index $t=0$ and t^+ = the maximum number of rounds available for negotiations.

1 Chance move

M: Select $i \in N$ at random and set *initiator*$=i$.

2 Initiation of proposal

M: Set $t=t+1$. If $t>t^+$, then set $(x^*,S^*,t) = (z,\emptyset,t^+)$ and go to 4 (default outcome). Otherwise, set $S^*=\emptyset$ and open the floor to the initiator, i.

P: The initiator i must choose just one of the following options:

- *pass* [optional: to $j \in N\backslash i$].

  [M: If j is specified, set *initiator*$=j$ and go to 2. Else, go to 1.]

- *propose* $x \in E_{i \cup R}$ to $R \subseteq N\backslash i$ (ordering R).

  [M: Set $S^*=S^* \cup i$ and $x^*=x$.]

3 Response to proposal

M: If $R=\emptyset$ go to 4. Otherwise, designate the first player in R as *responder*, j, and open the floor to j.

P: The responder, j, must choose just one of the following options:

- *reject* (x,S) [and assume the initiator position].
  [M: Set *initiator*$=j$. Go to 2.]
- *accept* (x,S).
  [M: Set $S^*=S^* \cup j$, $R=R \backslash j$. Go to 3.]

4 Final Decision

M: Announce final decision is (x^*,S^*,t). STOP.

Effectiveness under the bargaining procedure: Seizing the opportunity. Note that the mechanism adds the status quo option to the effective set of each coalition containing at least two players. This modification is necessitated by a complication that arises in implementing simple collective decision rules. Under the procedure it is possible for any set S of two or more players to monopolize the floor if some member i of S becomes the initiator. By one method, i passes to j, j passes to i, i to j, and so on, such that no other player gets a chance to assume the initiative. By a second method, i makes a proposal that includes j as first responder, j rejects and makes a proposal listing i as the first responder, i rejects to assume the initiative, and so on. Each assignment of the initiative in either sequence begins a new round. By these tactics, i and j could continue the game indefinitely or through the maximum number of rounds permitted by the rules, and thus ensure the continuation of the status quo. Moreover, if the rules limit the game to at most t^+ rounds, then i and j can exchange the floor in this way until *one* of them attains the initiativ for the ultimatum subgame to be played in round t^+: *"Take it or leave it."* Any such extended sequence in which a coalition S of two or more players monopolizes the floor, thereby enforcing the continuation of the status quo or consuming the available rounds until the ultimate subgame in round t^+ is reached, may be said to constitute a *filibuster* by S.

Filibusters must be taken into account when defining effectiveness for the bargaining game. I follow an approach similar to that employed by Elaine Bennett (1988) for another class of games by interpreting effectiveness of coalition S as the ability to *seize the opportunity* to enact an outcome if a member of S attains the initiator position. Precisely, for any nonempty subset $S \subseteq N$ of players, let $E_S \subseteq X$ constitute the alternatives that, under the rules, S can enact as the outcome of the game *if* some member i of S attains the initiative and makes a proposal to $S \backslash i$. Let E_S' denote the effective set of coalition S in the original representation of collective decision problem. If this problem is to be negotiated via the bargaining mechanism then, because of filibuster possibilities, the effective set of coalition S in the bargaining problem is $E_S = E_S' \cup z$ if S contains at least two members; otherwise, $E_S = E_S'$. A coalition containing at l ast two players is effective for the status quo in the bargaining game, and thus is not a losing coalition in the bargaining game, even though it might be a "losing coalition" in th simple collective decision game. On the other hand, if S belongs to the set of winning(W) or blocking(B) coalitions, as specified in the simple collective decision game, then S has the same effectiveness in the associated bargaining game: $E_S = E_S'$.

Although filibusters must be recognized thus in defining effectiveness in these bargaining games, they have no influence on the equilibrium solutions of the games analyzed here, because I restrict the analysis to games in which

(1) the rules place *no limit* on the number of rounds, hence there is no ultimatum subgame, and

(2) the default option, z, not only represents the status quo, which remains in effect until it is replaced by an alternative decision outcome, but also is *unattractive*, in the sense that for every player i there exists both a coalition C containing i and an alternative $x \epsilon E_C$ such that every member of C strictly pr fers x to z; thus no one benefits from infinitely long play.

At equilibrium, under these conditions, there is no incentive to filibuster.

1.3 Payoffs

Payoffs are defined over endpoints of the game, where each such outcome may be denoted as (x,S,t), indicating that the alternative x is enacted in round t by coalition S as the final decision, thus ending the game. Let the ordered set $T = \{0,1,2,...\}$ index the rounds in the negotiation process. Assume that the preferences of each player $i \epsilon N$ over the possible outcomes are represented by the strictly quasiconcave payoff function $h_i : X \rightarrow R$. Thus, payoffs are neither discount d nor subject to transaction costs, and the payoff to any player i for an outcome in which the alternative $x \epsilon X$ is enacted by any coalition $C \subseteq N$ in any round $t \epsilon T$ is given by $h_i(x,t) = u_i(x)$. Strict quasiconcavity of the payoff function implies that no indifference contour in X is "thick" or "flat."

The payoff function has other useful properties. Denote an upper contour set in X associated with u_i in either of the following two ways:

$R_i(x) = \{ y \epsilon X \mid u_i(y) \geq u_i(x) \}$, given any fixed $x \epsilon X$;

$R_i(q_i) = \{ y \epsilon X \mid u_i(y) \geq q_i \}$, given any fixed real number, $q_i \epsilon R$.

The interior of $R_i(x)$ [respectively, $R_i(q_i)$] relative to X constitutes the strict preference set, denoted by $P_i(x)$ [respectively, $P_i(q_i)$]. The assumptions concerning X and u_i are sufficient to guarantee that, for every $x \epsilon X$, $R_i(x)$ is nonempty, compact, and strictly convex. [A set V is *strictly convex* if, for all $x \epsilon V$, $y \epsilon V \backslash x$ and any λ such that $0<\lambda<1$, $w = [\lambda x + (1-\lambda)y]$ lies in the interior of V.] Moreover, u_i attains a maximum at a unique element of X, denoted as the bliss point β_i, and also attains a minimum in X. Without loss of g nerality, u_i may be scaled such that, for ev ry player i and ev ry $x \epsilon X$, $u_i(x)$ belongs to th closed interval $[0, u_i(\beta_i)]$.

Corresponding preference s ts in X for any non mpty subset $S \subseteq N$ of players may be defin d:

$R_S(x) = \cap_{i \in S} R_i(x)$, given any fixed $x \in X$;

$R_S(q) = \cap_{i \in S} R_i(q_i)$, given any fixed vector of n real numbers, $q \in R^n_+$.

The interior of these sets constitute the strict preference sets $P_S(x)$ and $P_S(q)$, respectively. Then the (weak) Pareto efficient subset of alternatives in X for any nonempty subset S of players may be defined simply as $\beta_S = \{ y \in X \mid P_S(y) = \emptyset \}$.

The assumptions imply that, for every nonempty subset S of players and $x \in X$, $R_S(x)$ is nonempty, compact, and strictly convex, whereas, for every vector q of n real numbers, $R_S(q)$ is compact and strictly convex (but can be empty). These properties facilitate the analysis. For example, they provide one justification for limiting attention to behavioral strategies that have finite carrier.

Since z has been assumed to represent an unattractive *status quo* that remains in effect until a different alternative is enacted, each player belongs to some winning coalition that seeks a decision in a finite number of rounds.

1.4 The Bargaining Game in Recursive Form

This game may be represented as a special instance of a general type of r cursive game of perfect information, $G = (N,Y,y^0,O,P,A,\pi,\tau,h)$, where

$N=\{1,..,i,..,n\}$ is a finite set of players.

Y is a set of nodes, partitioned as follows.

$O \subset Y$ is a set of decision outcomes representing alternative end points of the game.

$Y \setminus O$ is a set of nodes representing choice opportunities by Nature or any player.

$y^0 \in Y \setminus O$ is a distinguished node indicating the first move of the game.

$P = (P_0, P_1, \ldots, P_n)$ is a partition of $Y \setminus O$ such that P_0 denotes the moves assigned to Nature and $(\forall i \in N)$ $P_i = P_i^1 \cup P_i^2$, the union of positions in which player i is (1) initiator and (2) responder, respectively.

A is a choice function that assigns to every $y \in Y \setminus O$ a nonempty set $A(y) \subset Y$ of options.

π is a probability function such that $\pi[A(y)]$ governs any move $y \in P_0$,

$\tau: Y \rightarrow T = \{0,1,2,\ldots\}$ indexes the rounds such that $\tau(y^0) = 0$ and move $y \in Y$ occurs in round $t = \tau(y)$.

$h: O \rightarrow R^n$ is a function that assigns a payoff vector to every outcome $o \in O$.

Provided that the rules place no limit on the maximum length of the game, then the types of nodes in $Y \setminus O$ and the choices available at each such node are as follows.

$y \in P_0$ denotes a chance move by Nature in accordance with the simple lottery $\pi[A(y)]$, where $A(y) = \{p_i \mid i \in N, p_i \in P_i^1\}$ is the set of initiator positions and, for every $p_i \in A(y)$, $\pi(p_i) > 0$.

$p_i \in P_i^1$ denotes a node at which player $i \in N$ has the *initiative* to make a proposal.

The choice set $A(p_i)$ at any initiator node p_i consists of the following, mutually exclusiv options.

$y \in P_0$: pass (to random selection of initiator).

$p_j \in P^1_j$: pass the initiative to any specified player $j \in N \setminus i$.

$r_j \in P^2_j$: propose (x,C), where $C \subseteq N$ is an ordered set of players that includes i, $x \in E_C$, and j is the first player in $C \setminus i$.

$r_j = (x,C,S,j) \in P^2_j$ denotes a *responder* position of player j in relation to a proposal to enact the outcome (x,C). S includes the initiator of this proposal and any responder who has accepted it, and j is the first player in $C \setminus S$.

The choice set $A(r_j)$ at any responder node r_j consists of the following options.

$p_j \in P^1_j$: reject the proposal and assume the initiator position.

$r_k = (x,C,S \cup j,k) \in P^2_k$, if $S \cup j \neq C$, where k is the first player in $C \setminus (S \cup j)$: accept the proposal, so that the floor passes to the next responder, k.

$(x,C) \in O$ if $S \cup j = C$: accept the proposal and end the game.

An endpoint of the game may be denoted $(x,C,t) \in O$, indicating that x has been nacted in round t by coalition C; in this case each player $i \in N$ realizes the payoff $h_i(x,t)$. Given that payoffs are assumed to be independent of both the coalition enacting the outcome and the round, for convenience I represent the outcome (x,C,t) variously as (x,t), (x,C), or x, depending on the context.

Assume that all details defining the game are common knowledge.

2 NONCOOPERATIVE SOLUTION THEORY

Section 2 outlines some noncooperative solution theory for recursive bargaining games (Selten, 1981, §3), thus establishing foundations for the subsequent analys s in Sections 3 and 4 of strategies that constitute stationary equilibria in simple spatial games that are negotiated in accordance with the bargaining mechanism. Subsection 2.1 restricts the focus of this essay to behavioral strategies that are stationary, hence subgame consistent, and identifies a class of stationary strategi s that are path independent. 2.2 characterizes the expected payoff function when players adopt stationary behavioral strategies for playing such bargaining games. 2.3 concludes the section by defining the conditions under which stationary behavioral strategies constitute noncooperative equilibria in these games.

2.1 Behavioral Strategies

Local strategies. Define a *local strategy* b_y at any node $y \in Y \setminus O$ as a probability distribution over $A(y)$ with finite carrier. Such a local strategy is a pure strat gy only in the special case that some edge in $A(y)$ is assigned probability equal to one.

Also, recall that every move $y \epsilon P_0$ assigned to Nature is governed by the fixed probability function π given in the specification of the game in recursive form, such that $b_y = \pi[A(y)]$. To avoid measure-theoretic complications, the analysis is confined to local strategies that assign positive probability to a finite number of options. This restriction might seem problematic when y represents an initiator position, since, in selecting a proposal, the initiator must choose a point in the convex set X of decision alternatives. However, I shall prove in Section 3 that the initiator has only a small number of optimal proposals, given the constraints associated with payoff expectations of other players in the game.

Global strategies. A *global strategy*, denoted by b, is a function that assigns to every $y \epsilon Y \backslash O$ a local strategy, b_y.

Stationary strategies. The following analysis focuses exclusively on stationary strategies with finite carrier that are, in the following sense, stationary. Any behavioral strategy b is said to be *stationary* if, for every player $i \epsilon N$, $b_y = b_{y'}$ for every pair of moves y and y' in P_i such that either (1) y and y' are initiator positions of player i, or (2) y and y' are responder positions of player i such that $y = y' = (x, C, S, i)$. Note that the behavior prescribed by a stationary strategy at any move in round t of the game does not depend on the history of play through round t-1.

Subgame consistency. Any stationary strategy is *subgame consistent*, in that it prescribes the same behavior in any two subgames that maintain the players' identities and in all respects are isomorphic up to positive linear transformations of the payoffs. To be precise, let $G' = (N', Y', y^{0\prime}, O', P', A', \pi', \tau', h')$ and $G'' = (N'', Y'', y^{0\prime\prime}, O'', P'', A'', \pi'', \tau'', h'')$ denote any two subgames of the game G. Call the subgame G" *isomorphic* to G' if $N' = N'' := N$ and there exists a bijective homomorphism $f: G' \to G''$ such that, for every $y' \epsilon Y'$ and its image $f(y') = y'' \epsilon Y''$, all four of these conditions obtain: (1) for every $k \epsilon \{0, 1, \ldots, n\}$, $y' \epsilon P'_k$ iff $y'' \epsilon P''_k$, (2) $A''(y'') = A'(y')$, (3) $y' \epsilon P'_0$ implies $\pi''[A''(y'')] = \pi'[A'(y')]$, and (4) for every player $i \epsilon N$, outcome $y' \epsilon O'$ and image $y'' = f(y') \epsilon O''$, $h''_i(y'')$ is a positive linear transformation of $h'_i(y')$. Then define a global strategy b for game G to be *subgame consistent* if it possesses this property: for every pair of subgames G' and G" of G, if G" is isomorphic to G' then, for each decision node $y' \epsilon Y' \backslash (O' \cup P'_0)$ of G' and image $y'' = f(y') \epsilon Y''$ of G", $b_{y''} = b_{y'}$.

Path independence. Some of the analysis is restricted to strategies that are path independent. Note that the local strategy induced by a global stationary strategy b at any initiator position does not depend on the path through which this node is reached. On the other hand, the local strategy induced by a stationary strategy b at a responder position (x, C, S, r) can depend on S, hence on the history of prior play within the current round. Thus, a stationary strategy need not be path independent, in the following sense. Informally, a strategy b is said to be path independent if $b_y = b_{y'}$ for every player $j \epsilon N$ and for every pair of positions y and y' assigned to j such that the subgames beginning at y and y' are essentially equivalent, differing only in th path by which these subgames are reached. To be precise, define any global strategy b to be *path independent* if it is stationary and for every player $r \in N$ and every pair of responder positions $y=(x,C,S,r)$ and $y'=(x',C',S',r)$ assigned to r, $b_y = b_{y'}$ if both $x=x'$ and $C \backslash S = C' \backslash S'$. A path independent strategy is history-free at every choice position, whereas a stationary strategy need be history independent only at initiator positions, given the way in which responder positions are represented. Thus, path independence is a natural extension of stationarity. In particular, for any game G as defined earlier, let G' denote an associated game that differs from G only in that the description of each responder position (x,C,S,r) and outcome (x,C) in G is replaced, respectively, by $(x,C \backslash S,r)$ and (x) in G'. Then any global strategy b in game G is *path independent* if and only if b induces a strategy b' in G' that is stationary (and subgame consistent).

2.2 Expected Payoffs

In computations of expected payoffs, the expression (y,t) indicates that node $y \in Y$ occurs in round t of the game. Let $p[(x,t') \in O \mid (y,t),b,t^+]$ denote the conditional probability that the game ends with outcome (x,t'), $x \in X$, *given* that (1) the node (y,t) has been reached, (2) the global strategy b governs behavior in all subsequent play, and (3) the game ends in or before round t^+.

If the game is constrained by the rules to last at most a finite number t^+ of rounds, then, given that each local strategy has finite carrier, only a finite number of outcomes have positive probability, and the expected payoff to any player $j \in N$ at any node (y,t) is given by the expression

$$h_j(b \mid (y,t)) = \Sigma_{t' \in [t,t^+ +1]} \Sigma_{x \in X} \; p[(x,t') \in O \mid (y,t),b,t^+] \; h_j(x,t'),$$

where round $t^+ +1$ is included to account for the selection of the status quo outcome z by default.

On the other hand, if, as assumed in this essay, the rules place no constraint on the maximum length of the game, then $h_j(b \mid (y,t)) =$

$$p[\infty \mid (y,t),b] \; u_j(z) + \lim_{t^+ \to \infty} \Sigma_{t' \in [t,t^+]} \Sigma_{x \in X} \; p[(x,t') \mid (y,t),b,t^+] \; h_j(x,t'),$$

where $p[\infty \mid (y,t),b]$ denotes the probability that play will continue forever, given that (y,t) has been reached and that b governs behavior. Since each local strategy b_y has finite carrier, only a finite number of outcomes in X have positive probability. Moreover, for all $x \in X$ and $t' \in [t,t^+]$, $h_j(x,t') \geq 0$. Therefore, the sum across x and t' is nondecreasing in t^+. Also, the sum is bounded above by $u_j(\beta_j)$. Therefore, the limit exists and $h_j(b \mid (y,t))$ is well defined.

The assumptions (1) that X is convex, (2) that, for any player $j \in N$, $x \in X$ and round t, $h_j(x,t) = u_j(x)$, and (3) that $u_j(\cdot)$ is strictly quasiconcave, together imply that the expected payoff function is *strictly risk averse*: if two distinct outcomes x and y in X yield the same payoff to j, then j strictly prefers the "sure thing" outcome $w = \lambda x + (1-\lambda)y$ to the lottery that yields x with probability λ and otherwise yields y.

2.3 The Stationary Equilibrium Concept

For any player $j \in N$ and any move $y \in P_j$, define a local strategy b_y to be *optimal* at y if, for every $w \in A(y)$, $b_y(w)>0$ implies that $h_j(b|w) \geq h_j(b|v)$ for every $v \in A(y)$. A global strategy b is defined to be a *stationary equilibrium* if, to every node $y \in Y \setminus O$, b assigns a local strategy b_y that is optimal at y. A global strategy c_j is said to be a *deviation* by player j from b if the local strategies assigned by c_j and b differ only at one or more nodes assigned to player j. If b is a stationary equilibrium of the game, then, for every $j \in N$ and $y \in P_j$ and for every deviation c_j of j from b, $h_j(b|y) \geq h_j(c_j|y)$ [Selten, 1981: Thm. 1]. Also, any stationary equilibrium is optimal at each decision node; therefore it is *subgame perfect* (Selten, 1975).

3 STATIONARY EQUILIBRIA AND STABLE DEMANDS IN SIMPLE SPATIAL BARGAINING GAMES

This section investigates stationary equilibrium strategies and their relation to demand equilibria in simple collective decision games on a Euclidean set X of decision alternatives in which the process of negotiations is governed by the recursive bargaining procedure defined above. Recall that, by assumption, the rules impose no limit on the length of the game and that the status quo is unattractive. Subsection 3.1 establishes some preliminary results. 3.2 characterizes stationary equilibrium strategies under which each player adopts an "accommodating" local response strategy, in the sense that the responder accepts with certainty any proposal that yields a payoff no less than that to be expected if, instead, he rejected it to take the intiative. 3.3 defines a concept of demand equilibrium for spatial games and proves two theorems about the relationship between "semistable" demands and the payoff quotas associated with stationary equilibrium strategies in the noncooperative bargaining game. 3.4 discusses two examples.

3.1 Preliminary Results

This subsection identifies some basic properties of stationary equilibria in th se games. For any stationary strategy b and every player $j \in N$, define the *quota* of player j associated with b to be the expected payoff to j if j assumes the initiative, given that behavior is governed by b: $q_j(b) = h_j(b|p_j \in P^1_j)$. Th n $q(b) = (q_1(b),...,q_n(b))$ denotes the *quota vector* associated with b.

Lemma 1 (lower bound on responder's payoff). If b is any stationary equilibrium with associated quota vector q(b), then, for every play r $j \in N$ and resp nder position $r_j \in P^2_j$, $h_j(b|r_j) \geq q_j(b)$.

Proof. Suppose there xists some $j \epsilon N$ and $r_j \epsilon P_j^2$ such that $h_j(b|r_j) < q_j(b)$. Then b is not optimal at r_j, since j has an option $p_j \epsilon A(r_j) \cap P_j^1$ for which $h_j(b|r_j) < h_j(b|p_j) = q_j(b)$. Hence, b is *not* a stationary equilibrium. //

Lemma 2 (upper bound on equilibrium payoffs). Let b denote any stationary equilibrium with associated quota vector q(b). Then, for every player $j \epsilon N$ and every node $y \epsilon Y \backslash (O \cup P_j)$ such that $y \epsilon P^0$ is a chance move assigned to Nature or $y \epsilon P_k^1$ is an initiator position assigned to any other player $k \epsilon N \backslash j$, the expected payoff to j at y cannot exceed his quota: $h_j(b|y) \leq q_j(b)$.

Proof. If j has the initiative at position $p_j \epsilon P_j^1$, then j can pass to $y \epsilon P_i^1$ for any $i \epsilon N \backslash j$ or to $y \epsilon P_0 \backslash y^0$ (Nature). Since b is optimal at p_j, for any such y it must be true that $h_j(b|y) \leq h_j(b|p_j) = q_j(b)$. Also, j's expected payoff at the first move of the game, $y^0 \epsilon P_0$, is a convex combination of his expected payoffs across the initiator positions that may be selected at y^0. Therefore, $h_j(b|y^0) \leq q_j(b)$. //

Note that Lemma 2 applies to the game's starting point. Therefore, for any stationary equilibrium point, each player's expected payoff for participating in the game as a whole does not exceed his quota. The lemma implies also that there is a first-mover advantage: at equilibrium, the initiator's expected payoff equals his quota.

Lemma 3 (guaranteed acceptances). Let b be any stationary equilibrium with associated quota vector q(b). Then any proposal (x,C) by any initiator $i \epsilon N$ will be accepted with certainty if, for every responder $j \epsilon C \backslash i$, $u_j(x) > q_j(b)$.

Proof. Consider any proposal (x,C) and any responder position $r_j \epsilon P_j^2$ such that everyone in $S = C \backslash j$ has initiated or accepted the proposal. Thus, j is the last responder in C. Then, since b is optimal at r_j, j must accept the proposal, thereby ending the game. Therefore, by induction on responder positions, every player in $C \backslash i$ must accept the proposal. //

Lemma 4 (limited play). If b is a stationary equilibrium and default is unattractive, then the probability of infinitely long play equals zero.

Proof (Selten, personal communication). Suppose that there exists a node $y \epsilon Y$ in the game such that $p(\infty|y,b) \neq 0$. Then, in accordance with b, there must exist an infinite sequence *s* of rounds that is reached with positive probability from y such that *s* contains only a finite number of rounds in which the game ends with probability greater than zero (but less than one, by assumption). Since the number of such rounds is finite, ther must be a last such round, say t, in which the game can end in accordance with b. Now consider the subgame in the infinite sequence that begins with round t+1 and suppose that player i has the initiative. If the play continues into round t+2 in accordance with b, then each player $j \epsilon N$ expects the default payoff, $u_j(z)$. By assumption, default is unattractive, so, by definition, there exists a winning coalition C containing i and an alternative $x \epsilon E_C$ such that, for every player $j \epsilon C$, $u_j(x) > u_j(z)$. Thus, player i can propose the outcome (x,C) and, by Lemma 3, every responder must accept it, thereby ending the game. Therefore, b is not a stationary equilibrium. //

Lemma 5 (quota-feasible outcomes). Let b be a stationary equilibrium with associated quota vector q(b) and outcomes O(b). Then,

(1) for every player $i \epsilon N$, there exists an outcome $(x,C) \epsilon O(b)$ such that $u_i(x) \geq q_i(b)$, and

(2) for every outcome $(x,C) \epsilon O(b)$ such that $C \neq \emptyset$, and for every member i of C, $u_i(x) \geq q_i(b)$.

Proof. Part (1) follows immediately from the fact that $q_i(b)$ is a probability-weighted average of i's payoffs for outcomes in O(b).

The following proof of (2) is a restatement of one given by Selten(1981: s e the relevant portion of his proof of Lemma 3). Let s be any sequence of positions in Y that ends with the outcome $(x,C) \epsilon O(b)$ and consider any player $i \epsilon C$. Th re must be a position $y_i \epsilon P_i$ in s such that after y_i there is no initiator position p_k in s: either $y_i \epsilon P_i^1$, so that i proposes the outcome (x,C), or $y_i \epsilon P_i^2$, so that i is a responder to the proposal. In either case, Lemma 1 and the definition of $q_i(b)$ together imply $q_i(b) \leq h_i(b|y_i)$. Let θ denote the conditional probability, given that position y_i is reached and b governs behavior, that no responder after y_i in s rejects the proposal if i initiates or accepts it at y_i. Note that $\theta > 0$, since $(x,C) \ \epsilon \ O(b|y_i)$. Thus the proposal is rejected by one of these responders with probability $1-\theta$; in this event, Lemma 2 implies that player i's payoff cannot exceed $q_i(b)$. Hence, $q_i(b) \leq h_i(b|y_i) \leq \theta u_i(x) + (1-\theta)q_i(b)$. Given $\theta > 0$, this implies $q_i(b) \leq u_i(x)$. Since $(x,C) \epsilon O(b)$ and $i \epsilon C$ were selected arbitrarily, it follows that, for each $(x,C) \epsilon O(b)$ and $i \epsilon C$, $u_i(x) \geq q_i(b)$. //

Lemma 6 (efficient agreements). Let b denote any stationary equilibrium with quota vector q(b). For any *winning* coalition C and alternative $x \epsilon X$, the outcome (x,C) is assigned positive probability by b only if x is Pareto optimal for C.

Proof. Suppose that, in accordance with b, (1) player i initiates at position $p_i \epsilon P_i^1$ a proposal (x,C) such that $C \epsilon W$ and x is *not* Pareto optimal for C, and (2) every responder $j \epsilon C \backslash i$ agrees, thus ending the game. In this event, Lemma 1 impli s that, for every $j \epsilon C \backslash i$, $u_j(x) \geq q_j(b)$. Then, by th definitions of Pareto optimality and winning coalition, there exists an alternative $x' \epsilon X$ that C could enact that every member of C strictly prefers to x. Consid r a d viation c_i from b such that i proposes (x',C). Then, for every responder $j \in C\backslash i$ to i's proposal, $u_j(x') > q_j(b)$; hence, Lemma 3 impli s that the outcome (x',C) will occur with c rtainty if i proposes it. But then $h_i(c_i|p_i) > h_i(b|p_i)$, so b is not optimal at p_i. This contradicts the assumption that b is a stationary equilibrium. //

Lemma 7 (no slack). Let b be any stationary equilibrium with the associated quota vector $q(b)$. Then, for every coalition C, $E_C \cap P_C(q) = \emptyset$.

Proof. Suppose that b is a stationary global strategy with associated quota vector $q(b)$ but there exists a coalition $C \subseteq N$ and decision alternative $x \in E_C \cap P_C(q)$. Consider any member i of C and position $y \in P_i$ such that i has the initiative or, by rejecting any proposal, can assume it at position $p_i \in P_i^1$. Then there exists an outcome (x,C) that i can propose at p_i such that, for every $j \in C$, $u_j(x) > q_j(b)$. If i proposes the outcome (x,C), it will be accepted with certainty, by Lemma 3, and i expects to achieve the payoff $u_i(x) > q_i(b) = h_i(b|p_i)$. Therefore b is not optimal at y. //

3.2 Accommodating Responses and Extreme Proposals

Path independent strategies can exhibit either or both of the properties identified in this subsection. The theorem establishes a relationship between these properties at equilibrium.

Accommodating Responses. The local response strategies induced by one type of path independent global strategy satisfy

Property A (Accommodating Responses):

For every player $r \in N$ and responder position $y' \in (x,C,S,r) \in P_r^2$,

$$b_{y'}(p_r \in P_r^1) = \left\{ \begin{matrix} 0 \\ 1 \end{matrix} \right\} \text{ if } (\forall k \in C\backslash S)\ h_k(b|\text{accept}) \left\{ \begin{matrix} \geq \\ < \end{matrix} \right\} q_k(b).$$

If the global strategy b governing behavior exhibits property A, then a responder *rejects* any proposal that is generated in accordance with b if and only if h expects to achieve a strictly greater payoff by assuming the initiative. By induction on responder positions, it follows that, if the global strategy possesses property A, then a proposal by player i to enact any outcome (x,C) will be accepted as the decision outcome if and only if, for every responder k in $C\backslash i$, $u_k(x) \geq q_k(b)$.

Extreme Proposals. A path independent strategy can also exhibit a second property, in that it induces local initiation strategies that generate only proposals that are, in the sense to be defined, extreme. Given any stationary strategy b with associated quota $q(b)$, then for any player $i \in N$ and initiator position $p_i \in P_i^1$, let $F_i(q) = \{C \in W \cup B \mid i \in C,\ E_C \cap R_{C\backslash i}(q) \neq \emptyset\}$ denote the ffective coalitions for which i might generat an acceptable proposal. These are the coalitions to which i belongs that can enact an outcome such that the payoff to ev ryone $C\setminus i$ is a least as good as his quota. For any coalition $C \epsilon F_i(q)$, let x_c^* denote any decision alternative in X such that

$$u_i(x_c^*) = \underset{x \epsilon E_C \cap R_{C\setminus i}(q)}{\text{MAX}} u_i(x).$$

Next, define

$$F_i^*(q) = \{ C \epsilon F_i(q) \mid u_i(x_c^*) = \underset{C' \epsilon F_i(q)}{\text{MAX}} u_i(x_{c'}^*) \}, \text{ and}$$

$$A_i^*(q) = \{ r_j \epsilon P_j^2 \mid j \epsilon N\setminus i,\ r_j = (x,C,S,j),\ C \epsilon F_i^*(q),\ x = x_c^* \},$$

where, by definition, $A_i^*(q)$ constitutes the set of *extreme* (*maximal*) *proposals* by the initiator, i, given the quota vector q(b). Such proposals must exist:

Lemma 8 (existence, finiteness, and uniqueness of extreme proposals). Let b denote any stationary equilibrium with associated quota vector q(b). Then, for every player $i \epsilon N$ and initiator position $p_i \epsilon P_i^1$, the set of extreme proposals is nonempty, finite and unique.

Proof. Consider any player $i \epsilon N$ and any initiator position, $p_i \epsilon P_i^1$, assigned to i. By Lemma 2, if i passes or makes a proposal that is rejected, then his xpected payoff cannot exceed $q_i(b)$, whereas, by Lemma 5(2), if he proposes an outcome in $O(b)$ that is accepted by all responders, then his expected payoff is no less than $q_i(b)$. By Lemma 1 and the local optimality of any stationary equilibrium strategy, a proposal (x,C) by i will be accepted *only if*, for ev ry responder $j \epsilon C\setminus i$, $u_j(x) \geq q_j(b)$. Lemma 5 implies that, at equilibrium, such a proposal exists. Hence, the set of effective coalitions for which i has a feasible proposal, $F_i(q) = \{C \epsilon W \cup B \mid i \epsilon C,\ E_C \cap R_{C\setminus i}(q) \neq \emptyset\}$, is not empty.

For any $C \epsilon F_i(q)$, consider any decision alternative x_c^* in X such that

$$u_i(x_c^*) = \underset{x \epsilon E_C \cap R_{C\setminus i}(q)}{\text{MAX}} u_i(x).$$

If $C \epsilon B \cap F_i(q)$, then $E_C \cap R_{C\setminus i}(q) = z$, so $x_c^* = z$ is the default outcome. Now suppose $C \epsilon W \cap F_i(q)$, so that $E_C = X$, hence $E_C \cap R_{C\setminus i}(q) = R_{C\setminus i}(q)$. Then, since u_i is continuous and $R_{C\setminus i}(q)$ is nonempty and compact, u_i attains a maximum in $R_{C\setminus i}(q)$ at some point, x_c^*. Moreover, $u_i(\cdot)$ is strictly quasiconcave and $R_{C\setminus i}(q)$ is conv x, hence x_c^* is the unique maximal element of $R_{C\setminus i}(q)$. Thus for each $C \epsilon F_i(q)$ there is a unique point x_c^* such that the proposal (x_c^*,C) maximizes u_i. By definition, $F_i^*(q)$ contains each coalition C to which i belongs such that the associated point x_c^* maximizes $u_i(x_{c'}^*)$ across $C' \epsilon F_i(q)$. Since a unique point x_c^* is associat d with each coalition C in $F_i^*(q)$, and there is a finite number of permutations of $C\setminus i$ and of coalitions in $F_i^*(q)$, then, at equilibrium, the set $A_i^*(q)$ of extreme proposals by the initiator i is nonempty, finite and unique. //

Then one type of stationary strategy b with associated quota vect r q(b) satisfies

Property $\underline{E}$ (Extreme Proposals):
For very player $i \epsilon N$, initiative $y \epsilon P_i^1$, and option $y' \epsilon A_i(y)$,
$b(y') > 0$ implies $y' \epsilon A_i^*(q)$,
where $A_i^*(q)$ is the set of extreme proposals, given q.

Only extreme proposals are rational when all responders are accommodating:

Theorem 1 ($\underline{EA}$ equilibrium). Suppose that b is any stationary strategy that satisfies property $\underline{A}$. Then b is a stationary equilibrium *if and only if* b exhibits property $\underline{E}$.

Proof. *Sufficiency*. Let b denote any stationary strategy that satisfies property $\underline{A}$. Suppose that b does not possess property $\underline{E}$. Then there exists some $i \epsilon N$, $y \epsilon P_i^1$, and $y' \epsilon A_i(y)$ such that $b_y(y')>0$ but y' is not an element of $A_i^*(q)$. There are two possibilities. First, $A_i^*(q) \neq \emptyset$; hence b is not optimal at y. Second, $A_i^*(q) = \emptyset$; hence Lemma 8 applies. Thus, in either case, b is not a stationary equilibrium.

Necessity. Let b denote any stationary strategy that satisfies $\underline{A}$, and assume that b also exhibits property $\underline{E}$. Select arbitrarily any player $i \epsilon N$, option $y \epsilon P_i^1$, and alternative $y' \epsilon A_i(y)$ such that $b_y(y')>0$. By property $\underline{E}$, $y' = (x_c^*, C, S, k) \epsilon A_i^*(q|y)$. Consider any responder $r \epsilon C \setminus i$ to the proposal. Either r is the last responder in C or, in accordance with property $\underline{A}$, every responder in $C \setminus (S \cup i)$ will accept the proposal. In either case, by definition of the extreme proposals set, r expects a payoff of at least $q_r(b)$ if he conforms to the strategy b and thus accepts the proposal. If he rejects the proposal, then, by definition, r's expected payoff equals $q_r(b)$. Therefore, r has no incentive to deviate from b. Since r was chosen arbitrarily, this conclusion applies to every responder to the proposal, if it is offered. Therefore, the initiator i expects the proposal y' to be accepted, if he offers it at y. But this proposal was selected from the set $A_i^*(q|y)$ of proposals that maximize i's expected payoff at y, given that all responders are governed by b. Hence, i has no incentive to deviate from b. Therefore, b is a stationary equilibrium. //

Hereinafter, the term $\underline{EA}$ strategy refers to any stationary strategy that satisfies both properties $\underline{E}$ and $\underline{A}$.

Corollary 1 ($\underline{EA}$ quota points). Let b denote any $\underline{EA}$ strategy, with associated quota vector q(b). Then b is a stationary equilibrium such that, for every player $i \epsilon N$, initiative $y \epsilon P_i^1$, and proposal $(x, C, S, r) \epsilon A_i(y)$,
$b_y(x, C, S, r) > 0$ implies $u_i(x) = q_i(b)$.

Proof. This follows immediately from Theorem 1, Lemma 8, and the definitions of quota and xtreme proposals set. //

EA strategies have some special features. Every EA strategy is path independent and induces pure local response strategies (local initiator strategies can be pure or mixed). Also, any EA equilibrium strategy has shortness, in that, for every subgame that begins with an initiator position, it induces a proposal at that position that is accepted by all responders as the final decision outcome. At an EA equilibrium, the game ends in the first round. This type of equilibrium is based on a quite accommodating *pure strategy*: each responder accepts a proposal with certainty even in the knife-edge case in which he is indifferent between accepting the proposal or rejecting it to assume the initiative.

The next result demonstrates that, in any outcome under a stationary equilibrium that is enacted by a winning coalition, at least one responder must face this situation.

Lemma 9 (the knife's edge). Let b denote *any* stationary equilibrium strategy with associated quota vector q(b) and outcomes $O(b)$. For every winning coalition C and for every proposal (x,C,S,j) by any initiator $i\epsilon N$ to enact an outcome $(x,C)\epsilon O(b)$, there exists some responder $k\epsilon C\backslash i$ such that $u_k(x)=q_k(b)$.

Proof. Suppose that, in accordance with a stationary equilibrium b, some player i initiates the proposal (x,C,S,j) to enact an outcome $(x,C)\epsilon O(b)$ by a winning coalition, C. Assume also, contrary to the lemma's assertion, that for every $k\epsilon C\backslash i$, $u_k(x)\neq q_k(b)$. Then, by Lemma 1, $x\epsilon P_{C\backslash i}(q)$. There are two possibilities, depending on whether $x=\beta_i$.

First, suppose that $x\neq\beta_i$. Then, since X and $P_{C\backslash i}(q)$ are convex and the latter is open, there exists some $\lambda\epsilon(0,1)$ such that $w = [\lambda x+(1-\lambda)\beta_i] \epsilon P_{C\backslash i}(q)$. Then $u_i(w)>u_i(x)$, since u_i is strictly quasiconcave, and, by Lemma 5(2), $u_i(x)\geq q(b)$. Therefore, $u_i(w)>q_i(b)$. But then $P_C(q)\neq\emptyset$; hence, q(b) contains slack, in contradiction of Lemma 7.

Second, consider the remaining possibility: $x=\beta_i$. Select any responder $k\epsilon C\backslash i$ to the proposal by i to enact the outcome (β_i,C) and suppose that it is k's move to respond. Then k can reject the proposal, take the initiative, and propose the same outcome, (β_i,C), listing i as the first responder in $C\backslash k$. By Lemma 3, i is guaranteed that every responder in $C\backslash\{i\cup k\}$ will accept the proposal if i does. Thus, at this position, $h_i(b|r_i) = u_i(\beta_i)$. Since b is optimal at r_i and β_i is the unique point in X that maximizes u_i, k can be certain that, for every $(x',C')\epsilon O(b|r_i)$, $x'=\beta_i$. Therefore, $h_k(b|p_k)=q_k(b) \geq u_k(\beta_i)=u_k(x)$, contradicting the assumption that $q_k(b) < u_k(x)$. //

Corollary 1 indicates that, at an EA equilibrium, the initiator also operates at the knife's edge, proposing via a pure or properly mixed local strategy a point that yields exactly his quota. As demonstrated in the next section, this fact limits the usefulness of EA strategies in simple spatial games.

3.3 Noncooperative Implementation of Demand Equilibria

A Demand Equilibrium Concept. This subsection extends Albers (1975) basic concept of a demand equilibium in characteristic function games to simple spatial games without sidepayments. Although the translation of this concept to the spatial context is not straightforward (Albers, 1987; also see Bennett and Winer, 1983), in this essay I define a demand equilibrium for any simple spatial game as follows. A demand vector for a game is an n-tuple of real numbers, $d \epsilon R^n$, with the interpretation that that it is possible for coalition C to enact the feasible outcome (x,C) *only if* C is effective for x and, for every member i of C, $u_i(x) \geq d_i$: that is, $x \epsilon E_C \cap R_C(d)$. Let X(d) denote the alternative points in X that can be enacted, given the demands: $X(d) = \{x \epsilon X \mid (\exists\ C \subseteq N)\ x \epsilon E_C \cap R_C(d)\ \}$. For any player $i \epsilon N$ and demand vector d, let $F_i(d) = \{\ C \subseteq N \mid i \epsilon C,\ E_C \cap R_C(d) \neq \emptyset\ \}$. This set includes every coalition to which i belongs that is effective for an outcome that satisfies every member's demand. Given the demand vector d, player i is said to *depend* on player j if $F_i(d)$ is a proper subset of $F_j(d)$. Any demand vector d can be evaluated against these conditions:

(1) *Feasibility*. For every player $i \epsilon N$, $F_i(d)$ is not empty.

(2) *Maximality (no slack)*. For every coalition C, $E_C \cap P_C(d) = \emptyset$.

(3) *Independence*. For each $i \epsilon N$ there is no $j \epsilon N$ such that $F_i(d) \subset F_j(d)$.

Define any demand vector d to be *semistable* if it meets the first two of these conditions, and *stable* if it satisfies all three.

Two Results. I now prove two theorems about the noncooperative implementation of semistable demands. Let me say that any global stationary strategy b with associated quota vector q(b) *implements* the demand vector d if b is an equilibrium and d=q(b).

Theorem 2 (equilibrium quotas as semistable demands). Every stationary equilibrium b implements a semistable demand vector, d=q(b).

Proof. Immediate consequence of Lemmas 5(1) and 7. //

Theorem 3 (semistable demands implementable with accommodating responses). Let b denote any stationary global strategy exhibiting property A such that the associated quota vector q(b) constitutes a semistable demand vector, d. Then b is a stationary equilibrium if and only if this condition obtains:

for every option $x \epsilon X(d)$ and player $i \epsilon N$, $u_i(x) \leq d_i$.

Proof. *Sufficiency*. Choose any player $i \epsilon N$ and suppose that there exists a coalition $C \subseteq N$ and alternative $x^* \ \epsilon\ E_C \cap R_C(d)$ such that $u_i(x^*) > d_i = q_i(b)$. The effectiveness relation is superadditive. Therefore, at any initiator position $p_i \epsilon P_i^1$, i could make the extreme proposal $(x^*, i \cup C, i, r)$ and, by property A, it would be accepted by all responders. In this case, i's expected payoff equals

$u_i(x^*) > q_i(b)$. Hence, b is not an equilibrium.

Necessity. Suppose that for every player *i* and each alternative x that is feasible, given the demands, $u_i(x) \le d_i$. Since d is a semistable demand vector, by definition there exists for each player *i* some alternative x* and coalition C such that $x^* \epsilon E_C \cap R_C$ and $u_i(x^*) \ge d_i$. Hence $u_i(x^*) = d_i$. Consider any initiative $y \epsilon P_i$ of player *i*. Since $q_i(b)$ is a probability weighted average of *i*'s payoffs in $O(b|y)$ then for all $x' \epsilon O(b|y)$, $u_i(x') = q_i(b)$. Hence, *i* has no option that would yield him a greater payoff, and thus no incentive to deviate from b at y. By property A, every responder achieves at least his quota in any proposal that he accepts, in accordance with b, and thus has no incentive to deviate from b. Therefore, b is an equilibrium. //

All examples presented in the sequel but the first demonstrate that the condition displayed in the last line of Theorem 3 need not be satisfied in a spatial game. Therefore, these results contrast sharply with those obtained by Selten (1981, Theorem 2; also see Winter, 1989), who proved for one-stage characteristic function games not only that the quota vector associated with any stationary equilibrium is a semistable demand vector, but also that every semistable demand vector can be implemented by a pure strategy equilibrium that exhibits properties E and A.

3.4 Examples

All examples in this essay share these features. A point is to be selected from $X \subset R^2$ by a set N of five players governed by simple majority rule. Each player $i \epsilon N$ seeks to minimize the Euclidean distance of the decision point in X from his bliss point, β_i, and is indifferent among points that are equidistant from β_i. Thus, the arc in X of any circle centered on β_i is an indifference class of player *i*. The default outcome is unattractive and may be ignored.

Stable demands for a symmetric two-insiders game. Consider first the game plotted in Figure 1, and the stable demand vector, d, represented by the circles in the figure. Diamonds in the figure indicate blisspoints and the centers of the three squares constitute the set X(d) of solution points that are consistent with the stable demands. No element of X(d) lies inside any of these circles, so no solution point exceeds any player's demand. Thus, by Theorem 3, this stable demand vector can be implemented in the bargaining game via an EA stationary equilibrium.

Stable demands for Game F: The Fox. The opposite conclusion obtains if I modify the first example to create the simple majority game shown in Figure 2. Again, exactly three points in this game are feasible, given the stable demands plotted in the figure. Yet, by Theorem 3. this stable demand vector cannot b implemented by a stationary equlibrium strategy that satisfies property A, because two of the three solution points exceed some player's demand.

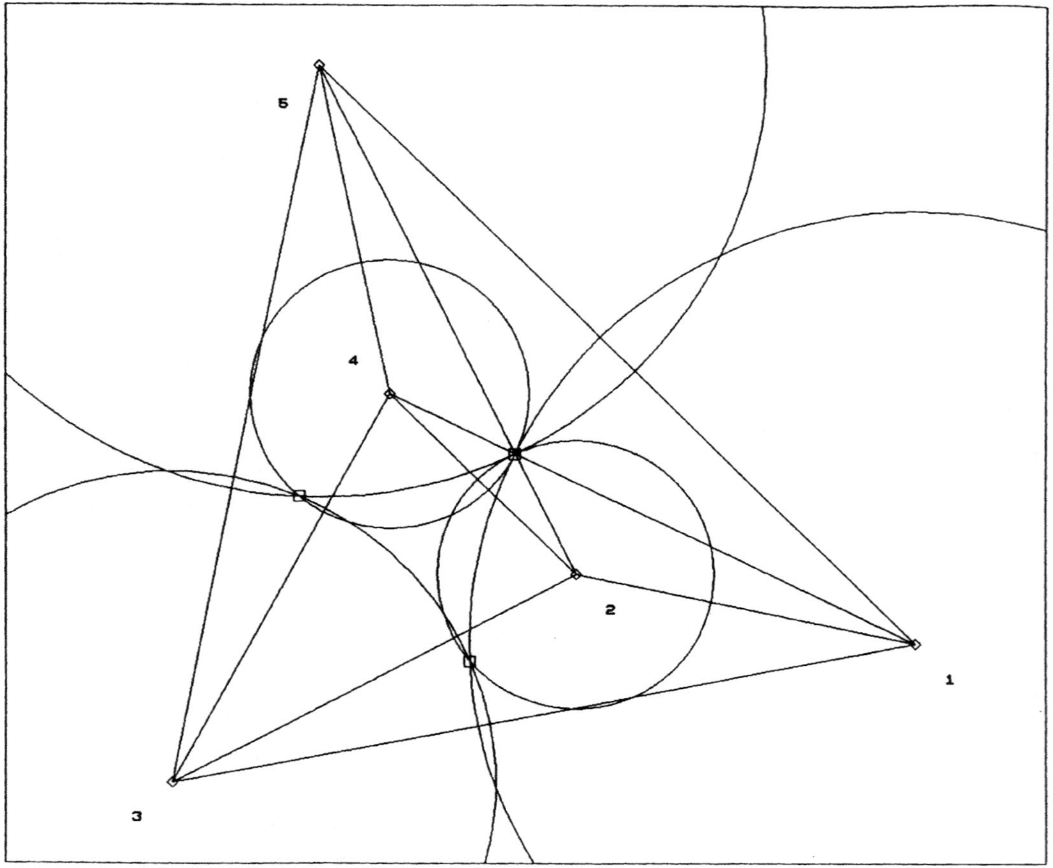

FIGURE 1. STABLE DEMANDS IMPLEMENTED BY STATIONARY EQUILIBRIUM WITH ACCOMMODATING RESPONDERS

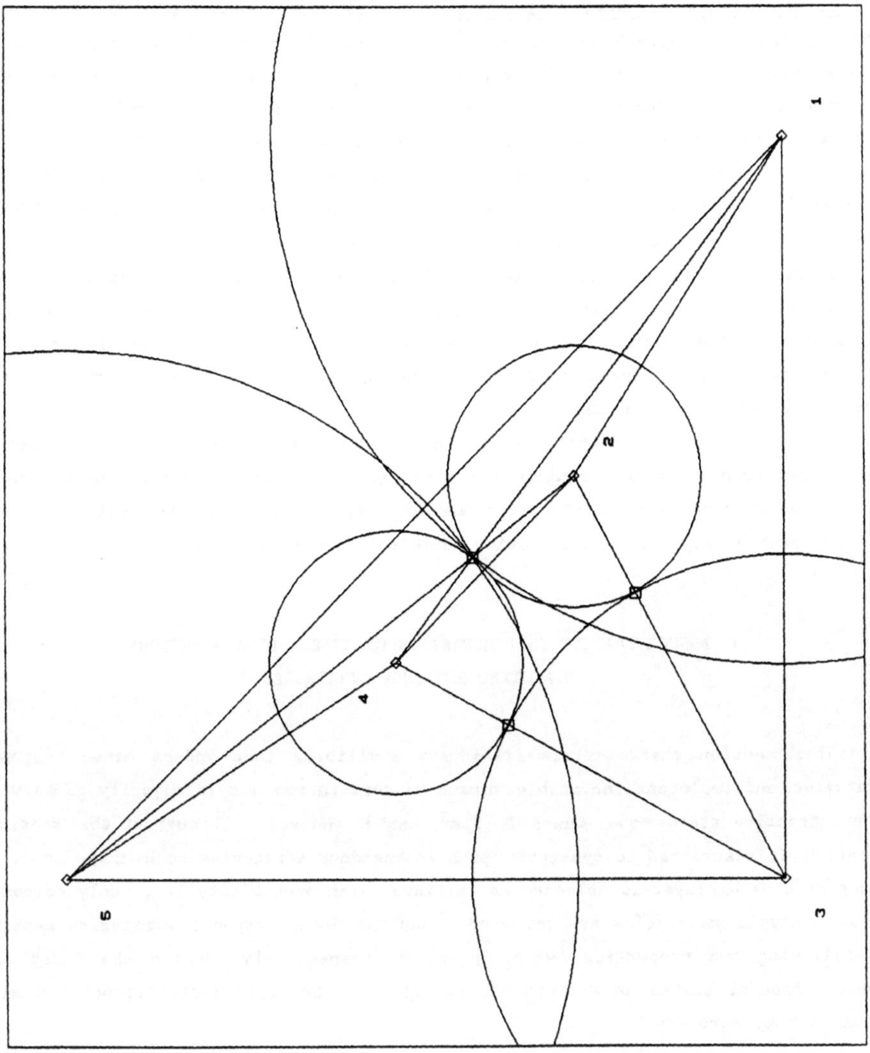

FIGURE 2. STABLE DEMANDS IN GAME F: THE FOX

Stable demands for Game S: The Star. The perfectly symmetric, simple majority game depicted in Figure 3 has an obvious solution: the stable demand vector plotted in the figure. This is the only demand vector for the game that is both semistable and symmetric. It happens to correspond in this particular example to another cooperative equilibrium concept called the competitive solution (McKelvey, Ordeshook, and Winer, 1978; Forman and Laing, 1982) if only three-person coalitions are considered, but there are other games in which these two concepts do not agree. Given this stable demand vector d, exactly the five decision alternatives indicated by squares in the figure are feasible, and they may be indexed in such a way that, for each player $i \epsilon N$ and $x_i \epsilon X(d) = (x_1, \ldots, x_5)$, $u_i(x_i) > d_i$. Therefore, by Theorem 3, the stable demand vector for this game cannot be implemented via a stationary strategy that satisfies property A. The same conclusion applies for another example given in the next section, Game H.

Thus, Theorem 3 implies that the stable demand vector in Games F, S, and H can be implemented by a stationary equilibrium strategy only if it instructs some responder who is indifferent between accepting and rejecting a proposal to reject it with positive probability. The next section addresses such strategies.

4 NONCOOPERATIVE IMPLEMENTATION OF STABLE DEMAND VECTORS VIA MIXED RESPONSE STRATEGIES

This section characterizes stationary equilibria that induce mixed response strategies and implement the stable demand vectors in two simple majority games with an unattractive status quo: Games S (Star) and H (House). Throughout the section, attention is restricted to symmetric path independent strategies such that, at very chance move each player is selected as initiator with probability 1/5, only outcomes based on simple majorities are proposed, and the local response strategies satisfy the following two properties, where j and k, respectively, denote the first and second responder listed in a proposal, and q(b) is the quota vector associated with the stationary strategy b.

Property F:

$$j \text{ rejects with probability} \begin{Bmatrix} 0 \\ f_j \\ 1 \end{Bmatrix} \text{ if } h_j(b|accept) \begin{Bmatrix} > \\ = \\ < \end{Bmatrix} q_j(b), \text{ where } f_j \epsilon [0,1].$$

Property S:

$$k \text{ rejects with probability} \begin{Bmatrix} 0 \\ s_k \\ 1 \end{Bmatrix} \text{ if } h_k(b|accept) \begin{Bmatrix} > \\ = \\ < \end{Bmatrix} q_k(b), \text{ where } s_k \epsilon [0,1].$$

Since only three-person coalitions are considered, allow me to adopt the following convention for identifying a strategy's properties. Let the term EFS strategy refer to a stationary strategy that induces a local strategy for the initiator, first responder, and second r sponder, r spectively, that satisfies property E, F, and S. Also, A will be insert d in place of F or S to denote the

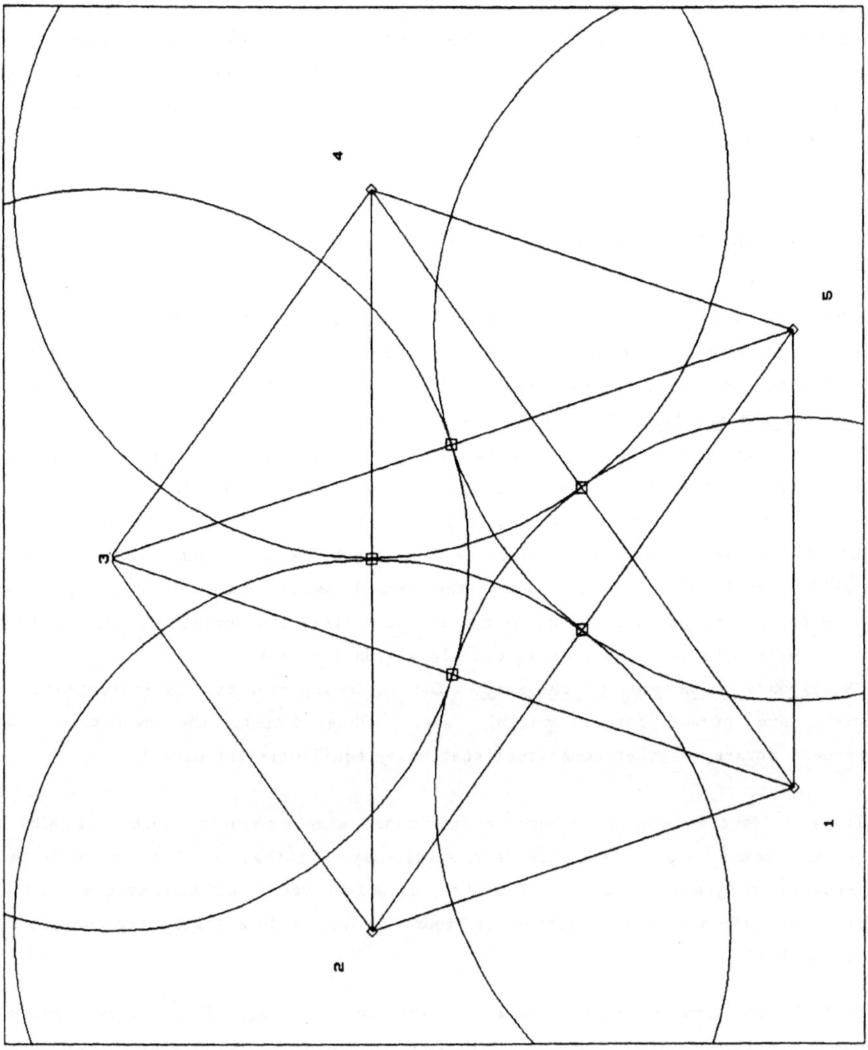

FIGURE 3. STABLE DEMANDS IN GAME 5: THE STAR

special case in which the corresponding r sponder, r, accepts with certainty any proposal for which the expected payoff equals his quota ($f_r=0$ or $s_r=0$, respectively). By this convention, for example, EAS describes a stationary strategy under which the initiator makes extreme proposals, and, in the first and second responder's strategies, $f_j=0$ and $s_k \geq 0$.

4.1 Implementing Stable Demands in Game S: The Star

Given the complete symmetry of Game S as depicted in Figure 3, assume that all players adopt identical strategies, and that every initiator's strategy is uniformly mixed over the proposals in the support set of b at each of his initiator positions.

Scaling the payoffs. Exactly five alternatives are feasible in this game, given the stable demand vector. Recall that each of these points is indexed such that player i strictly prefers the solution point x_i to the other four that are consistent with the demands. Assume that the utility function over the decision alternatives X of each player $i \in N$ is linear in distance from his bliss point and, without loss of generality, scale this function (and the demand vector) such that $u_i(x_i)=1$ and $u_i(x_j)=0$ for any solution point x_j that does not satisfy i's demand. Then, for every player i, $u_i(x_j):=q$ [$\approx 0.545^+$] if x_j exactly yields i's demand..

Stationary strategies in the Star. The following results, as indicated by an asterisk, are proved for a special case. They delimit the symmetric, path independent strategies that constitute stationary equilibria for Game S.

Corollary 2* (quota demands). For the Star under simple majority rule, let q(b) be the quota vector associated with any stationary strategy b that is both path independent and symmetric, and let d denote the unique vector of demands that is both symmetric and semistable, as plotted in Figure 3. Then b is a stationary equilibrium only if q(b)=d.

> Proof. By the first part of Theorem 2, every stationary equilibrium b implements a demand vector d=q(b) that is semistable. Since b is symmetric and path independent, then q(b) is symmetric. But d is the *unique* semistable and symmetric demand vector for the Star. Therefore, the symmetric and path independent strategy b is a stationary equilibrium for the Star *only if* q(b)=d. //

Lemma 10* (initiator's strict advantage). In the Star under simple majority rule, given that default is unattractive, let b denote any stationary strategy equilibrium that is both symmetric and path independent. Then — for every player $i \in N$ and every initiative $p_i \in P_i^1$ of i, and for every position y such that either $y \in P^0$ or $y \in P_j^1$ for some $j \in N \setminus i$ — $h_i(b|p_i) > h_i(b|y)$.

Proof. Since b is an equilibrium and default is unattractive, we may ignore the possibility of infinite play, by Lemma 4. Without loss of generality, given all the symmetry, consider the situation of player 1. Let b_{ki} denote the probability that alternative x_k is enacted as the outcome, given that the game has reached the inititiator position p_i assigned to player i and that b governs behavior. Then, by symmetry and the way payoff functions are scaled, $h_1(b|p_1) = b_{11} + 2qb_{21} = q$. Therefore

$$b_{11} = q(1-2b_{21}) \qquad [1]$$

If player 2 has the initiative, then the expected payoff to 1 is

$$h_1(b|p_2) = b_{12} + q(b_{22} + b_{52})$$

By symmetry, this may be rewritten

$$h_1(b|p_2) = b_{21} + q(b_{11} + b_{31})$$
$$= b_{21} + q[b_{11} + .5(1-b_{11}-2b_{21})]$$

Substituting by [1] in this expression and combining terms yields

$$h_1(b|p_2) = b_{21}[1-q(1+q)] + q(1-q)/2$$

Therefore

$$h_1(b|p) < q \text{ iff } b_{21} < q(1-q)/2[1-q(1+q)] \approx .789^+ \qquad [2]$$

By symmetry, $b_{51}=b_{21}$, and $\Sigma_k b_{k1}=1$. Hence $b_{21} \leq 1/2$ and the strict inequality in [2] obtains. Therefore, $h_1(b|p_2)$ [hence, by symmetry, $h_1(b|p_5)$] is strictly less than 1's quota.

The proof for p_3 (hence p_4) is similar:

$$h_1(b|p_3) = b_{13} + q(b_{23} + b_{53}) = b_{31} + q(b_{21} + b_{31})$$
$$= -b_{21} + .5(1+q)(1-b_{11})$$

Using [1] to substitute for b_{11} in this expression and simplifying yields

$$h_1(b|p_3) = -b_{21}[1-q(1+q)] + (1+q^2)/2, \text{ hence}$$
$$h_1(b|p_3) < q \text{ iff } -b_{21}[1-q(1+q)] + (1+q^2)/2 < q$$

Simplifying yields

$$h_1(b|p_3) < q \text{ iff } b_{21} > [1-q(2+q)]/2[1-q(1+q)] \qquad [3]$$

The numerator of the fraction in [3] is negative, while the denominator is strictly positive; thus, since $b_{21} \geq 0$, $h_1(b|p_3) < q$.

Finally, at any move assigned $y \epsilon P^0$ assigned to Nature, there is positive probability that some $j \neq 1$ will be selected as initiator, and $h_1(b|p_j) < q$ for all $j \epsilon N\backslash 1$. Therefore, $h_1(b|y) < q$. //

Lemma 11* (never propose the worst). For the Star under simple majority rule, let b denote any stationary strategy that is both symmetric and path independent and $q(b)= d$ be the stable demand vector. For any player $i \epsilon N$ and any initiator position $p_i \epsilon P_i^1$, if b assigns positive probability at p_i to any proposal by i to enact a point $x \epsilon X(d)$ such that $u_i(x) < d_i$, then b is *not* an equilibrium.

Proof. Suppose that at p_i player i proposes that such a point x be enacted by coalition C. Then, lest Lemma 5(2) be contradicted, some responder $j \epsilon C\backslash i$ must reject the proposal, h nce, by Lemma 10^*, the expected payoff to i for making the proposal is less than q. Th r fore, b is not optimal at p_i. //

Extreme proposals in the Star. The next result identifies a stationary equilibrium instructing each initiator to propose his favorite solution point.

Proposition 1^*(EFA equilibrium in the Star). In the simple majority rule Star with unattractive status quo, let d be the stable demand vector and, for each player $i \in N$, let x_i denote player i's favorite demand-feasible point. Assume that b is a symmetric global stationary strategy with associated quota vector q(b)=d that satisfies properties E, F and S. Then b is a stationary equilibrium if and only if $s=0$ and $f=f^*$, where $f^* \in (0,1)$ is unique ($f^* \approx 0.8297$).

Proof. *Necessity*. Given any probability $f \geq 0$, assume that $s>0$. Without loss of generality, suppose that b is an equilibrium that implements d and assume that, in accordance with b, player 1 exercises the initiative at p_1 by proposing (x_1, C). Since $x_1 \neq z$, $C \in W$. If $C \cap (3,4) \neq \emptyset$ then the proposal would be rejected, hence, by Lemma 10^*, b would not be an equilibrium. Without loss of generality, given the symmetry, suppose now that 1 proposes the outcome $(x_1, 152)$. Note that $u_2(x_1)=q$, so 2 is indifferent to accepting this proposal, thus ending the game, or rejecting it to take the initiative. Therefore, in accordance with b, 2 rejects the proposal with probability s, so the expected payoff to player 5 for accepting the proposal is

$$h_5(b|\text{accept}) = (1-s)\, u_5(x_1) + s\, h_5(b|p_2) = (1-s)q + s\, h_5(b|p_2)$$

If $s>0$, then the right hand side is strictly less than q, by Lemma 10^*. Thus, 5 will reject the proposal and assume the initiative; hence, again by Lemma 10^*, the proposal is not optimal for player 1 at p_1, contradicting the assumption that b is a stationary equilibrium. Therefore, b is a stationary equilibrium only if $s=0$. Also, given $s=0$ and $u_1(x_1)>d_1$, $f=0$ implies that b is not an equilibrium, by Theorem 3. Therefore, if b is an equilibrium then $s=0$ and $f>0$.

Now assume $s=0$ and $f>0$. Given the way payoffs are scaled and that the game and global strategy b are both symmetric, then the expected payoff to 1 if he has the initiative is

$$h_1(b|p_1) = b_{11}+q(b_{21}+b_{51}) = b_{11}+2qb_{21}$$

Hence, since $h_1(b|p_1)=d_i=q$,

$$b_{11}+2qb_{21} = q \qquad [4]$$

In accordance with the Markov process induced by b,

$$b_{11} = .5f(b_{12}+b_{15}) + (1-f)\text{ , and} \qquad [5]$$

$$b_{21} = .5f(b_{22}+b_{25}) \qquad [6]$$

Since b is symmetric, we may rewrite [5] as

$$b_{11} = fb_{21} + (1-f) \qquad [5']$$

and [6] as

$b_{21} = .5f(b_{11}+b_{31})$, hence

$b_{21} = .5f[b_{11}+.5(1-b_{11}-2b_{21})]$

Rearranging terms in the last equation yields

$$b_{11} = [2b_{21}(2+f)/f] - 1 \qquad [6']$$

which, given $f>0$, is well defined. Solving [5'] and [6'] simultaneously,

$fb_{21} + (1-f) = [2b_{21}(2+f)/f] - 1$, hence

$$b_{21} = f(2-f)/[4+f(2-f)] \qquad [7]$$

where the denominator is greater than zero, given that f is a probability. Substituting this expression for b_{21} in [5'] yields

$$b_{11} = [f^2(2-f)/[4+f(2-f)]] + (1-f) \qquad [8]$$

Substituting [7] and [8] into [4] yields the quadratic

$$-(1+q)f^2 - 2(1-q)f + 4(1-q) = 0 \qquad [9]$$

Given that f is a probability, [9] has a unique solution:

$f^* = .8297161245$

Therefore, b is a stationary equilibrium in the Star only if $s=0$ and $f=f^*$.

Sufficiency. Now suppose that behavior is governed by b and that $s=0$ and $f=f^*$. Without loss of generality, consider again the proposal $(x_1, 152)$. The above calculations have established that, by making this proposal, 1 expects to attain his quota. No other proposal can yield him a greater payoff. Hence 1 has no incentive to deviate from b. Given the opportunity 2 will accept th proposal, and both 5 and 2 will attain their quotas. Therefore 2 has no incentive to deviate from b. On the other hand, if 5 rejects and assumes the initiative, then, again, 5 expects to achieve only his quota. Thus no player in this coalition has incentive to deviate from b and these players were chosen arbitrarily. Therefore, b is a stationary equilibrium. //

The preceeding results have begun to delimit the symmetric, path independent, stationary strategies that implement the stable demand vector for the Star. Lemma 11* eliminates strategies that induce any initiator to propose a solution point that falls short of his demand. Proposition 1* then identifies a very special, knif -edge case ($s=0$ and $f=f^*$) in which the stable demand vector for the Star is implemented by a strategy that induces the initiator to make an extreme proposal.

Modest proposals. The stable demand vector in this game can be implemented also via a global strategy that induces a local strategy at every initiator position such that positive probability is assigned at that position only to proposals that, if accepted, would yield the initiator's quota exactly. Such a strategy b is defined by the following characteristic.

Property Q:

For every player $i \in N$, initiative $y \in P_i^1$, and ption $y' \in A_i(y)$ at y, $b_y(y')>0$ implies that y' is a proposal of an outcome (x,C) satisfying $u_i(x)=q_i(b)$.

The next results pertain to the implementation of the stable demand vector for the Star via a strategy that satisfies properties Q, F, and S.

Proposition 2* (QFS implementation of stable demands in the Star). For the Star under simple majority rule with an unattractive status quo, let d denote the stable demand vector and let b denote any symmetric, stationary global strategy that seeks only a bare majority and exhibits properties Q, F, and S. Then b implements the stable demand vector d *only if*, for every proposal that is induced by b to enact an outcome (x,*ijk*), either of these conditions obtain:

(a) $f>0$, $s=0$, $u_j(x)>d_j$ and $u_k(x)=d_k$, or

(b) $f=0$, $s>0$, $u_j(x)=d_j$ and $u_k(x)>d_k$,

where *j* and *k*, respectively, denote the first and second responders to the proposal.

Proof. Suppose b is a stationary and symmetric QFS equilibrium that implements the stable demand vector d. For any $i\epsilon N$, consider any proposal by *i* to enact an outcome (x, *ijk*), in accordance with b. By Theorem 2 and the definition of a stable demand vector, x satisfies both responders' demands. Since b is a stationary, symmetric equilibrium that satisfies Q and implements d, x exceeds one responder's demand and satisfies the other's exactly. By Theorem 3, at least one responder $r\epsilon\{j,k\}$, if indifferent, must reject the proposal with strictly positive probability. However, if the proposal is rejected with positive probability, then, by Lemma 10*, the initiator's expected payoff does not satisfy his demand, contradicting the assumption that b is a stationary equilibrium that implements the stable demand vector. Therefore, the indifferent responder must accept with certainty, and the responder who, if indifferent, would reject the proposal with positive probability, must be offered more than his demand. //

Therefore, if b is a symmetric QFS strategy that implements the stable demand vector for the Star, then b must be a QAS or QFA stationary equilibrium, as characterized by Proposition 2*. Note that, in accordance with these strategies, the responder who rejects an offer with positive probability when indifferent is offered a solution point that exceeds his quota. The next two results deal with these two types of strategies, thus completing the characterization of symmetric, path independent equilibria that implement the stable demand vector for this game.

Proposition 3* (QFA implementation of stable demands in the Star). For the Star, let b denote a QFS stationary global strategy such that the associated quota vector q(b) equals the stable demand vector, d, and b not only satisfies condition (a) in Proposition 2* but also is uniformly mixed over the two proposals that satisfy this condition. Then b is a stationary equilibrium if and only if $f \geq 2(1-q)$ [$\approx .909^+$].

Proof. Without loss of generality, given the symmetries, suppose that, at initiative $p_1 \epsilon P_1^1$, player 1 proposes the outcome $(x_2, 123)$. As second responder, 3 would accept this proposal with certainty since $s=0$ and $u_3(x_3)=q$. Then 2 also must accept with certainty, since $u_2[\text{accept } (x_2, 123)] = u_2(x_2)=1 > q$. In this situation, neither responder has incentive to deviate from b. Therefore, the expected payoff to 1 for making this proposal is $u_1(x_2)=q$.

To determine whether 1 has reason to deviate from b at p_1 note that only the outcome x_1 could increase 1's payoff, given the others' demands. Again without loss of generality, given symmetry, suppose that c_1 is a global strategy in which player 1 deviates unilaterally from b at position p_1 by proposing the outcome $(x_1, 152)$. Since $s=0$ and $u_2(x_1)=q$, player 2 would accept this proposal with certainty. Therefore, $u_5[\text{accept } (x_1, 152)] = u_5(x_1) = q$, hence 5 is indifferent and, in accordance with b, rejects the proposal with probability $f>0$. If 5 rejects this proposal to take the initiative, then, in accordance with b, 5 proposes $(x_1, 512)$ with probability 1/2 and otherwise proposes $(x_4, 543)$. Player 1 has no incentive to reject the first proposal since 2 would accept it, thus giving 1 his favorite solution point. The second proposal also would be accepted with certainty, in accordance with b. Therefore, the expected payoff to 1 for deviating from b at p_1 by proposing $(x_1, 152)$ is

$$h_1(c_1|p_1) = (1-f)u_1(x_1) + (f/2)[u_1(x_1)+u_1(x_4)] = 1-(f/2).$$

Thus 1 has no incentive to deviate unilaterally from b, hence b is a stationary equilibrium, iff

$$h_1(c_1|p_1)=1-(f/2) \le h_1(b|p_1)=q$$

or, equivalently, $f \ge 2(1-q)$ [$\approx 0.909^+$]. //

Thus, the QFA strategy b implements the stable demands in the Star if and only if it satisfies a rather severe restriction.

The next result indicates that the stable demands in this game can be implemented under considerably less restrictive conditions.

Proposition 4* (QAS implementation of stable demands in the Star). For the Star, let b denote a QFS stationary global strategy such that the associated quota vector q(b) equals the stable demand vector, d, and b both satisfies condition (b) in Proposition 2* and is uniformly mixed over the two proposals that satisfy this conditon. Then b is a stationary equilibrium.

Proof. Without loss of generality, given the symmetries, suppose that 1 has the initiative at p_1 and proposes the outcome $(x_2, 132)$. Since $u_2(x_2)= 1>q$, 2 has no incentive to deviate from b as second responder to this proposal, and would accept with certainty. Consequently, at 3's move as first responder to this proposal, $u_3[\text{accept } (x_2, 132)]=u_3(x_2)=q$. Hence, 3 has no incentive to deviate unilaterally from b and would accept the proposal with certainty. Therefore, $u_1[\text{propose } (x_2, 132)]=q$. Only the outcome x_1 would yield 1 a great r payoff.

Without loss of generality, given symmetry, suppose 1 deviates from b at p_1 by proposing the outcom $(x_1, 152)$. Since $u_2(x_1)=q$, 2 would reject this proposal with probability $s>0$, in accordance with b. If 2 rejected this proposal, he would propose $(x_1, 251)$ with probability 1/2 and, otherwise, $(x_3, 243)$. Either of these proposals would be accepted with certainty. As a result, by accepting 1's proposal as first responder, player 5 expects to achieve

$$u_5[\text{accept } (x_1, 152)] = (1-s)u_5(x_1) + 0.5s[u_5(x_1)+u_5(x_3)] = (1-0.5s)q < q.$$

Therefore, 5 would reject it, then propose $(x_1, 521)$ with probability 1/2 and, otherwise, $(x_4, 534)$; either proposal would be accepted. Hence, in deviating from b at p_1 by proposing $(x_1, 152)$, 1 expected payoff is $0.5[u_1(x_1)+u_1(x_4)] = 1/2 < q$. Neither the initiator nor any responder has incentive to deviate from b. //

In sum, these propositions indicate three types of stationary equilibria that implement the stable demand vector in the Star. However, the unique EFA equilibrium of Proposition 1^* based on extreme proposals requires that a very stringent condition be satisfied by the responders' strategies. In this sense, the equilibria based on modest proposals — the QFA and QAS strategies, respectively, of Propositions 3^* and 4^* — provide a more direct way of implementing the stable demand vector in this game. For this reason I shall now investigate whether analogous strategies can be constructed to implement the stable demands for the House.

4.2 Implementing Stable Demands in Game H: The House

The simple majority game depicted in Figure 4, called the House, is selected for attention here because of its importance to the previous literature on simple spatial games (e.g., McKelvey, Ordeshook and Winer, 1978; Laing and Olmsted, 1978). Again the alternatives that are feasible, given the stable demands shown in the figure, happen to agree with the competitive solution of this game. To implement this demand equilibrium noncooperatively, I shall construct QFS strategies for the House that pres rve the symmetries among players and seek to enact the demand-feasible outcomes via three-person majorities. It is apparent by inspection of the game that players 1 and 5 are in symmetric roles, as are players 2 and 4. The linear payoff functions across the solution points are normalized as in Table 1.

	DEMAND-FEASIBLE POINTS				
PLAYER	x_1	x_2	x_3	x_4	x_5
1:	1	q_1	0	w_1	q_1
2:	q_2	1	q_2	w_2	0
3:	0	q_3	1	q_3	0
4:	0	w_4-w_2	q_4-q_2	1	q_4-q_2
5:	q_5-q_1	w_5-w_1	0	q_5-q_1	1

Note: $q_1 \approx .535^-$, $w_1 \approx .137^-$, $q_2 \approx .669^-$, $w_2 \approx .193^+$, $q_3 \approx .503^-$.

Table 1. Normalized Payoffs of Solution Points for Game H

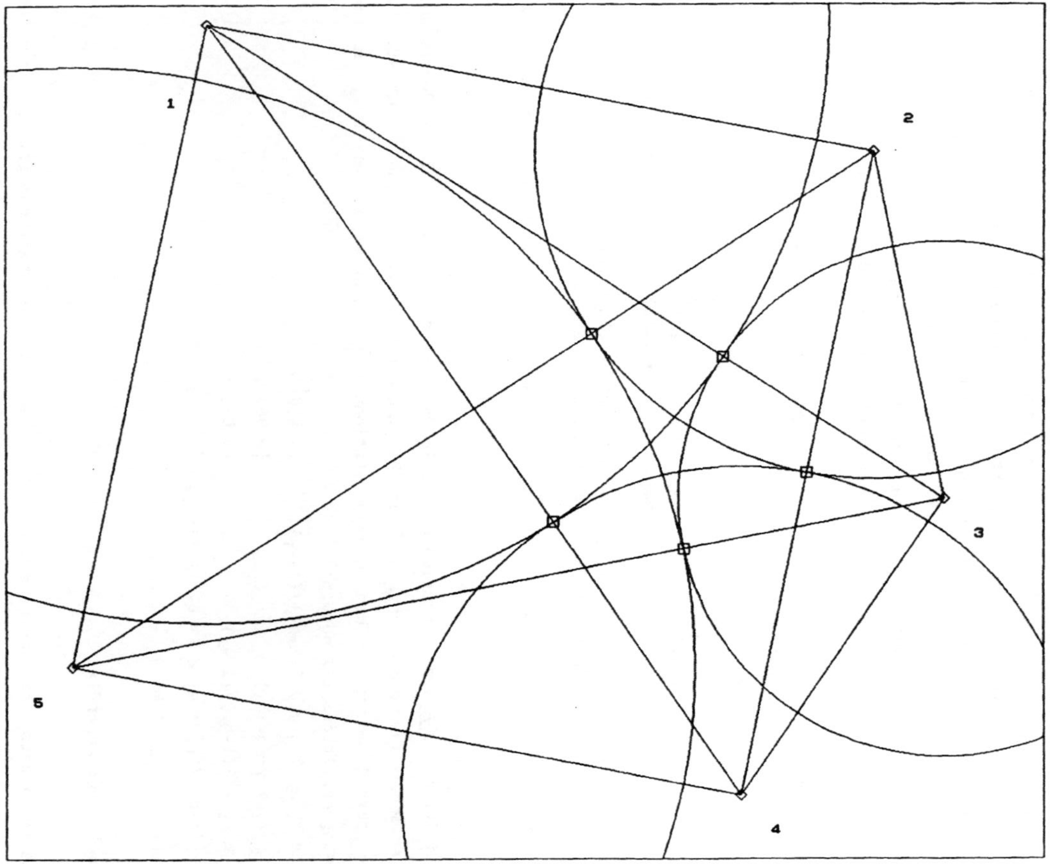

FIGURE 4. STABLE DEMANDS IN GAME H: THE HOUSE

QFA strategies for the House. Given a proposal of any outcome (x, ijk), assume that the local *response* stategies are: the second responder $k \epsilon N$ rejects the proposal with probability $s=0$ if he is indifferent, and the first responder $j \epsilon N$ rejects with probability $f_k > 0$ if he is indifferent, where, by symmetry, $f_1 = f_5$ and $f_2 = f_4$. Also stipulate that the symmetric QFA global strategies at issue induce local strategies at ach *initiative* to propose solution outcomes as specified in Table 2.

INITIATOR:	PROPOSES OUTCOME:	WITH PROBABILITY:	OTHERWISE PROPOSES OUTCOME:
1	$(x_2, 123)$	m_1	$(x_5, 154)$
2	$(x_3, 234)$	m_2	$(x_1, 215)$
3	$(x_2, 321)$	1/2	$(x_4, 345)$
4	$(x_3, 432)$	$m_4 = m_2$	$(x_5, 451)$
5	$(x_4, 543)$	$m_5 = m_1$	$(x_1, 512)$

Table 2. Modest Proposals Induced by a Symmetric QFA Strategy to Enact Three-Person Solution Outcomes in Game H

Given these specifications, the proof given in the appendix constructs this result.

Proposition 5* (QFA implementation of stable demands in the House). Let b denote any such symmetric QFA global strategy for the House with associated quotas $q(b) = d$, the stable demand vector. Then b is a stationary equilibrium if and only if the following conditions all obtain:

$$1 \geq f_1 \geq (1-q_2) + [(1-q_1)/(1-w_1)] \qquad [\approx .870^+]$$

$$1 \geq f_2 \geq 2-(q_1+q_3) \qquad [\approx .963^-]$$

$$1 \geq f_3 \geq 2(1-q_2)/(1-w_2) \qquad [\approx .821^-]$$

$$1-[(1-q_2)/f_1] \geq m_1 \geq (1-q_1)/f_1(1-w_1)$$

$$1-[(1-q_3)/f_2] \geq m_2 \geq (1-q_1)/f_2$$

Proof. See appendix.

If the responders' strategies satisfy the first three of these conditions, then the last two inequalities imply, to three places, that $m_1 \epsilon (.539, .668)$ and $m_2 \epsilon (.465, .503)$. Thus, at such a QFA equilibrium, the first responders reject with high probability any proposal to which they are indifferent, while, except for player 3, each initiator's strategy tends to be biased in favor of proposing a specific solution point that yields his demand: e.g., player 1 leans in favor of proposing x_2, whereas player 2 tends to favor x_1.

QAS strategies for the House. Now assume instead that, given a proposal of any outcome (x, ijk), the local *response* stategies are these: the first responder, j, r jects the proposal with probability $f_j = 0$ if indifferent, and the second responder, k, r j cts with probability $s_k > 0$ if indiff rent, wh re, by symmetry, $s_1 = s_5$ and $s_2 = s_4$.

Also stipulate that the symmetric QAS global strategies at issue induce modest proposals at each *initiative* in accordance with the local strategies specified in Table 3.

INITIATOR:	PROPOSES OUTCOME:	WITH PROBABILITY:	OTHERWISE PROPOSES OUTCOME:
1	$(x_2, 132)$	m_1	$(x_5, 145)$
2	$(x_3, 243)$	m_2	$(x_1, 251)$
3	$(x_2, 312)$	1/2	$(x_4, 354)$
4	$(x_3, 423)$	$m_4 = m_2$	$(x_5, 415)$
5	$(x_4, 534)$	$m_5 = m_1$	$(x_1, 521)$

Table 3. Modest Proposals Induced by a Symmetric QAS Strategy to Enact Three-Person Solution Outcomes in Game H

Given these specifications, the following result is proved in the appendix.

Proposition 6^* (QAS implementation of stable demands in the House). Let b denote any such symmetric QAS global strategy for the House with associated quotas q(b)=d, the stable demand vector. Then b is a stationary equilibrium if and only if both of the following conditions obtain:

$q_2 \geq m_1 \geq (1-q_1)/(1-w_1)$ and

$q_3 \geq m_2 \geq (1-q_1)$

[where $q_2 \approx .669^-$, $(1-q_1)/(1-w_1) \approx .539^+$, $q_3 \approx .503^-$, and $(1-q_1) \approx .465^+$].

Proof. See the appendix.

Note that the two conditions on initiator strategies identified by Proposition 6^* are identical to those of Proposition 5^* if $f_1 = f_2 = 1$. Moreover, both types of equilibria induce initiator strategies that are biased in the same way: player 1 clearly leans towards proposing that x_2 be enacted, whereas player 2, tends to favor proposing x_1. These equilibria also have implications for the probability with which each outcome is reached. For example, if at every chance node each player is chosen as initiator with probability 1/5, and if, in accordance with the QAS equilibrium strategy b of Proposition 6^*, $m_1 = .6$ and $m_2 = .5$, then $b(x_1|y) = b(x_5|y) = .18$, $b(x_2|y) = b(x_4|y) = .22$, and $b(x_3|y) = .20$. Perhaps most importantly, both types of QFS equilibria provide rather natural strategies that implement the demand equilibrium in this game.

4.3 Implementing Stable Demand Vectors via Pure Response Strategies

Moreover, QFS equilibria include as special cases global strategies that induce *pure* response strategies at each responder position in the game. For example, Propositions 3^* and 5^*, respectively, indicate for Game S and H that there is a QFS strategy that implements the stable demand v ctor such that the local response strategy of very player $i \epsilon N$ satisfies $f_i=1$ and $s_i=0$. The stable demand vector for Game F, displayed in Figur 2, also can be implement d by a QFS strat gy that induces at each responder position a pure local strategy. It remains an open question as to whether QFS strategies implement the stable demand vector in other games.

5 CONCLUSIONS

The foregoing results demonstrate that the noncooperative approach used by Selten to analyze bargaining in characteristic function games extends naturally to simple spatial games without sidepayments. The results indicate that Albers' cooperative solution, the demand equilibrium, can be implemented in a direct way via path independent equilibrium strategies that govern each player's negotiations in a corresponding noncooperative bargaining game that is represented in extensive form. Th se noncooperative equilibria add precision to the characterization of not only the cooperative solution but also the process through which this solution is reached in rational play. For example, the equilibrium strategies of Propositions 5^* and 6^* specify what choices would be made at each decision node in the game tree and yield immediate implications about the probabilities with which various outcomes of the game will be reached. These features illustrate some important advantages to be gained from analyzing a bargaining problem as a noncooperative game. In this way one may hope to rest cooperative solution theory on first principles pertaining to strategies employed by rational players to reach a cooperative agreement, rather than on arbitrary dicta about "reasonable" properties that the cooperative solution should satisfy, or on informal appeals to intuition about the process through which the cooperative solution is reached. One hopes that it will be possible to *derive* the cooperative solution by solving the noncooperative bargaining game, but this requires advances in the theory about how rational players select among alternative noncooperative equilibria (Harsanyi and Selten, 1988). At the v ry least, noncooperative equilibrium analysis can be used to determine whether or not a cooperative solution concept is consistent with a detailed analysis of the bargaining process. It also provides a crisp benchmark for identifying ways in which cognitively limited human beings systematically deviate in their bargaining from perfectly rational play.

APPENDIX

Proof of Proposition 5^*. To avoid unnecessary repetition, note that, by conforming to b, each responder will accept any proposal that is consistent with b, thereby attaining a payoff that satisfies his demand (quota). A responder has incentive to deviate unilaterally from b if and only if he exp cts to achieve a greater payoff by rejecting the prop sal and then, as *initiator*, making a proposal that deviates from b. Therefore I need to consider only unilateral deviations from b that might tempt the *initiator*, under the assumption that no responder (*qua* responder) has incentive to deviate from b. Each such deviation will be examined and necessary and sufficient conditions constructed so that the initiator has no incentive to deviate in this way from b. Without loss of generality, since b is stationary and symmetric, I consider one initiative each for players 1, 2, and 3.

Initiator-1. Player 1 has two deviations from b at his initiative p_1 that might tempt him.

First, suppose that 1 proposes $(x_1,125)$. Since 5 would accept, 2 is indifferent and, in accordance with b, rejects the proposal with probability $f_2>0$. Given the initiative, 2 would propose $(x_3,234)$ with probability m_2 and, otherwise, $(x_1,215)$; 2's proposal would be accepted. Therefore the expected payoff to 1 for making this proposal is

$$h_1[\text{propose } (x_1,125)] = (1-f_2)u_1(x_1) + f_2[m_2u_1(x_3)+(1-m_2)u_1(x_1)]$$
$$= (1-f_2)+f_2(1-m_2) = 1-f_2m_2$$

Then $h_1[\text{propose } (x_1,125)] \le q_1$, hence 1 has no incentive to deviate from b in this way, if and only if $f_2m_2 \ge (1-q_1)$.

Second, assume that 1 proposes $(x_1,152)$. Since 2 would accept and $u_5(x_1)=q_5$, 5 would reject 1's proposal with probability $f_5=f_1>0$. Given the initiative, 5 would propose $(x_4,543)$ with probability $m_5=m_1$ and, otherwise, $(x_1,512)$; either proposal would be accepted. Hence

$$h_1[\text{propose } (x_1,152)] = (1-f_1)u_1(x_1) + f_1[m_1u_1(x_4)+(1-m_1)u_1(x_1)]$$
$$= (1-f_1)+f_1[m_1w_1+(1-m_1)] = 1-f_1m_1(1-w_1)$$

Then $h_1[\text{propose } (x_1,125)] \le q_1$ if and only if $f_1m_1 \ge (1-q_1)/(1-w_1)$.

Therefore, 1 has no incentive to deviate in either way from b at his initiative p_1 if and only if both $f_1m_1 \ge (1-q_1)/(1-w_1)$ and $f_2m_2 \ge (1-q_1)$. By symmetry, an analogous conclusion applies to any initiative of player 5.

Initiator-2. There are two deviations that might tempt 2 at $p_2\epsilon P_2^1$.

First, suppose 2 deviates from b by proposing $(x_2,213)$. In accordance with b, the second responder would accept, yielding the first responder, 1, his quota. Hence 1 would reject the proposal with probability $f_1>0$ and, given the initiative, would propose $(x_2,123)$ with probability m_1 and, otherwise, $(x_5,154)$; 1's proposal would be accepted. Hence

$$h_2[\text{propose } (x_2,213)] = (1-f_1)u_2(x_2) + f_1[m_1u_2(x_2)+(1-m_1)u_2(x_5)]$$
$$= (1-f_1)+f_1m_1 = 1-f_1(1-m_1)$$

Consequently, $h_2[\text{propose } (x_2,213)] \le q_2$ if and only if $f_1(1-m_1)] \ge (1-q_2)$.

Second, assume that 2 proposes $(x_2,231)$. Since 1 would accept, yielding 3 his quota, 3 would reject this proposal with probablity $f_3>0$. Given the initiative, 3 would propose $(x_2,321)$ or $(x_4,345)$, each with probability 1/2, and the proposal would be accepted. H nce

$$h_2[\text{propose } (x_2,231)] = (1-f_3)u_2(x_2) + (f_3/2)[u_2(x_2)+u_2(x_4)]$$

$$= (1-f_3)+(f_3/2)(1+w_2) = 1-(f_3/2)(1-w_2)$$

Then h_2[propose $(x_2,231)$] $\leq$ q_2 if and only if $f_3 \geq 2(1-q_2)/(1-w_2)$.

Th refore, 2 has no incentive to deviate from b at p_2 if and only if $f_1(1-m_1) \geq (1-q_2)$ and $f_3 \geq 2(1-q_2)/(1-w_2)$. By symmetry, the analogous conclusion obtains when 4 has the initiative.

Initiator–3. Although 3 also has two tempting deviations from b at his initiative, they are symmetric. Therefore, without loss of generality, assume that 3 deviates from b by proposing $(x_3,324)$. Since 4 would accept this proposal and 2 is indifferent, 2 would reject the proposal with probability $f_2>0$. Given the initiative, 2 would propose $(x_3,234)$ with probability m_2, $(x_1,215)$ otherwise, and the proposal would be accepted. Hence

$$h_3[\text{propose } (x_3,324)] = (1-f_2)u_3(x_3) + f_2[m_2u_3(x_3)+(1-m_2)u_3(x_1)]$$
$$= (1-f_2)+f_2m_2 = 1-f_2(1-m_2)$$

Then h_3[propose $(x_3,324)$] $\leq q_3$ if and only if $f_2(1-m_2) \geq (1-q_3)$.

Initiator–$i\epsilon N$. Therefore, it follows from the above results for the various initiators that b is a stationary equilibrium if and only if the following system of inequalities is satisfied:

$$f_1m_1 \geq (1-q_1)/(1-w_1) \text{ and } f_1(1-m_1) \geq (1-q_2) \qquad [5.1]$$
$$f_2m_2 \geq (1-q_1) \text{ and } f_2(1-m_2) \geq (1-q_3) \qquad [5.2]$$
$$f_3 \geq 2(1-q_2)/(1-w_2) \qquad [5.3]$$

Conditions [5.1] and [5.2] may be replaced by their respective solutions:

$f_1 \geq (1-q_2) + [(1-q_1)/(1-w_1)]$ and $1-[(1-q_2)/f_1] \geq m_1 \geq (1-q_1)/f_1(1-w_1)$

$f_2 \geq 2-(q_1+q_3)$ and $1-[(1-q_3)/f_2 \geq m_2 \geq (1-q_1)/f_2$. //

<u>Proof of Proposition 6*</u>. As in the proof of Proposition 5*, only deviations from b by initiators 1, 2 and 3 need be analyzed to construct conditions that are necessary and sufficient for equilibrium.

Initiator–1. Player 1's only temptation for a deviation from b is to propose that his favorite solution point, x_1, be enacted. Given the others' demands, 1 has only two such deviant proposals.

First, suppose 1 deviates from b by proposing $(x_1,125)$. As second responder, player 5 would be indifferent between accepting this proposal or rejecting it to take the initiative. Hence, in accordance with b, 5 would reject the proposal with probability $s_5=s_1>0$ and, as initiator, propose instead the outcome $(x_4,534)$ with probability $m_5=m_1$ and, otherwise, propose $(x_1,521)$. Either proposal would be accepted. Therefore, by accepting 1's proposal as first responder, player 2 expects the payoff

$$h_2[\text{accept } (x_1,125)] = (1-s_5)u_2(x_1) + s_5[m_1u_2(x_4) + (1-m_1)u_2(x_1)]$$
$$= (1-s_5)q_2 + s_5[m_1w_2 + (1-m_1)q_2]$$

Cons quently, since $s_5>0$ and $q_2>w_2>0$, h_2[accept $(x_1,125)$] $< q_2$, hence 2 rejects this proposal, iff $m_1>0$. Given $m_1>0$, 2 would reject this proposal, then propose $(x_3,243)$ with probability m_2 and, otherwise, $(x_1,251)$. Either proposal would be accepted. Thus, at 1's initiative,

$$h_1[\text{propose } (x_1,125)] = m_2u_1(x_3) + (1-m_2)u_1(x_1) = 1-m_2$$

Therefore, 1 has no incentive to deviate from b in this manner if and only if both $m_1>0$ and $1-m_2 \le q_1$.

Second, suppose that 1 deviates at his initiative p_1 by proposing $(x_1,152)$. Player 2 would reject this proposal with probability $s_2>0$ and, given the initiative, propose $(x_3,243)$ with probability m_2 and, otherwise, $(x_1,251)$. Either of 2's counterproposals would be accepted. Knowing this, by accepting 1's proposal as first responder, 5 expects the payoff

$$h_5[\text{accept } (x_1,152)] = (1-s_2)u_5(x_1) + s_2[m_2u_5(x_3) + (1-m_2)u_5(x_1)]$$
$$= (1-s_2)q_5 + s_2(1-m_2)q_5 = (1-s_2m_2)q_5$$

Thus, since $s_2>0$, player 5, in accordance with b, would reject this offer if and only if $m_2>0$. Given the initiative, 5 would then propose $(x_4,534)$ with probability $m_5=m_1$ and, otherwise, $(x_1,521)$. Either proposal would be accepted. Consequently, the expected payoff to 1 for deviating from b in this way is

$$h_1[\text{propose } (x_1,152)] = m_1u_1(x_4) + (1-m_1)u_1(x_1) = m_1w_1 + (1-m_1)$$
$$= 1-m_1(1-w_1)$$
$$\le q_1 \text{ iff } m_1 \ge (1-q_1)/(1-w_1).$$

Therefore, 1 has no incentive to make this proposal iff m_1 satisfies the last condition.

Combining these results, when player 1 exercises the initiative, neither any responder nor 1 has incentive to deviate from b if and only if the following conditions both obtain: $m_1 \ge (1-q_1)/(1-w_1)$ and $m_2 \ge (1-q_1)$. By symmetry, these same conditions together are necessary and sufficient for b to be an equilibrium at any initiator position assigned to player 5.

Initiator-2. Player 2 has two deviant proposals that might be tempting when he has the initiative.

First, suppose 2 deviates from b by proposing $(x_2,213)$. Player 3 would reject this proposal with probability $s_3>0$ and, given the initiative, would propose $(x_2,312)$ with probability 1/2 and, otherwise, $(x_4,354)$. Either proposal would be accepted. Therefore, as first responder to 2's proposal, player 1 expects

$$h_1[\text{accept } (x_2,213)] = (1-s_3)u_1(x_2) + (s_3/2)[u_1(x_2)+u_1(x_4)]$$
$$= (1-s_3)q_1 + (s_3/2)(q_1+w_1)$$
$$< q_1 \text{ since } s_3>0 \text{ and } w_1< q_1.$$

Therefore 1 would reject 2's proposal and, instead, propose $(x_2,132)$ with probability m_1 and, otherwise, $(x_5,145)$; 1's proposal would be accepted. Hence,

$$h_2[\text{propose } (x_2,213)] = m_1u_2(x_2) + (1-m_1)u_2(x_5) = m_1$$

and thus 2 has no incentive to deviate in this way if and only if $m_1 \le q_2$. Assume hereinafter that m_1 does not exceed this upper bound.

Second, suppose 2 deviates from b at p_2 by proposing $(x_2,231)$. Player 1 would reject this proposal with probability $s_1>0$ and, given the initiative, would propose $(x_2,132)$ with probability m_1 and, otherwise $(x_5,145)$. Either proposal would be accepted. Therefore,

$$h_3[\text{accept }(x_2,231)] = (1-s_1)u_3(x_2) + s_1[m_1u_3(x_2) + (1-m_1)u_3(x_5)]$$
$$= (1-s_1)q_3 + s_1m_1q_3$$
$$< q_3 \text{ since } s_1>0 \text{ and } m_1 \le q_2 < 1.$$

Therefore 3 will reject 2's offer and propose $(x_2,312)$ or $(x_4,354)$, each with probability 1/2. Either proposal would be accepted. Hence,

$$h_2[\text{propose }(x_2,231)] = [u_2(x_2)+u_2(x_4)]/2 = (1+w_2)/2 \approx .596^+ < q_2$$

and 2 has disincentive to deviate from b by proposing $(x_2,213)$.

Thus, 2 has no incentive to deviate from b at p_2 in any way iff $m_1 \le q_2$. By symmetry, the analogous conclusion applies at any initiator position assigned to player 4.

Initiator=3. As initiator, player 3 also may be tempted by two deviations from b, but these cases are symmetric. Therefore, without loss of generality, suppose that player 3 deviates by proposing $(x_3,324)$. Player 4 would reject this proposal with probability $s_4=s_2$ and, given the initiative, would propose $(x_3,423)$ with probability $m_4=m_2$, $(x_5,415)$ otherwise, and the proposal would be accept d. Hence,

$$h_2[\text{accept }(x_3,324)] = (1-s_2)u_2(x_3) + s_2[m_2u_2(x_3) + (1-m_2)u_2(x_5)]$$
$$= [1-s_2(1-m_2)]q_2$$
$$< q_2 \text{ iff } m_2<1, \text{ given that } 0<s_2.$$

Therefore, provided that $m_2<1$, player 2 would reject 3's proposal and, given the initiative, propose $(x_3,243)$ with probability m_2 and, otherwise, $(x_1,251)$. Either proposal would be accepted. Consequently,

$$h_3[\text{propose }(x_3,324)] = m_2u_3(x_3) + (1-m_2)u_3(x_1) = m_2.$$

Thus player 3 has no incentive as initiator to deviate from b iff $m_2 \le q_3$.

Initiator=$i\epsilon N$. Taken together, the results obtained when each player njoys the initiative imply that the specified symmetric QAS strategy for the House is an equilibrium if and only if it always induces mixed local initiator strategies that satisfy both $(1-q_1)/(1-w_1) \le m_1 \le q_2$ and $(1-q_1) \le m_2 \le q_3$. //

REFERENCES

Albers, Wulf (1975). Zwei Lösungskonzepte für kooperative Mehrpersonenspiele, die auf Anspruchsniveaus der Spieler basieren. *OR-Verfahren* 21:1-13.

___________(1987). Aspirations and aspiration adjustment in location games. Working Paper 159, Institut für Mathematische Wirtschaftsforschung, University of Bielefeld.

Bennett, Elaine (1988). Unique opportunities and the A-stable set: A cooperative and a noncooperative model of bargaining in NTU games. Working paper 10, Zentrum fur interdisziplinare Forschung, University of Bielefeld.

________and Mark Winer (1983). The bargaining equilibrium for spatial games. Draft presented at the annual meetings of the Public Choice Society, Savannah, GA, March 24-26.

Forman, Robin, and James D. Laing (1982). Metastability and solid solutions of cooperative games. *Mathematical Social Science* 2: 397-420.

Harsanyi, John, and Reinhard Selten (1988). *A General Theory of Equilibrium Selection in Games*. Cambridg , MA: MIT Pr ss.

Laing, James D., Sampei Nakabayashi, and Benjamin Slotznick (1983). Winners, blockers, and the status quo: Simple collective decision games and the core. *Public Choice* 40:317-321.

______and Scott Olmsted (1978). An experimental and game theoretic study of committees. In Peter C. Ordeshook, ed., *Game Theory and Political Science*. New York: NYU Press, 215-281.

McKelvey, Richard D., Peter C. Ordeshook, and Mark Winer (1978). The competitive solution for n-person games without transferable utility, with an application to committee games. *American Political Science Review* 72: 599-615.

Selten, Reinhard (1975). Reexamination of the perfectness concept for equilibrium points in extensive games. *International Journal of Game Theory* 4: 22-55.

_______(1981). A noncooperative model of characteristic-function bargaining. In V. Boehm and H. Nachthanp, eds., *Essays in Game Theory and Mathematical Economics in Honor of Oskar Morgenstern*. Wissenschaftsverlag Bibliographisches Institut Mannheim, Wien - Zürich, 131-151.

Winter, Eyal (1989). An axiomatization of the stable and semistable demand vectors by the reduced game property. Projektbereich A, Discussion Paper A-254, University of Bonn.

DEMAND COMMITMENT BARGAINING:

- The Case of Apex Games[1] -

by

Elaine Bennett and Eric van Damme

Abstract

An apex game is a bargaining situation in which there is one major (apex) player and n "minor" players. The only profitable coalitions contain either the apex player and any one of the minor players or else all of the minor players. The demand commitment model is a bargaining procedure, i.e. an extensive form game. This paper investigates the payoffs that result (as subgame perfect outcomes) for apex games when players use the demand commitment bargaining procedure. We show that whenever the apex player has the first move he forms a coalition with a minor player and obtains the fraction $(n-1)/n$ of the coalition's value while his (minor-player) partner obtains the remaining $1/n$. When a minor player has the first move he either forms a coalition with the apex player (and obtains $1/n$) or else forms a coalition with all of the remaining minor players. When this minor-player coalition forms there are many subgame perfect payoff distributions. A refinement of subgame perfection is proposed and is shown to select a unique payoff distribution ($1/n$ for each minor player) for the minor-player coalition.

1 Introduction

In this paper we present a noncooperative model for bargaining in characteristic function games, and explore its implications for the class of apex games. An *apex game* is a bargaining situation

[1]Both authors are grateful to the Center for Interdisciplinary Research at the University of Bielefeld for its support and hospitality. They thank Bill Zame for helpful discussions and for suggestions that improved the presentation. The first author is also grateful to the CentER for Economic Research at Tilburg University for its support and hospitality and to the National Science Foundation grant for its support through grant SES-8706631.

in which there is one major (apex) player and n minor players; only the coalitions consisting of the apex player and any one of the minor players, and the coalition consisting of all of the minor players are profitable, and these coalitions have equal value. (It is convenient, and involves no loss of generality, to normalize payoffs so that the value of each of these coalitions is n, the number of minor players.) In our model, bargaining takes place by means of a procedure which we call *demand commitment*.[2] Each player in turn may set a price (a payoff demand, expressed in utility terms) for his participation in any coalition. Having set his price a player can form a coalition if his partners in the coalition have already named their prices and the coalition can afford these prices. The game terminates as soon as one coalition is formed; players not belonging to the "successful" coalition receive a payoff of zero.

Our purpose in this paper is to investigate the effects of the existence of alternative coalitions on bargaining outcomes. The demand commitment model is well-suited to this purpose: In this model, each player must make a trade-off between the higher payoff he might obtain by setting a higher price and the possibility of pricing himself out of the market entirely. We focus on apex games since they represent the simplest situations in which competition for partners is relevant.[3]

Considerable attention has been focused on providing extensive form models for bargaining in characteristic function games. The seminal work in this direction is that of Nash (1953) on the two-person simple bargaining problem. Nash argued for the importance of modeling the bargaining process explicitly by means of an extensive form game, and viewed cooperative and noncooperative approaches as complementary. The cooperative theory avoids any specification of the bargaining process, hence, it achieves great generality, is tractable and easy to apply. However, it is typically difficult to assess (the reasonableness of) the intuitions (axioms) underlying a cooperative solution, or the range of situations in which it should apply, without having a specific bargaining procedure in mind. A cooperative solution is in serious doubt if it is not compatible with some sensible noncooperative bargaining procedure. Conversely, a noncooperative bargaining procedure is not likely to be sensible if its outcomes are not supported by (the intuition of) some cooperative solution. Seminal papers on extensive form models for situations with many players and many potential coalitions (i.e., characteristic function games) are those of Harsanyi (1971) and Selten (1981). These papers exemplify the insight that can be gained by reinterpreting

[2]The demand commitment procedure was first suggested by Reinhard Selten. We are grateful to him stimulating us to carry out the research described in this paper.

[3]An apex game can be viewed as a unanimity game played among the minor players augmented with a single outside option for each player (that of forming a coalition with the apex player).

cooperative solution concepts as equilibria of noncooperative games. As is the case in many other models where time pressure plays no role, the models of Harsanyi and Selten are plagued with a mulitude of equilibria; Harsanyi and Selten use equilibrium selection to argue for "sensible" outcomes.

More recent papers in this area include those of Binmore (1985) and Chatterjee et al (1990). Both papers can be viewed as offspring of a mating between Rubinstein's noncooperative model of the two-person simple bargaining problem and Selten's noncooperative model of characteristic function games. Rubinstein (1982) describes a model in which two players alternate in making offers on the division of their total payoff, until agreement is reached. Pressure to reach agreement comes because players discount future payoffs. Selten (1981) describes a model in which many players bargain over which coalition to form and the division of payoffs within the coalitions. In Selten's model (the proposal-making model) players (sequentially) propose coalitions and feasible divisions of the payoffs obtainable in these coalitions. Binmore's "telephone bargaining" model and Chatterjee et al's model explore the effects of time pressure in the proposal-making model. In Section 7 we discuss the results obtained by Chatterjee et al, for now it suffices to remark that Binmore (1985) argues that an instability is built into this model since it imposes constraints that players would like to violate.

In our model and Binmore's (1985) "market demand" model players make demands and not proposals. This distinction is important because commitment to a demand means that bargaining takes place in many different coalitions simultaneously (a player's price is a demand for payoff which is uniform across *all* of his potential coalitions while a proposal specifies payoffs for only one coalition). The importance of this difference can be seen by contrasting the outcomes of Binmore's "telephone bargaining" and his "market demand" model.

Our model is also different in a second important way: competition among coalitions for players and not time pressure is the driving force in determining the division of payoffs and consequently in determining which coalitions are going to form. See Section 7 for further discussion of this issue.

Demand commitment games are extensive form games with perfect information. As is well-known, finite extensive form games with perfect information that have different payoffs for each player at different terminal nodes, have unique subgame perfect equilibria. Although demand commitment games are games with perfect information and finite horizons, they are not finite since each player's payoff demand can be any nonnegative real number. Moreover, for each player there will be many terminal nodes that yield the player the same payoff. As a consequence, it will typically be the case that players are faced with choices among which *they* (but not the other players) are indifferent and in such cases multiplicity of subgame perfect equilibria may result.

We show that such multiplicity does not arise when the apex player moves first or when there are only two minor players: in the first instance the apex player forms a coalition with one of the minor players and, in the second instance, the first two players to move — whether or not one of them is the apex player — form a coalition. In either case successful[4] minor players each obtain a payoff of 1 and the apex player, when successful, obtains a payoff of $n-1$. In cases in which there are at least three minor players and a minor player moves first, however, one can generate a continuum of subgame perfect equilibrium outcomes by varying players' conjectures about the choices other players will make when faced with choices among which they are indifferent.

In our view, not all conjectures about how players will resolve their indifferences are equally convincing. We introduce, in Section 5, a refinement of subgame perfection, called ***credible subgame perfect equilibrium*** which is based on the idea that players strategically exploit their indifferences. We assume that a player i can credibly threaten player j to resolve his indifferences in a particular way, if carrying out the threat will not reduce i's payoff, will reduce j's payoff and if j's best response to the threat will increase i's payoff.

In Section 6 we show that allowing credible threats drastically reduces the number of equilibrium outcomes. The only outcomes resulting from credible subgame perfect strategies are: formation of the coalition of all minor players with each minor player obtaining a payoff of 1 and formation of a coalition consisting of the apex player and one of the minor players with the apex player obtaining a payoff of $n-1$ and the minor player obtaining a payoff of 1.

Section 7 concludes the paper with a discussion the relationship between our model and other cooperative and noncooperative models for apex games.

2 Apex Games and Demand Commitment

Let $N_0 = \{0, 1, \ldots, n\}$ be the set of players. We call the players in $N = \{1, \ldots, n\}$ the *minor* players and call player 0 the *apex* player. The characteristic function of the apex game is given by

[4] "Successful" means that the player is a member of the coalition that forms.

$$v(S) = \begin{cases} n \text{ if } S = \{0, j\} \text{ with } j \in N \text{ or } S = N \\ 0 \text{ otherwise} \end{cases}$$

In demand commitment games play proceeds by each player, in turn, setting a price (a demand for payoff in utility terms) for his coalitional participation. Having set his price a player can form a coalition if his partners in the coalition have named their prices and the coalition can afford these prices. If the player doesn't form a coalition, he selects, as the player to have the next move, any player who hasn't announced his price. Before stating the rules more formally, we introduce the following notation.

Let p_j denote the price demanded by player j. The coalition S is *feasible* for i if $v(S) = n$ (i.e., S is a profitable coalition), S contains player i, every player in $j \in S$ has already announced his price, and $\sum_{j \in S} p_j \leq v(S)$.

The demand commitment game is played according to the following rules. Nature randomly selects a player from N_0 to move first. This player announces a demand (a nonnegative real number) which is then made known to all players and selects a player (any player who has not yet moved) to move next. When it is player i's turn to move, he announces a demand and either forms a feasible coalition or selects a player to move next. (If every other player has already moved and the last player to announce a demand does not form a coalition, the game ends and each player obtains a payoff of 0.) If a feasible coalition does form the game ends: each player in the coalition is paid his demand while players who are left out obtain nothing.

The demand commitment game is a finite length extensive form game with perfect information and continuous action spaces. (We prefer to work with the continuum rather than with discrete money units to avoid making additional case distinctions.) The solution concept we will employ is subgame perfect equilibrium in pure strategies, i.e. we will be looking for a pure strategy profile that induces a Nash equilibrium in every subgame. As this concept has by now become a standard element in the economist's toolkit there is no need to be more formal at this point. There is just one special feature of SPE in extensive form games with continuous action spaces of which the reader should be aware. This is the fact that in such a game not every SPE of a subgame can be extended to an SPE of the overall game because ties cannot always be broken in an arbitrary way. Requiring a strategy combination to be an SPE of the overall game may constrain a player to choose one action (consistent with one SPE of the subgame) rather than another action (consistent with a different SPE of the subgame) in situations where the player making the decision is indifferent between the two actions, this in order to ensure that players moving earlier in the game indeed have best responses. The simple analysis of the 3-person 3-cakes problem in Section 3 suffices to acquaint the reader with this peculiar property of SPEa in continuum games. For general results on SPEa in continuum games the reader is referred to Hellwig et al (1990).

3 The Three Player / Three Equal Cakes Problem

We begin with the simple case of an apex game with two minor player to familiarize the reader with the basic steps of the general argument. The apex game with two minor players is special because the apex player is not in a distinguished role — any two players can form a profitable coalition. In the literature, such bargaining situations have become known as 3-player/3-cakes problems (see for instance Binmore (1985)). Since apex games with two minor players are entirely symmetric, we deviate from our general notation and refer to the players simply as 1, 2 and 3. For this case the characteristic function is given by

$$v(S) = \begin{cases} 2 & \text{if } |S| = 2 \\ 0 & \text{otherwise.} \end{cases}$$

We will show that the game has a unique[5] SPE and that this SPE results in the outcome where the two first moving players divide their cake equally.

The SPE is found by backwards induction, so w.l.o.g. let us assume that the players 1 and 2 have already moved and that they have demanded prices p_1, resp. p_2. The optimal choice for player 3 is to form a coalition with the player that asked the lowest price, at least when $\min(p_1, p_2) < 2$. If both players have demanded the same price, he may pick either one.

Next, consider the decision problem faced by player 2 after player 1 has demanded p_1. Assume $p_1 > 0$. If player 2 accepts[6] the demand of player 1 his payoff is $2 - p_1$. If player 2 rejects this demand, then he can (in SPE) count on the cooperation of player 3 only if he demands $p_2 \leq p_1$,

[5]There is a recurring nonuniqueness which we ignore and will continue to ignore: the nonuniqueness which occurs when a player must choose among otherwise indistinguishable players. In this case the first player must select one of two indistinguishable players, in subsequent sections the apex player must select one among several indistinguishable minor players.

[6]By "player i accepts the demand of player j" we mean that player i accepts the residual payoff in the coalition $\{i, j\}$ as his price and forms the coalition $\{i, j\}$.

hence accepting p_1 is strictly optimal if $2 - p_1 > p_1$, or $p_1 < 1$. If $p_1 > 1$, then player 2 is sure that player 3 will accept his demand if $p_2 < p_1$, hence player 2 can almost (but not quite) guarantee p_1. We see that, if it would be the case that in the subgame where both player 1 and player 2 have demanded p_1 with $p_1 > 1$, player 3 would form the coalition with player 1, then player 2 would not have a best response after player 1 has demanded $p_1 > 1$. Hence, to guarantee existence of an SPE, player 3 has to break ties in the favor of player 2 if player 1 demands more than 1. Consequently, in any SPE, if player 1 demands $p_1 > 1$, then player 2 puts $p_2 = p_1$ and player 3 forms the coalition with player 2. By the same argument one sees that the subgame with $p_1 = 1$ admits two SPEa: player 2 accepts the demand of player 1 or player 2 matches player 1's demand and 3 forms the coalition with 2.

Finally, consider player 1's decision problem. The above analysis has shown that player 1 will end up with zero if he demands $p_1 > 1$ and that any demand $p_1 < 1$ will be accepted by player 2. Again we have that player 1 does not have a best response unless player 2 breaks the tie in 1's favor if $p_1 = 1$. Hence, player 2 should break the ties in this way, and we have shown

Theorem 1 . *For the 3-person apex game, the demand commitment model has a unique SPE. The outcome of this equilibrium is that the two first moving players agree on an equal split of their cake.*

4 The n-player Apex Game – Subgame Perfect Equilibria

In this section we derive two main results for apex games with at least 4 players

(i) if the apex player starts the game there is a unique SPE outcome: the apex player demands $n - 1$ from some minor player and the latter accepts, (i.e. he demands 1 and forms the coalition with the apex player)

(ii) if a minor player starts the game there are infinitely many SPE outcomes and the first moving minor player may obtain any payoff between 1 and n (hence, he may actually obtain the entire cake).

We proceed by proving a series of lemmas about SPEa in the various continuation games that may arise. It will be notationally convenient (and without loss of generality) to assume that the minor players have to move in the order $1, 2, \ldots, n$. (Alternatively one simply doesn't fix the players' names in advance, player i is the ith moving minor player.) For a vector $p = (p_1, \ldots, p_{i-1}) \in \mathbf{R}_+^{i-1}$

with $i \leq n$ it is also convenient to write

$$P_i = \begin{cases} 0 & \text{if } i = 1 \\ \sum_{j=1}^{i-1} p_j & \text{if } i > 1 \end{cases} \tag{4.1}$$

$$p^i = \begin{cases} \infty & \text{if } i = 1 \\ \min_{j<i} p_j & \text{if } i > 1 \end{cases} \tag{4.2}$$

$$r_i(p) = (n - P_i)/(n - i + 1) \tag{4.3}$$

Note that $r_i(p)$ is the per capita remainder in the coalition N of all minor players when player i has to move. p^i is the minimal demand of all minor players that moved before minor player i: If i wants the cooperation of the apex player then he can demand at most p^i. Intuitively it is therefore clear that player i's optimal decision will depend on p^i and $r_i(p)$. We will analyse only those subgames with $0 \leq p_0 \leq n$, $0 \leq p_i \leq n$ (for all $i \in N$) and $r_n(p) \geq 0$, this to avoid having to make uninteresting case distinctions. The reader can easily verify that subgames that do not satisfy these restrictions will not be reached by any SPE.

Lemma 1 . *Assume that the apex player (player 0) has already moved and that he demanded p_0. Consider the subgame starting with player i after each player $j < i$ has demanded p_j. Then*

(i) If $r_i(p) < n - p_0$, then in the unique SPE of the subgame player i forms the coalition with the apex player.

(ii) If $r_i(p) > n - p_0$, then in the unique SPE of the subgame player i asks for $n - P_i - (n - p_0)(n - i)$ and the coalition N forms with each remaining minor player $j > i$ asking for $n - p_0$.

(iii) If $r_i(p) = n - p_0$, then there are exactly two SPEa in the subgame: Player i demands $p_i = n - p_0$ and either

(a) forms the coalition with the apex player or

(b) selects another minor player and every remaining minor player demands $p_j = n - p_0$ and the minor-player coalition, N, forms.

Proof. The proof is by induction with respect to i in N. The result is obviously true for $i = n$, so assume the statements have all been proved for $j > i$ and consider player i's decision problem.

(i) Suppose $r_i(p) < n - p_0$. Then player i can guarantee $n - p_0$ by forming the coalition with player 0. Assume player i asks $p_i \geq n - p_0$ and gives the move to $i+1$. Then $r_{i+1}(p) < n - p_0$ so that $i + 1$ will form the coalition with 0. Hence, the coalition N is suboptimal for player i, player i's unique optimal action is to form the coalition with player 0.

(ii) Suppose $r_i(p) > n - p_0$. If player i asks for p_i with $n - p_0 < p_i < n - P_i - (n - p_0)(n - i)$, then $r_{i+1}(p) > n - p_0$ and, by induction, the coalition N will form with player i getting the payoff p_i. To ensure that player i has a best response, the players $j > i$ also have to continue with the equilibrium from *(iii)b* if player i demands $p_i = n - P_i - (n - p_0)(n - i)$. Hence, the unique equilibrium is as described.

(iii) Suppose $r_i(p) = n - p_0$. Player i gets payoff $n - p_0$ by forming the coalition with player 0. Player i cannot get more from the coalition N since if $p_i > n - p_0$, then $r_{i+1}(p) < n - p_0$ and $i + 1$ forms the coalition with 0. Hence, there is an SPE in which i accepts the demand of player 0. On the other hand, the induction step and *(iii)b* guarantee that there exists an SPE in which N is formed if player i demands $p_i = n - p_0$ and gives the move to $i + 1$. It follows that there are two SPEa. □

Corollary 1 . *If the apex player starts the game there is a unique SPE outcome: the apex player demands $p_0 = n - 1$ from a minor player and the latter accepts.*

Proof. This follows immediately from Lemma 1 together with the fact that the first moving minor player should, in equilibrium, resolve ties in favor of player 0. □

Before moving to the most interesting case where player i has the choice between calling the apex player or calling player $i+1$, let us consider the behavior of the apex player when he is called.

Lemma 2 . *Assume that player i gives the move to player 0. Then in the subgame starting with the move of player 0*

(i) If $i = n$ or if $p^{i+1} < r_{i+1}(p)$, then in any SPE player 0 forms a coalition with some player $j \leq i$ with $p_j = p^{i+1}$.

(ii) If $i < n$ and $p^{i+1} > r_{i+1}(p)$, then in the unique SPE, player 0 rejects all previous demands, instead he asks for $p_0 = n - r_{i+1}(p)$ and gives the move to player $i+1$ who accepts.

(iii) If $i < n$ and $p^{i+1} = r_{i+1}(p)$, then there are multiple equilibria that correspond to those of the cases (i) *and* (ii).

Proof. The assertion clearly holds if player 0 is the last one to move (that is, $i = n$), so assume $i < n$. Player 0 is guaranteed $n - p^{i+1}$ by forming a coalition with some player $j \leq i$. (Recall from (4.4) that p^i is the minimum price of any player *preceeding* player i.) If $p^{i+1} < r_{i+1}(p)$ and player 0 demands $p_0 \geq n - p^{i+1}$ from player $i+1$, then $r_{i+1}(p) > n - p_0$ so that (by Lemma 1) the coalition N will be formed. This proves (i). To prove (ii) one notices that, if $p^{i+1} > r_{i+1}(p)$, then player 0 is guaranteed of the cooperation of player $i+1$ as long as he asks $p_0 < n - r_{i+1}(p)$. (Of course, one also invokes the usual tie-breaking argument.) The proof of (iii) is a combination of the above arguments. □

Lemma 3 shows that if the players preceeding i have made modest demands (so that the per capita remainder exceeds the lowest previous price), the minor-player coalition will recessarily form. Lemma 4 on the other hand shows that, if one or more of the preceeding players has been "greedy" (so that the per capita remainder is less than the lowest previous price), then (in some SPE continuation) player i may form a coalition with the apex player.

Lemma 3 . *Consider a subgame where i has to move and where the apex player has not yet moved. Then, if $r_i(p) > p^i$, in the unique SPE player i demands $p_i = (n-i)(r_i(p) - p^i) + r_i(p)$, each player $j > i$ demands p^i and the minor-player coalition N is formed.*

Proof. Induction w.r.t. i. The statement clearly holds if $i = n$ (with the obvious modification that n forms the coalition N). Assume $i < n$ and that the statement has already been proved for $j > i$. If player i demands p_i with $p^i < p_i < (n-i)(r_i(p) - p^i) + r_i(p)$, then $r_{i+1}(p) > p^{i+1}$ so that, by the induction hypothesis i will end up with p_i. Hence, player i can guarantee $(n-i)(r_i(p) - p^i) + r_i(p)$ by cooperating with N. On the other hand, player 0 will accept i's demand only if $p_i \leq p^i$. Clearly, i's unique optimal action is to cooperate with N and to demand the highest possible price that doesn't jeopardize the formation of that coalition. □

Lemma 4 . *Consider a subgame where player i has to move and where the apex player has not yet moved. Then if $r_i(p) \leq p^i$, there exists an SPE where player i calls the apex player who forms the coalition with i. In this SPE player i demands $p_i = p^i$ if $i = n$ and $p_i = r_i(p)$ if $i < n$.*

Proof. Induction w.r.t. i. The statement is obviously true for $i = n$. Consider player $i < n$ and suppose that the statement has already been proved for $j > i$. Also assume $r_i(p) \leq p^i$. If i demands p_i with $p_i < r_i(p)$, then $p_i = p^{i+1} < r_{i+1}(p)$ and Lemma 2 guarantees that in this case the apex player accepts the demand of player i. Hence, by the usual argument, player i can guarantee $r_i(p)$ from the apex player. Lemma 2 also shows that the apex player will reject i's demand if $p_i > r_i(p)$. Assume i demands $p_i > r_i(p)$ from $i+1$. Then $r_{i+1}(p) < p^{i+1}$, so that by the induction hypothesis there exists an SPE continuation where $i+1$ calls the apex player. If player $i+1$ chooses the latter continuation for each $p_i > r_i(p)$, then it is optimal for player i to put $p_i = r_i(p)$ and to call the apex player. □

The condition from Lemma 4 is obviously satisfied for the first moving minor player, hence

Corollary 2 . *If player 1 starts the game there exists an SPE where player 1 demands $p_1 = 1$ and calls the apex player who accepts player 1's demand.*

Note that Lemma 3 implies that player 1 can also enforce that the coalition N is formed by demanding p_1 slightly less than 1. Hence, player 1 has at least a payoff of 1 in any SPE. In fact, the previous Lemmas (together with the usual tie breaking arguments) imply that there exists an SPE in which each minor player i demands $p_i = 1$ and in which the coalition N of all minor players is formed. In this SPE, player 2 threatens to call the apex player (i.e. to play the SPE from Lemma 4) as soon as $p_1 > 1$. Note, however, that Lemma 4 does not imply that in a subgame with $p_1 > 1$, player 2 necessarily calls the apex player. Indeed, also in subgames with $r_i(p) \leq p^i$ there exist SPEa in which the coalition N is formed, hence, player 1 may obtain more than 1 in some SPE. In fact it is easy to see that player 1 can obtain the entire cake. Namely, suppose player 1 demands $p_1 = n$. Then player 2, as well as any other remaining minor player, knows that his payoff will be zero anyhow. (The payoff is zero in the coalition N, but the apex player will exploit this fact and demand the entire cake for himself as well when given the move.) Facing this fait accompli, one may as well accept it and agree to the formation of N. The following corollary, describing the worst and best payoffs player 1 can receive in the minor-player coalition, summarizes the above discussion.

Corollary 3 .

(a) *If player 1 starts the game there is an SPE where player 1 demands $p_1 = 1$, every other minor player demands 1 and the minor-player coalition forms.*

(b) If player 1 starts the game there is an SPE in which player 1 demands $p_1 = n$ every other minor player demands 0 and the minor-player coalition forms.

The previous lemmas also allow us to describe the set of all SPE payoffs that can result in the demand commitment game if the coalition of all minor players is formed.

Theorem 2 . *If player 1 starts the game, then there exists an SPE in which the minor-player coalition N forms and agrees on the payoff vector $p \in \mathbf{R}^n_+$ if and only if p satisfies*

$$\sum_{i=1}^{n} p_i = n \tag{4.4}$$

$$0 \leq r_i(p) \leq p_i \qquad \textit{(all } i\textit{)} \tag{4.5}$$

$$p_{n-1} = r_{n-1}(p) \tag{4.6}$$

Proof. (Necessity.) Condition (4.4) is obvious. Since each player i can guarantee a payoff zero (by putting $p_i = 0$) we must have $r_i(p) \geq 0$. Lemma 3 implies that if there exists an SPE with payoffs p, then $r_i(p) \leq p^i$ for all $i = 2, \ldots, n$. Namely, if $r_i(p) > p^i$, then player $i - 1$ can increase his demand without jeopardizing the formation of the coalition N. Therefore, for $i = 1, \ldots, n-1$, we must have

$$p_i \geq p^{i+1} \geq r_{i+1}(p) = r_i(p) - \frac{p_i - r_i(p)}{n - i} \tag{4.7}$$

which shows that $p_i \geq r_i(p)$ for $i = 1, \ldots, n - 1$. Condition (4.4) already implies that $p_n \geq r_n(p)$. We finally must have $p_{n-1} = r_{n-1}(p)$ since otherwise (by (4.7)) $r_n(p) < p^n$ and player n prefers to form the coalition with the apex player.

(Sufficiency.) Assume $\bar{p}$ satisfies the conditions (4.4) – (4.6) and consider the following strategy combination:

For the apex player: Play in accordance with the strategies from Lemma 2, breaking ties in favor of that minor player who moved last.

For a minor player $i \in N$:

(i) If the apex player has already moved, play in accordance with Lemma 1, breaking ties in favor of the apex player.

(ii) If the apex player has not yet moved

a) if each $j < i$ demanded $\bar{p}_j$, demand $\bar{p}_i$ and call player $i+1$ (if you are $i = n$, form the coalition N in this case);

b) if you are in a subgame covered by Lemma 3, continue with the SPE described in that Lemma.

c) In all cases not covered by a) or b), play in accordance with the SPE from Lemma 4.

By construction, this strategy profile constitutes an SPE for each subgame that starts with a move of the apex player or that is covered by the cases (ii) a) or (ii) c). Hence, it remains to be verified that along the equilibrium path no profitable deviations are possible i.e. it suffices to check that the strategies form a Nash equilibrium. Now note that since $\bar{p}$ satisfies (4.4) – (4.6) we have that

$$r_i(\bar{p}) \leq \bar{p} \qquad \text{(all } i\text{)} \qquad \text{and} \qquad r_n(\bar{p}) = \bar{p}^n \tag{4.8}$$

so that definitely player n cannot profitably deviate if all other players conform. If player $i < n$ is the first minor player to deviate to a demand $p_i > \bar{p}_i$, then Lemma 2 and (ii) c) guarantee that the coalition $\{0, i+1\}$ will be formed after this deviation. Hence, no player can profit by deviating unilaterally and we have an SPE. □

5 Credible Subgame Perfect Equilibria

In the previous section we found a plethora of equilibria with a corresponding continuum of payoff divisions in the minor-player coalition. However the strategies that support many of these equilibria rest on logic that seems uncompelling. In this section, we formalize this intuition as a refinement of subgame perfect equilibrium which, for lack of a better name, we call *credible* subgame perfect equilibrium (CSPE). In the next section, we show that credibility leads to a drastic reduction in the number of equilibria.

To motivate our refinement, consider the following game.

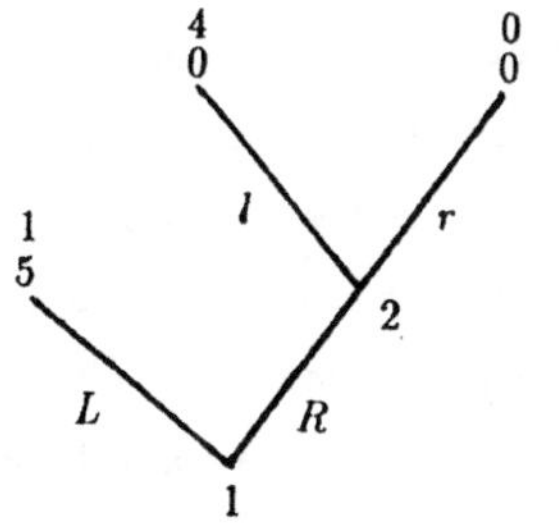

When it is his turn to move, player 2 is indifferent between l and r. Player 1's optimal strategy therefore depends on his belief about how player will play in the face of this indifference. Harsanyi and Selten (1998) argue that a rational player should randomize equally among all alternatives over which he is indifferent. If player 1 believes that player 2 will choose l and r with equal probability, then player 1 should choose R, and player 2 will obtain a payoff of 0. However, if player 1 believes that player 2 will play r, then player 1 will choose L and player 2 will obtain a payoff of 5. The threat by player 2 to play r is completely rational, since player 2 is indifferent between l and r. Moreover, the threat, if believed, yields player 2 a higher payoff. In our view, therefore, player 2 will indeed threaten to play r and player 1 has every reason to believe that player 2 will carry out his threat, so player 1 should choose L.

The logic above is quite different from that underlying forward induction arguments. The usual forward induction argument (see, for example Van Damme (1989)) is that player 1, by his action, can indicate his desire to play a particular subgame perfect equilibrium in the subgame that follows this action. In demand commitment games, as well as the game above, the forward induction logic is not compelling: although player 1 may indicate his desire to play a particular subgame perfect strategy combination, he has no means of enforcing this strategy combination since he has no further moves in the game.

An intuition similar to ours led Leininger (1986) to define a refinement of subgame perfection called ***strategic equilibrium***. However his formalization of this intuition is different from ours. One manifestation of this difference is that every credible subgame perfect equilibrium is self-consistent while strategic equilibria need not be. That is, a credible subgame perfect equilibrium necessarily induces a credible subgame perfect equilibrium in every subgame; a strategic equilibrium need not induce a strategic equilibrium in every subgame.[7]

We now turn to the formal description. Some terminology first. We restrict attention to the class of demand commitment games but the definition of credible subgame perfect equilibria can be readily extended to the class of all extensive form games with perfect information. Let Γ be a demand commitment game based on an $(n+1)$-person apex game. Let Γ be a subgame of Γ and denote its length, i.e. the maximum number of moves on a path in Γ, by $l(\Gamma)$. Let σ be a strategy combination in Γ. If τ_i is an alternative strategy of player i in Γ then $\sigma\backslash\tau_i$ denotes the strategy profile in which all players play in accordance with σ except player i who plays τ_i. The subgame γ is said to be *consistent* with σ if γ is reached if σ is played. We say that τ_i' is an ε-*best reply* of player i to σ in Γ if $h_i(\sigma\backslash\tau_i') \geq \sup_{\tau_i} h_i(\sigma\backslash\tau_i) - \varepsilon$, where h_i is player i's payoff in Γ.

We define credible threats and credible subgame perfect equilibria (CSPE) by induction on the length of the game. If $l(\Gamma) = 1$, then there are no credible threats so every SPE is a CSPE. If CSPEa have been defined in all subgames of length $l(\Gamma) - 1$, then we define credible threats in Γ as follows:

Definition 1. Credible Threats.
Let σ be an SPE of a subgame Γ with $l(\Gamma) > 1$. Let j be a player who has to move when σ is played. The strategy τ_j of player j is a *credible threat* of player j against i at σ if the following three conditions are satisfied

(a) Ex post indifference: $h_j(\sigma\backslash\tau_j) = h_j(\sigma)$

(b) Ex ante improvement: There exists $\varepsilon > 0$ such that for every ε-best reply σ_i' of player i against $\sigma\backslash\tau_j$ we have $h_j(\sigma\backslash\tau_j\backslash\sigma_i') > h_j(\sigma)$.

(c) Credibility: For each choice σ_i' of player i, the strategy combination $\sigma\backslash\tau_j\backslash\sigma_i'$ induces a CSPE in the subgame that follows σ_i'.

Condition (a) requires that if j is called upon to carry out his threat, his is no worse off than by following the equilibrium strategy. Condition (b) captures the idea that i's best response must improve j's payoff. Since, however, players' demands are continuous variables, player i may not have an exact best response against j's threat. Hence, condition (b) requires that every 'almost best' response of player i to j's threat improves j's payoff. Definition 1 assumes that a player k $(k \notin \{i,j\})$ will not deviate from σ. Condition (c) from the definition guarantees that this

[7] An example is available from the authors upon request.

assumption is justified: No matter what player i will do, player j's threat results in the players continuing with a credible equilibrium. Hence, condition (c) formalizes the idea that credible threats are accompanied by credible promises.

A credible subgame perfect equilibrium (CSPE) is an SPE in which there are no credible threats. Formally:

Definition 2. Credible subgame perfect equilibrium (CSPE).

(i) Initialization: Every SPE of Γ is a CSPE if $l(\Gamma) = 1$.

(ii) Induction: let Γ be a subgame with $l(\Gamma) > 1$ in which player i has the first turn to move. An SPE σ of Γ is a CSPE if

(a) The strategy combination σ_γ that σ induces in γ is a CSPE of γ for each proper subgame γ of Γ, and

(b) no player j who has a move on the path of σ has a credible threat against σ.

6 Credible SPEa of Demand Commitment Games

We next investigate the extent to which the credibility requirement reduces the set of SPEa in apex games. We show that only two types of outcomes survive: coalitions consisting of the apex player and a minor player, with payoffs of $n-1$ and 1 respectively, and the coalition of all minor players, with a payoff of 1 for each minor player.

In the demand commitment model a threat of player i against player j takes the form: "If you don't reduce your demand — thereby allowing me to obtain a higher payoff — I will not form a coalition with you." Such a threat is credible exactly when player i can in fact obtain the same payoff in a coalition without player j, while player j cannot obtain the same payoff in a coalition without player i.

We first consider subgames that have a unique SPE. The next lemma shows that for such a subgame the SPE is credible.

Lemma 5 . *If* Γ *admits a unique SPE* σ, *then* σ *is a CSPE of* Γ.

Proof. The proof is by induction with respect to the length of Γ and closely follows the arguments given in Section 4. In particular it uses the fact that the player moving first in such a subgame Γ can, by lowering his demand slightly force the others to accept. Hence, there can be no credible threats. For example, the minor player 1 has no credible threat against the apex player if the latter demands $p_0 < 1$. Consequently, the apex player can counter the threat that player 1 will form the coalition N if $p_0 = 1$ by demanding slightly less and giving the move to another minor player. Hence, player 1 does not have a credible threat if $p_0 = 1$. We leave further details to the reader. □

Hence, we may concentrate on the interesting subgames with multiple SPEa. We first consider subgames starting with player $i = n$ and in which the apex player still has to move.

Lemma 6 . *Assume that the apex player has not yet moved and consider a subgame Γ starting with player $i = n$. Then any SPE of Γ is a CSPE.*

Proof. The results from the previous section imply that if $r_n(p) \neq p^n$ there is a unique SPE of the subgame Γ, hence, this is a CSPE by Lemma 5. Assume $r_n(p) = p^n$ so that there are two SPE. The one in which N is formed is a CSPE since no player is moving after n. The SPE in which player n demands $r_n(p)$ from player 0 is a CSPE since the apex player's threat to form a coalition with $j \neq n$, can be countered by forming the coalition N. □

The next lemma describes two CSPEa for subgames starting with player $i < n$ in which the apex player has not yet moved.

Lemma 7 . *Assume the apex player has not yet moved and consider a subgame Γ starting with a move of a player $i \in N$ with $i \neq n$ and $r_i(p) \leq p^i$. Then the following two outcomes can be sustained by CSPEa of Γ.*

(i) Player i demands $p_i = r_i(p)$ and gives the move to the apex player who accepts this demand.

(ii) Each player $j \geq i$ demands $p_j = r_i(p)$ and the coalition N is formed.

Proof. Statement (i) follows immediately from Lemmas 2, 3 and 5: By demanding p_i slightly less than $r_i(p)$, player i forces any player to whom he gives the move to accept. Statement (ii) follows in a similar way by using an induction argument: If $j \in N\backslash\{i\}$ threatens not to accept player i's demand of $r_i(p)$, player i can counter by forming the coalition $\{0, i\}$. □

The next Lemma shows that the subgames described in Lemma 7 have no other CSPE outcomes.

Lemma 8 . *Assume the apex player has not yet moved and consider a subgame* Γ *starting with a move of player* $i \in N$ *with* $i \neq n$ *and* $r_i(p) \leq p^i$. *Then in each CSPE of* Γ *player* i *demands* $p_i = r_i(p)$.

Proof. Induction with respect to i. The results from the previous section imply that for $i = n-1$ or in case $r_i(p) = p^i$ the statement holds even for any SPE. Therefore, let $i \leq n-2$, assume $r_i(p) < p^i$ and consider an SPE in which player i has a payoff more than $r_i(p)$. Hence, i demands $p_i > r_i(p)$ and gives the move to $i+1$. Let $0 < \eta < p_i - r_i(p)$. Player $i+1$ has the following credible threat against this SPE: If you demand $p_i' \leq r_i(p) + \eta$ then we continue with the CSPE from Lemma 7 (ii), if you demand $p_i' > r_i(p) + \eta$, then we continue with the CSPE from Lemma 7 (i). For an appropriate value of ε this threat indeed satisfies the conditions from Definition 1 (if ε is small enough, then in any ε-best response, player i still gives the move to player $i+1$ and the latter ex ante gains at least $(p_i - r_i(p) - \eta)/n$), hence the SPE is not credible. □

The following theorem summarizes the results obtained in this section.

Theorem 3 . *In the demand commitment game the following and only the following outcomes can be sustained by credible subgame perfect equilibria:*

(i) the apex player starts the game, demands $p_0 = n-1$ *and forms a coalition with a minor player.*

(ii) a minor player starts the game, demands 1 and forms the coalition with the apex player.

(iii) a minor player starts the game, demands 1 and calls on a minor player, who demands 1 and calls on another minor player ... and forms the minor-player coalition N.

A natural question is whether one can give additional arguments in favor of, or to dismiss, either of the outcomes described in Theorem 3 (ii), (iii). At first it seems that the minor player might prefer the coalition with the apex player since the apex player will accept for certain while attempting to form the coalition of all minor players seems more risky. (The apex player accepts $n-1$ for certain, because if he rejects he can demand at most $n-1$ from any other minor player, and risks the possibility that this player chooses the 'wrong' continuation.) Upon closer inspection this argument is not valid. Consider the decision situation faced by player i in N when each minor player $j < i$ has demanded $p_j = 1$ and the apex player has not yet moved. Clearly, if $i = n$, then

i will prefer to form N: By forming N i has 1 for sure, if he gives the move to the apex player, he has to compete for this player's favor with the other minor players. Continuing inductively, we see that each player $i > 1$ will prefer to form the coalition N to avoid the competition with the minor player that already moved. Hence, player 1 not only knows that the apex player would accept his demand $p_1 = 1$, he also knows that, if he demands $p_1 = 1$ and gives the move to player 2, then the coalition N is formed for sure. Hence, player 1 is indifferent and both equilibria are viable.

7 Discussion

Next we relate the results of our model to those of other cooperative and noncooperative models.

In the early 1960's Davis and Maschler polled a number of prominent game theorists about the 4-minor player version of the apex game asking them what they thought should be the division of payoff between the apex player and a minor player. The replies are reported in Davis and Maschler (1965). Not surprisingly, different people made different suggestions but the majority of responses favored the payoff division (3,1). The discussion is quite interesting and reveals several intuitions that one might have about the game. Interesting is also the fact that two of the experts stress the importance of the extensive form. Martin Shubik demanded more information about the rules of the game, he did not want to commit himself using the argument that "one cannot predict anything on a game given solely in terms of the characteristic function form". Lloyd Shapley noted that "A good deal depends here on the extensive form of the game; i.e. on whether the game is actually presented to the players as a pure coalition-forming exercise, or whether there is a structure of moves and strategies which just happen to yield the indicated characteristic function. The passage from extensive (or normal) form to characteristic function form is not without pitfalls; its validity depends to some extent on the nature of the solution-concept that is applied to the characteristic function."

Two concepts discussed in Davis and Maschler that do not yield the (3,1)-division are the bargaining set and the kernel. The bargaining set of Aumann and Maschler (1964) admits any division of payoff from (2,2) to (3,1); the kernel of Davis and Maschler (1965) admits only the payoff vector (2,2).[8].

[8]The bargaining set and kernel were designed to analyze payoff divisions for situations with a fixed coalition structure, and not for situations with endogenous coalition formation. See Section 13 of Aumann and Drèze (1974), and Section 4 of Bennett and Zame (1988) for further discussion of this point.

The cooperative solution closest in spirit to our noncooperative model is that of bargaining aspirations. Bargaining aspirations were introduced independently by several authors — under a variety of names — the first was Albers (1974). Later is was recognized that bargaining aspirations are any extension of the bargaining set of Aumann and Maschler to the aspiration solution space (see Bennett and Zame (1988)). Formally, a vector $p = (p_0, p_1, \ldots, p_n)$ is an aspiration if: (a) there is no payoff leftover in any coalition after its members are paid their prices ($\sum_{i \in S} p_i = v(S)$), and (b) no player is priced out of the market (for player i there is a coalition $S(i)$ containing i with $\sum_{i \in S(i)} p_i = v(S(i))$).

In the context of apex games, an *objection* of player i against player j takes the form: "If you don't lower your price — and thereby allow me to raise mine — I will not form a coalition with you." This objection is *justified* when player i can in fact obtain the same price in a coalition without player j, while player j cannot obtain his price in a coalition without player i. A vector p is a bargaining aspiration if it is an aspiration and there are no justified objections against it. Clearly similar intuitions lie behind bargaining aspirations and credible threats; so it may not be surprising that CSPE prices are related to bargaining aspirations. For apex games with n minor players $(n-1, 1, \ldots, 1)$, the only price vector consistent with CSPE strategies, is also the unique bargaining aspiration.[9]

Chatterjee et al (1990) consider an alternative noncooperative bargaining model for the class of TU games. Their model extends those of Rubinstein (1982) and Selten (1981). In contrast to our model, in which each player announces a price, in their model, players make, accept, or reject proposals. (A proposal $< S, x >$ consists of a feasible coalition S together with a payoff division x for that coalition.) For the class of apex games, the rules are as follows. If there are no proposals on the table, the player who has to move makes a proposal $< S, x >$, where S is a coalition containing the given player. If there is a proposal on the table, the player who has to move may accept or reject it. If he accepts it, the proposal remains on the table and the move passes to the next player (the order of play within coalitions is given exogenously). If he rejects it, the proposal vanishes from the table and he must make a new proposal. The game terminates as soon as a proposal $< S, x >$ has been accepted by all players in S. Players discount payoffs by

[9]Binmore (1985) presents a multilateral Nash bargaining model as the cooperative solution concept which supports his noncooperative model. A multilateral Nash bargaining model (similar in spirit but different in details) can also be shown to support this noncooperative model. See Bennett (1990) for an overview of multilateral bargaining models.

a factor of $(1-\varepsilon)$ for each rejection that has occurred, so if r is the number of rejections, then each player $j \in S$ receives $(1-\varepsilon)^r x_j$; players not in S receive 0.[10]

Chatterjee et al look for SPE of this game in stationary strategies. Let $\bar{x}_p = n/(2-\varepsilon)$, let $\bar{x}_r = n - \bar{x}_p$ and write $\bar{x} = (\bar{x}_p, \bar{x}_r)$. Then it is easily seen that the following strategies constitute such an SPE.

For player 0: Once 'matched' with minor player i, never leave this player; propose $< \{0,i\}, \bar{x} >$, and accept any proposal x that gives you at least $\bar{x}_r$.

For player $i \in N$: Propose $< \{0,i\}, \bar{x} >$, accept any proposal of player 0 that gives you at least $\bar{x}_r$; reject any proposal to form the coalition N unless you are the last, or next to last one to move in this coalition and your acceptance guarantees a payoff of at least $\bar{x}_r$.

For small ε, these strategies lead to the formation of a two person coalition (i.e., a coalition of the apex player and one minor player); as ε tends to 0, the payoff within this coalition tends to equal division. Thus, even when there are 1,000 minor players, the apex player does not fully exploit his bargaining power: the apex player and his minor partner each obtain 500.

This odd outcome seems to result from the requirement that players use stationary strategies. The apex player, ignoring the presence of other minor players to whom he might switch, remains with the minor player to whom he is initially matched. This strategy is sensible for the apex player only because the minor players also ignore the presence of other minor players, so the apex player cannot gain from switching. Put another way, the minor players "refuse to learn" during the game, and there is nothing the apex player can do to teach them; this severely limits the bargaining power of the apex player.

Chatterjee et al do not provide a convincing motivation for the assumption of stationarity. (They just note that without it, not much can be said: in strictly superadditive games with at least 3 players, any individually rational, efficient allocation can be generated by a SPE, for ε small enough.) In our opinion, stationarity is a strong behavioral assumption, and is not justified. Furthermore, there appears to be no convincing alternative way (yet) to select among the infinity of nonstationary equilibria. In the demand commitment model, there are also an infinity

[10] Equivalently, we could assume that players do not discount payoffs, but interpret ε as the probability the game ends following a rejection.

of SPE, but there is a relatively straightforward (and in our opinion convincing) way to reduce to multiplicity. We find the way in which the outcomes depend on the number of minor players in the demand commitment model to be considerably more satisfying.

We expect that the demand commitment game, or natural variations of it, can give interesting insights for other classes of cooperative games as well. Of course, the analysis in the present paper depends on the assumption that demands cannot be renegotiated. In general it may be more natural to allow for such renegotiation, that is to allow multiple bargaining rounds with players in each round having the opportunity to quote a new demand. Also in each new round one may want to select the player moving first in that round at random, this to avoid monopoly power of the player moving first. We plan to study such multiple-round demand commitment games in the future.

References

Albers, W. (1974). "Zwei Lösungskonzepte für Kooperative Mehrpersonspiele, die auf Anspruchsnivaus der Spieler Basieren", *OR-Verfahren*, **21**, 1-13.

Aumann, R. and J. Drèze (1974). "Cooperative Games with Coalition Structures", *International Journal of Game Theory*, **3**, 217-237.

Aumann, R. and M. Maschler (1964). "The Bargaining Set for Cooperative Games", in *Advances in Game Theory (Annals of Math. Studies 52*, Eds. M. Dresher,L. Shapley and A.W. Tucker, Princeton University Press, Princeton.

Bennett, E. (1986, revised 1990). "Multilateral Bargaining Problems", Economic Department, University of California at Los Angeles WP594.

Bennett, E. (1990a). "Three Approaches to Bargaining in NTU Games", *Strategic Bargaining*, Ed. R. Selten, Springer-Verlag, Berlin.

Bennett, E. and W. Zame (1988). "Bargaining in Cooperative Games" *International Journal of Game Theory*, **17**, 279-300.

Binmore, K. (1985). "Bargaining and Coalitions", *Game Theoretic Models of Bargaining*, Ed. Alvin E. Roth, Cambridge University Press.

Chatterjee, K., B. Dutta, D. Ray and K. Sengupta (1990). "A Non-Cooperative Theory of Cooperative Bargaining", *Mimeo*, Penn State University.

Davis, M. and M. Maschler (1965). "The Kernel of a Cooperative Game", *Naval Research Logistic Quarterly*, **12**, 223-259.

Harsanyi, J. (1971). "An Equilibrium-Point Interpretation of Stable Sets and a Proposed Alternative Definition", *Management Science*, **20**, 1422-1495.

Harsanyi, J. and R. Selten (1988). *A Theory of Equilibrium Selection in Games*, MIT Press, Cambridge.

Hellwig, M., W. Leininger, P. Reny and A. Robson (1990). "Subgame Perfect Equilibrium in Continuous Games of Perfect Information: An Elementary Approach to Existence and to Approximation by Discrete Games", *Journal of Economic Theory*, forthcoming.

Leininger, W. (1986). "Strategic Equilibrium in Games with Perfect Recall", *Technical Memorandum*, Bell Laboratories.

Nash, J. (1950). "The Bargaining Problem", *Econometrica*, **18**, 155-162.

Nash, J. (1953). "Two-Person Cooperative Games", *Econometrica*, **21**, 128-140.

Rubinstein, A. (1982). "Perfect Equilibrium in a Bargaining Model", *Econometrica*, **50**, 97-108.

Selten, R. (1981). "A Noncooperative Model of Characteristic Function Bargaining", *Essays in Game Theory and Mathematical Economics in Honor of Oskar Morgenstern*, Eds. V. Böhm and H.H. Nachtkamp, 131-151.

Van Damme, E. (1989). "Stable Equilibria and Forward Induction", *Journal of Economic Theory*, **48**, 476-496.

PROMINENCE, COMPETITION, LEARNING, AND THE GENERATION OF OFFERS IN COMPUTER-AIDED EXPERIMENTAL SPATIAL GAMES

Wulf Albers and James D. Laing

Collective decision making by people who, as humans, are merely quasirational is subject to influences of substantively irrelevant features of the decision environment. This paper analyzes data from a computer-aided laboratory study of decision making in spatially represented, majority rule games to identify influences of prominent features of the problem environment on the processes and outcomes of decision, and the way in which these influences change through competition and learning.*

The discussion is organized as follows. Section 1 describes the class of spatial majority decision problems at issue and the laboratory situation. Section 2 demonstrates that the laboratory design induces highly motivated play. Section 3 analyzes the exactness with which players select a decision alternative in formulating their proposals. Section 4 presents the paper's main results on the effects of prominence on bargaining processes and decision outcomes. Section 5 draws some general conclusions. Appendix A specifies precisely the formal rules that govern these decisions in the laboratory situation. Appendix B presents a translation into English of the laboratory instructions.

1 SPATIAL DECISION PROBLEMS

Consider the following type of decision problem. A set N of five players must choose, by simple majority rule, one element from a set $X \cup z$ of decision alternatives, where z denotes the default outcome, the status quo. In principle, the status quo, z, becomes the decision outcome if it is enacted by a simple majority, or if no alternative in X is enacted by a majority, so that z becomes the outcome by d fault. The set X of decision alternatives in this situation is represented by a grid of integer-valued points embedded in a two-dimensional Euclidean space. Each player $i \epsilon N$

*We wish to express our gratitude to YAN LAU, who programmed the laboratory software, to INGO KROHN and BERND HELLMICH, who helped to establish the negotiations laboratory in Bielefeld, to BERND HELLMICH, BIRGITT KONIETZNY, SABINE KOST, BRITTA GRÜNDEL, and JÖRGE NAEVE, who assisted in running the experiments, and to LEIF ALBERS, who computed the initial results and typed the first draft of the paper. Also, we gladly acknowledge support provided by Deutsche Forschungsgemeinschaft, th National Science Foundation (Grant SES87-09476), the Wharton School, and the R s arch Fund of the University of Pennsylvania.

is assigned a payoff function (defined later in this paper) such that his payoff decreases linearly with the Euclidean distance in X of the decision point from that player's blisspoint, $\beta_i \epsilon X$. Everyone's payoff function, up to the addition of a positive constant, is common knowledge. Players with different blisspoints have a conflict of interests, so compromises must be made if they are to agree on a decision outcome. Thus, each geometric pattern (graph) of five blisspoints, which we shall call a *configuration*, represents a distinct five-person decision problem. The players' task is to select one decision alternative by simple majority rule through a process that is governed in minute detail by this formal procedure.

1.1 A Formal Mechanism for Negotiations

We characterize somewhat informally here the formal mechanism governing the negotiations. A precise specification is given in Appendix A (ALBERS and LAING, 1988). The mechanism identifies two phases in the negotiation process: (A) the *tentative decision* (or *selection*) *phase*, and (B) the *final decision* (or *verification*) *phase*. If the selection phase produces a tentative agreement (*x*,S) by any majority coalition S to enact decision alternative *x*, then that agreement must survive the verification phase if it is to become the final decision outcome. The verification phase is included to permit players to "test the waters" by exploring *tentative* decisions before they commit themselves to a final decision outcome.

The mechanism incorporates well-defined rules specifying at any stage in the bargaining process such details as (1) who has the floor, (2) what messages are in order, and (3) when a final decision outcome has been reached. Only the player who has the floor may send a message. The player who has the floor occupies just one of two roles or *positions*: *initiator* or *responder*. The set of options that are available to a player in either of these positions depends on whether the negotiations are in the selection or verification phase.

In the selection phase the player occupying the floor has these options. The *initiator*, player *i*, can choose either to (1) *pass* the initiative to a chance selection of the next initiator or pass the initiative to a designated other player, $j \epsilon N \backslash i$ or to (2) make a *proposal* (*x*,S), in which $i \epsilon S$ both offers an alternative $x \epsilon X$ that S can enact, and specifies the order in which players in $S \backslash i$ are queued for responding to the proposal. In the *responder* position that follows a proposal (*x*,S) from *i*, player $r \epsilon S \backslash i$ (as *responder)* can choose either to (1) *accept* the proposal, or to (2) *reject* (thus cancelling) the proposal and thereby assume the initiator position. If the proposal (*x*,S) is accepted by all responders, then the process moves to the verification phase with (*x*,S) as the tentative decision.

In the verification phase for the tentative decision (*x*,S), each player will be given the initiative no fewer than a prespecified number of times, a^*:-2, before the tentative agre ment is nacted as the final decision outcome. In this phase, the options available at each position ar modified as follows. The initiator can make a counterproposal or choose to pass, but in the verification phase he can pass only to a chance selection of the next initiator from the set of players who have not yet occupied the initiator position a^* times in the verification phase for the current tentative decision. This same chance selection is reached if any responder to a counterproposal in the verification phase rejects the counterproposal. If any member of the coalition S forming the current tentative agreement either initiates or accepts a counterproposal during the verification phase, then the tentative agreement is cancelled and the selection phase is restarted to process the counterproposal. On the other hand, in the verification phase for the tentative decision (x,S), once every player has the initiative a^* times and no member of S has cancelled the tentative decision by either making or accepting a counterproposal, then (x,S) is verified as the final decision.

The maximimum number of rounds to be allowed in the negotiation process can be set in the mechanism via the parameter t^*, such that if no final decision is enacted by the end of round number t^*, then the default outcome z is imposed as the final result. By definition, a new *round* in the negotiation process begins if (1) a player assumes the initiator position in the selection phase, or if (2) a member of the coalition forming the current tentative agreement either makes or accepts a new proposal in the verification phase. [In the latter case, the software implementation of this mechanism discussed next labels (relabels, if necessary) the new proposal as the first message in the new round.] In our laboratory situation, $t^*:=999$, so that the maximum number of rounds permitted by the rules is practically unlimited.

1.2 Collective Decision Process Software

This mechanism provides a logic for a software system named Committee (LAING and LAU, 1988) that was used in the laboratory situation to govern the bargaining process. In this system, players communicate in accordance with the bargaining mechanism via a network (NETBIOS calls) of 80286-based personal computers, each equipped with a hard disk, Enhanced Color Graphics, Microsoft Windows, and a mouse. This software implements simple collective decision games in which the set of $n=5$ players must select an outcome from a 1400x1200 (horizontal x vertical) coordinate system of spatially represented decision alternatives in accordance with the mechanism and m/n majority rule ($m > n/2$), such that any coalition with at least m members can enact any alternative in X that it chooses, once a member of the coalition gains access to the floor. (In our laboratory situation, $m:=3$.) This system controls the network in accordance with the bargaining mechanism. All messages are sent via the network. At each player's station the system presents a graphical display of the decision problem and th location of every player's blisspoint, plots and records the default option, each proposal, and the tentative decision, maintains a current log of the messages sent in the bargaining on this problem, and identifies both the round and the name of the player who currently has the floor. It also provides a variety of graphical aids that the player may use. Thus the player can plot or "unplot": points; everyone's circle through a selected point; or straight lines connecting the blisspoints of any or every pair of players. The screen display can be zoomed from the "big picture" (Zoom 0) for a magnified view of the subset of decision alternatives centered on any selected point in the grid. Two levels of magnification (Zoom 1 or 2) are available. Also, a window displays the coordinates of the current location of the mouse-controlled cursor in the grid and the distance in grid units of this point from the player's blisspoint. The player can choose to display also in this window the distances of the cursor point from the other four players' blisspoints. We shall present additional details about these features of the laboratory situation when they are needed in the subsequent discussions. For a more complete description, see the laboratory instructions in Appendix B.

1.3 The Laboratory Study

In the spring of 1988 we conducted a series of laboratory studies of five-person groups negotiating within this environment in accordance with the process mechanism and simple majority rule. The subjects were 30 students of business administration and economics at the University of Bielefeld. Each subject participated in at least two, and at most four, laboratory sessions.

Group assignments. In assigning subjects to groups for each session we used this system. We divided the 30 subjects into two sets (A and B) of 15. We partitioned each of these sets into 3 five-person groups (indexed as A1, A2, and A3; B1, B2, and B3) for each subject's first laboratory session. For the second sessions, we again partitioned each subject set, A and B, to form six different groups (A11, A12, and A13; B11, B12, and B13). Then a set C of 15 volunteers from A and B was partitioned into three five-person groups (C1, C2, C3) for a third session, and, again, for a fourth session, but only two of the fourth sessions (groups C11 and C12) were run. We made these assignments such that no member of any given group had participated in a previous session with more than one other member of the current group. Thus, altogether, we conducted 17 laboratory sessions. A session typically lasted 6-8 hours, but two lasted 12 hours or more.[1] On average, decision number 1 in the first session took 3 and 1/4 hours; therefore, to some extent, the players were already experienced when they began decision number 2. Clearly, the players who

---

[1]These are the approximate durations of the laboratory sessions, in hours: A1:12 1/2, A2:7, A3:7 3/4, B1:12, B2:7 1/4, B3:8 3/4, A11:7, A12:6 1/2, A13:7 1/2, B11:6, B12:6, B13:5 3/4, C1:6, C2:8 1/2, C3:6, C11:7 1/2, C12:6. Times for the first sessions include about 2 hours required for the laboratory instructions and practice with the various tools available at the subject's station. These preliminaries consumed appreciably less time in subsequent sessions.

participated in a third or fourth session wer highly exp rienced in this situation, having logged as much as 30 hours or mor in th lab by th time they had finished. Thus, the design permits us to observe increasingly sophisticated bargaining as players gain experience in the situation.

Payoffs. We provided substantial monetary incentives in this situation to induce highly motivated play. Although similar in certain respects to the "nonmonetary duplicate method" of KORMENDI and PLOTT (1980), our method of calculating payoffs "at the margin" involves considerable sums of money. Specifically, it was common knowledge that each person's payoff would be computed by the following schem . Let player i be any member of the group. Suppose that, for the decision probl m in which the blisspoints are arranged in configuration c, the group selects as the majority decision outcome any particular alternative $x \epsilon X$ at a Euclidean distance of $d_i(x)$ grid units from i's blisspoint. The payoff to player i is computed on the margin by comparing $d_i(x)$ to the mean distance, denoted D_{ic}, from i's blisspoint to the points chosen in configuration c by all other groups that do not contain player i. Then the increment to i's payoff(in DM) for any decision outcome x equals 0.20 $[D_{ic} - d_i(x)]$. Thus, to every player in the group facing any of these decision problems, every distance unit in the 1200 x 1400 grid is worth, at the margin, DM 0.20 (roughly equivalent to \$0.11 at an exchange rate that then averaged about DM 1.8 per US dollar). This means that, by belonging to the coalition that enacts the outcome in a given decision problem, a player could easily increase his payoff by DM 50 (about \$28) or more. Subject knew that the most successful player in the study could win about DM 300 ("...perhaps even DM 400 ...") overall, but that no one's total payoff over two laboratory sessions would fall below DM 40. After explaining these details to subjects, we gave this advice: "So adjust yourself to a payoff between DM 40 and DM 400, and try to maximize your payoff by optimal behavior." Also, if the session exceeded 6 1/2 hours, we paid subjects in addition for overtime at DM 10 per hour. Omitting complications introduced by overtime and the lower bound, the sch me implies that the payoff (in DM) to group member i in any decision problem p facing the group is given by the function $u_i{:}X \to R$, such that $u_i(x) = a_{ip} - 0.2\ d_i(x)$ for every decision alternative x, where $a_{ip} > 0$ is unknown to players but is independent of i's actions.

Payoff results. The total payoffs actually earned in the study by th thirty subjects ranged from a low of DM 55 (for only two sessions), to a high of DM 544 (including DM 65 for overtime). In the first two laboratory sessions, the total payoff associated with the decision points for two of the thirty subjects fell below the DM 40 minimum. For four of the fifteen players in set C, the total payoffs won through decision outcomes after the second session also fell below the lower bound. But no player dropped below this bound in both pairs of sessions. We asked the players to contact the experimentor if they felt that their losses were so substantial that they were not sufficiently motivated to play the subs quent games. No player did so. In fact, in ignorance of the results being obtained in other groups, it is difficult to estimate one's own success in comparison to the others'. For this reason, the incremental payoffs associated with each decision problem, calculated at DM 0.20 per unit, appear to be the primary motivation arising from this payoff scheme. This and, perhaps, other features of the laboratory design induced highly motivated play.

Other features of the laboratory games. Within a given session, the group played a sequence of distinct spatial games each governed by the mechanism and simple-majority rule. All negotiations were communicated publicly via the network. The default option in each game was set at a distance of 2000 grid units from each players blisspoint, and thus was very unattractive. The maximum permissible length of the game on any decision problem was virtually unlimited (t^*:=999). Each game was played, in fact, until an outcome was reached. Within any such game, each player sat in a separate room and was identified by a color name that changed across games. [In assigning subjects to blisspoints in the various configuration and selecting players who were first given the floor to begin negotiations in the various decision problems, we used a rotation scheme to ensure that each player was placed with equal frequencies in both advantageous and disadvantageous positions.] Thus, the laboratory design permits us to model each of the decision problems as a distinct game. Altogether, we recorded data about the processes and outcomes of 103 decisions and 7 different configurations representing distinct spatial games.

2 INTENSITY OF BARGAINING

This section discusses some general features of the multilateral bargaining processes that we observed in this laboratory situation. It begins with a preliminary overview of the data, then reports some results pertaining to the length of the bargaining process.

2.1 Points Proposed in Seven Spatial Games

The data. Figures 1 through 7 plot all the points that players proposed in this laboratory study in the seven distinct spatial games. These 1730 points, and the coalitions to which they were proposed, constitute the primary data for this paper. Let us make some preliminary observations about these figures.

Seven games. The names of these games — Games A, F("Fox"), H("House"), M, W, B("Bear"), and S("Star") — reflect the geometry of the configuration of five blisspoints shown in the respective figure, with this exception. Game B is dubbed the Bear, in recognition of the stubborn resistance it offers to precis math matical solution for a typ of equilibrium from the th ory of cooperative gam s (LAING and

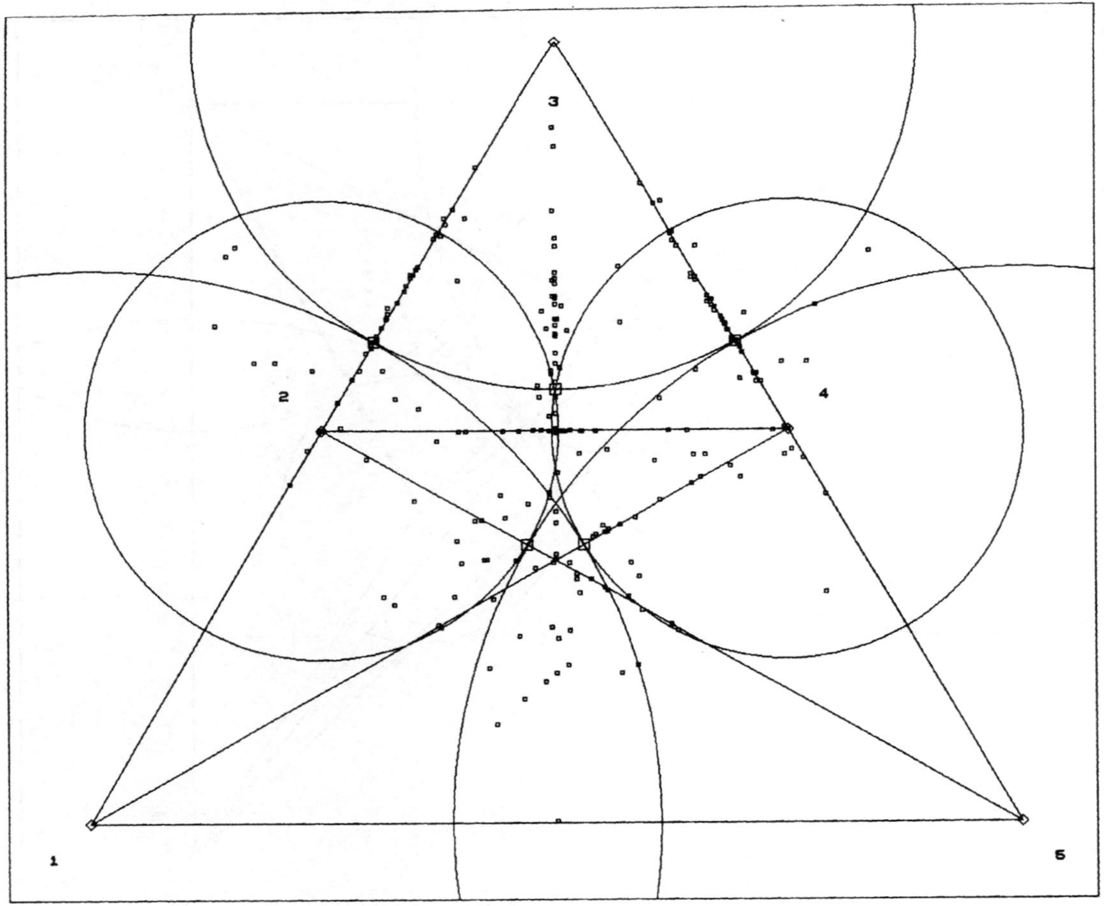

FIGURE 1. GAME A: STABLE DEMANDS AND ALL(N=34) POINTS PROPOSED.

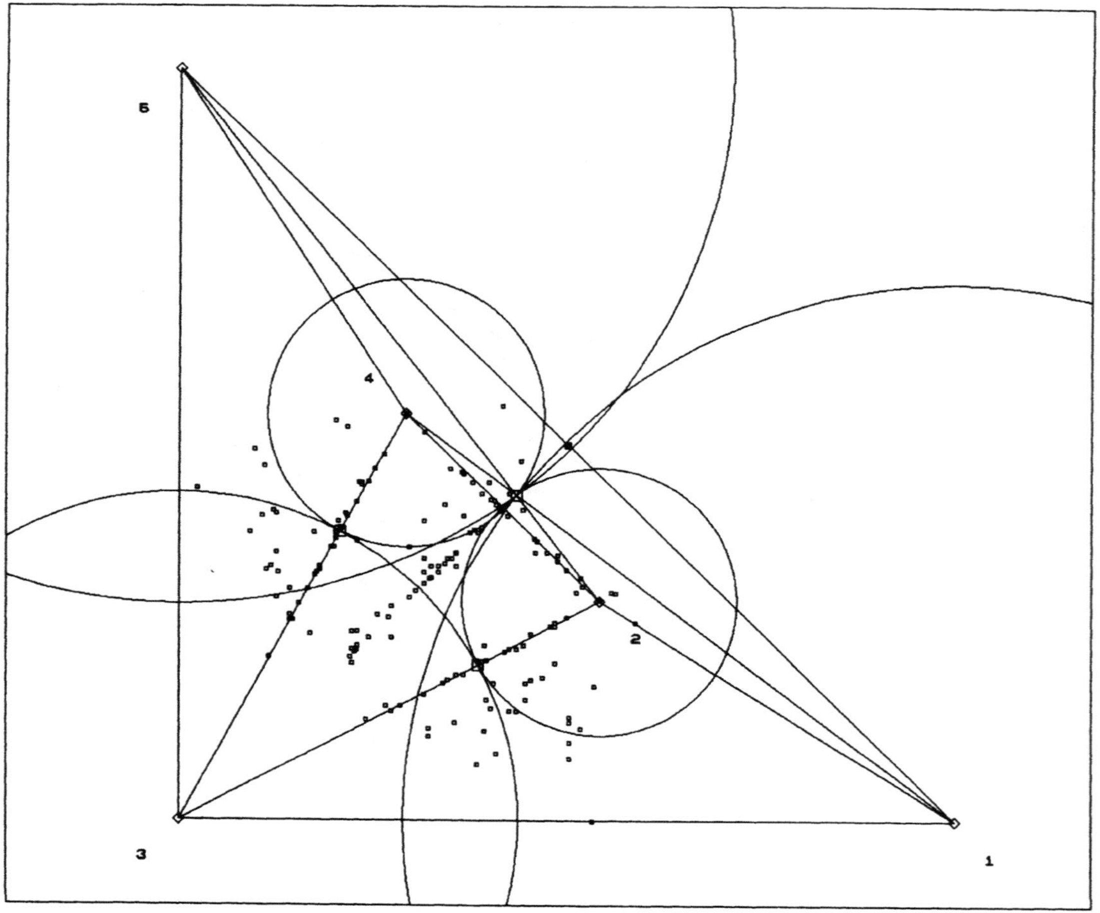

FIGURE 2. GAME F: TABLE DEMANDS AND ALL(N=219) POINTS PROPOSED.

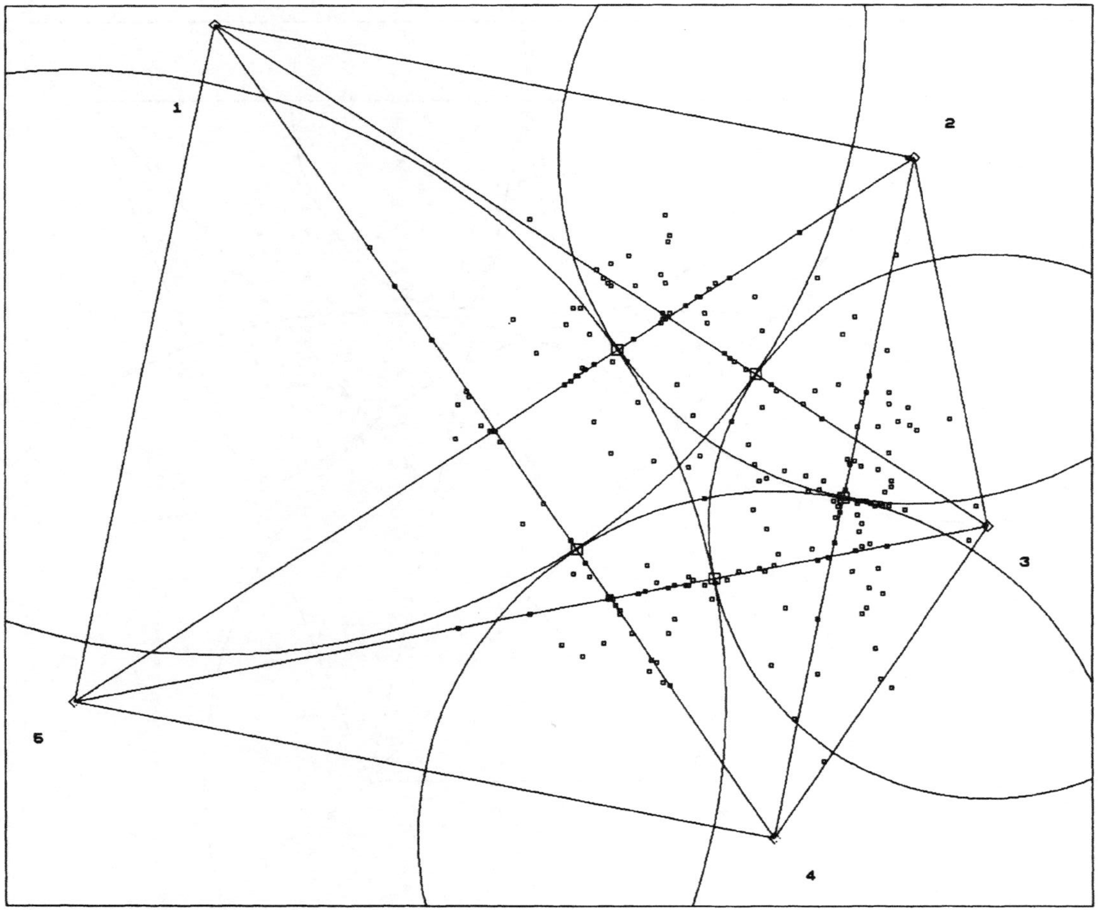

FIGURE 3. GAME H: STABLE DEMANDS AND ALL(N=384) POINTS PROPOSED.

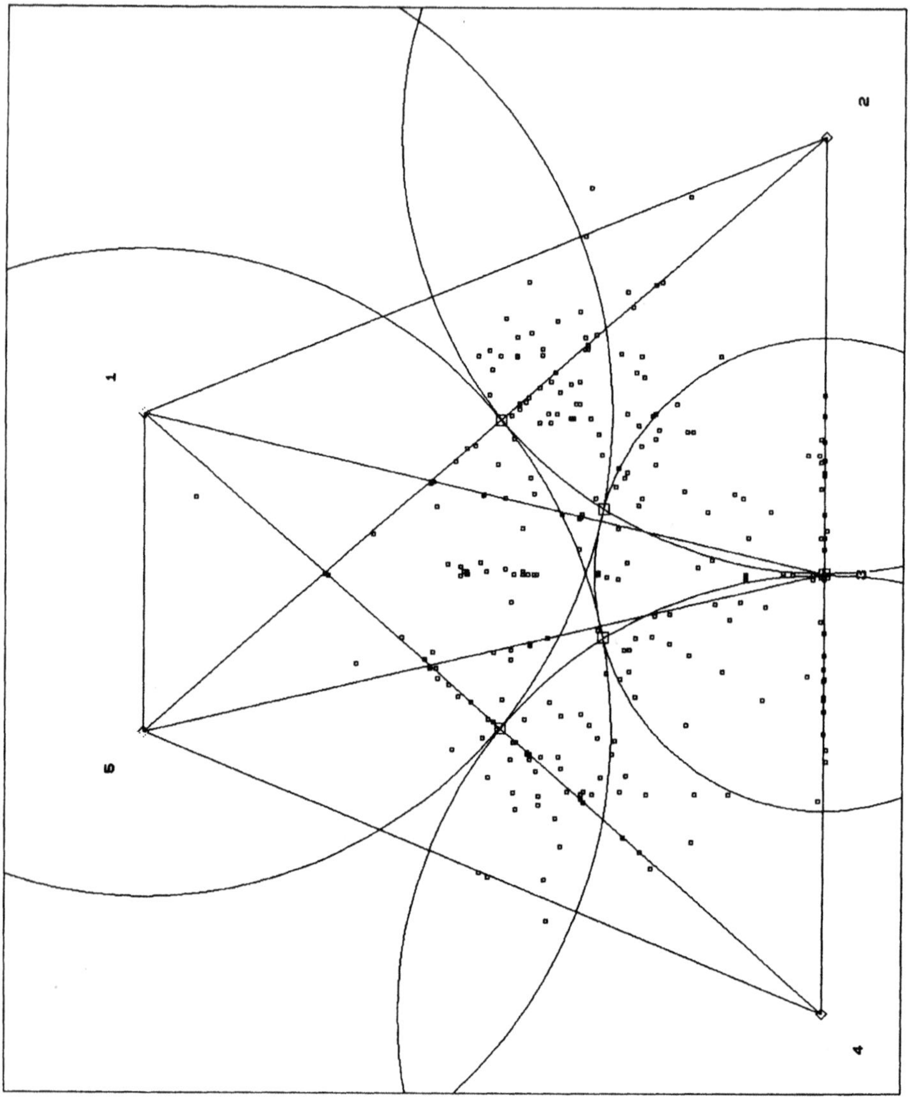

FIGURE 4. GAME M: STABLE DEMANDS AND ALL(N=336) POINTS PR POSED.

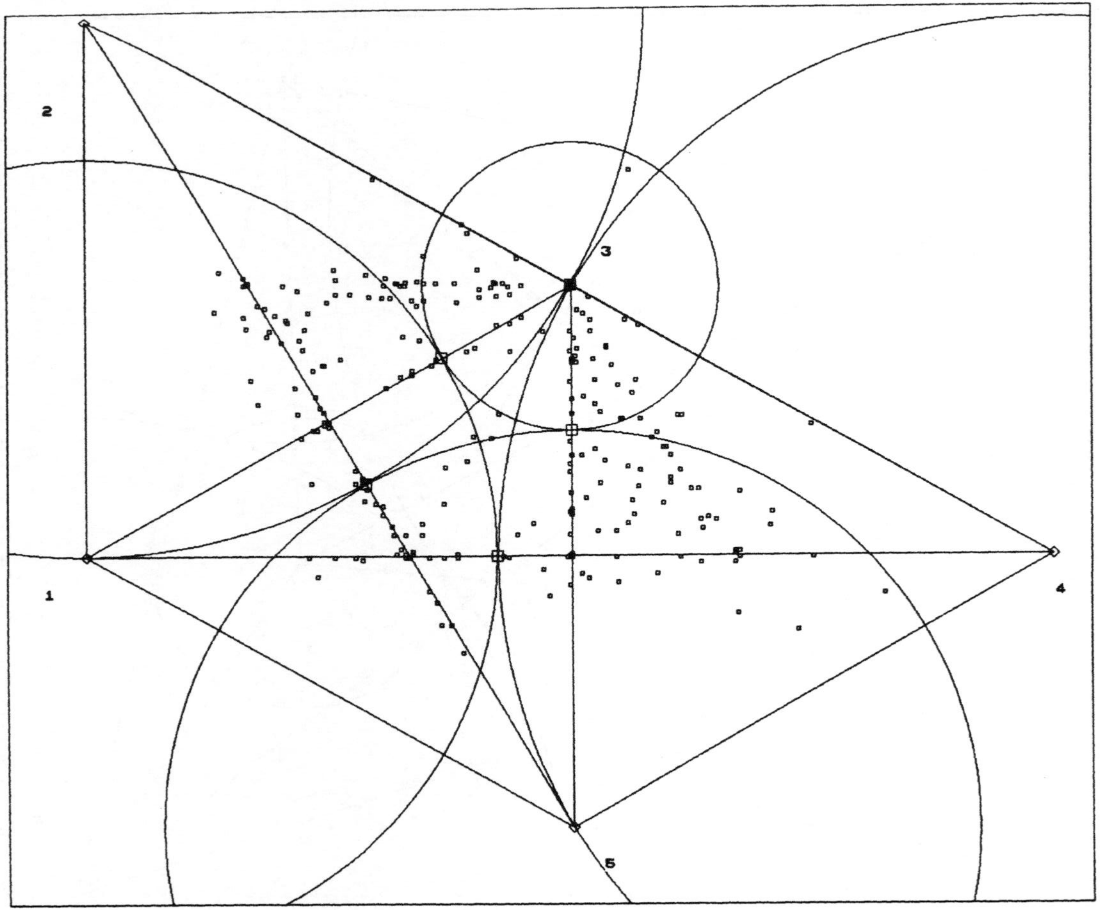

FIGURE 5. GAME W: TABLE DEMAND AND ALL(N=34) POINT PROPOSED.

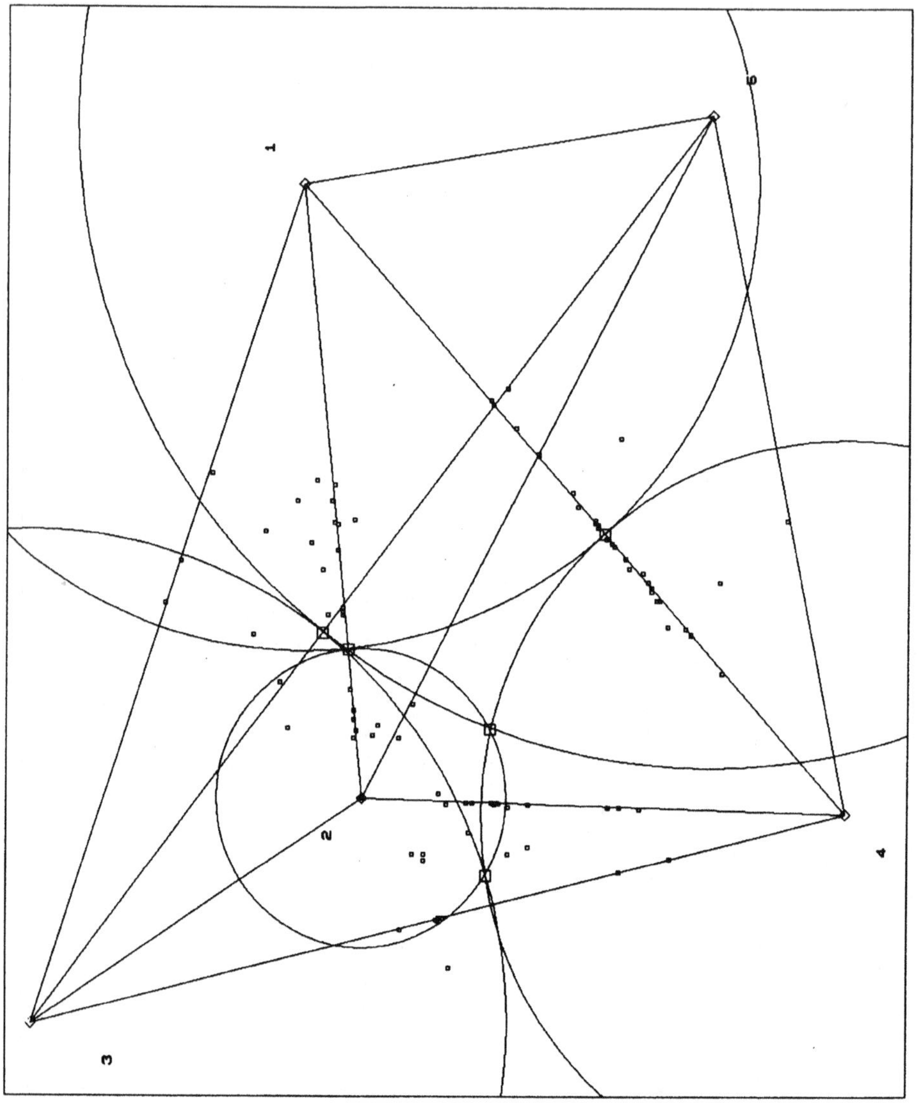

FIGURE 6. GAME B: STABLE DEMANDS AND ALL(N=104) POINTS PROPOSED.

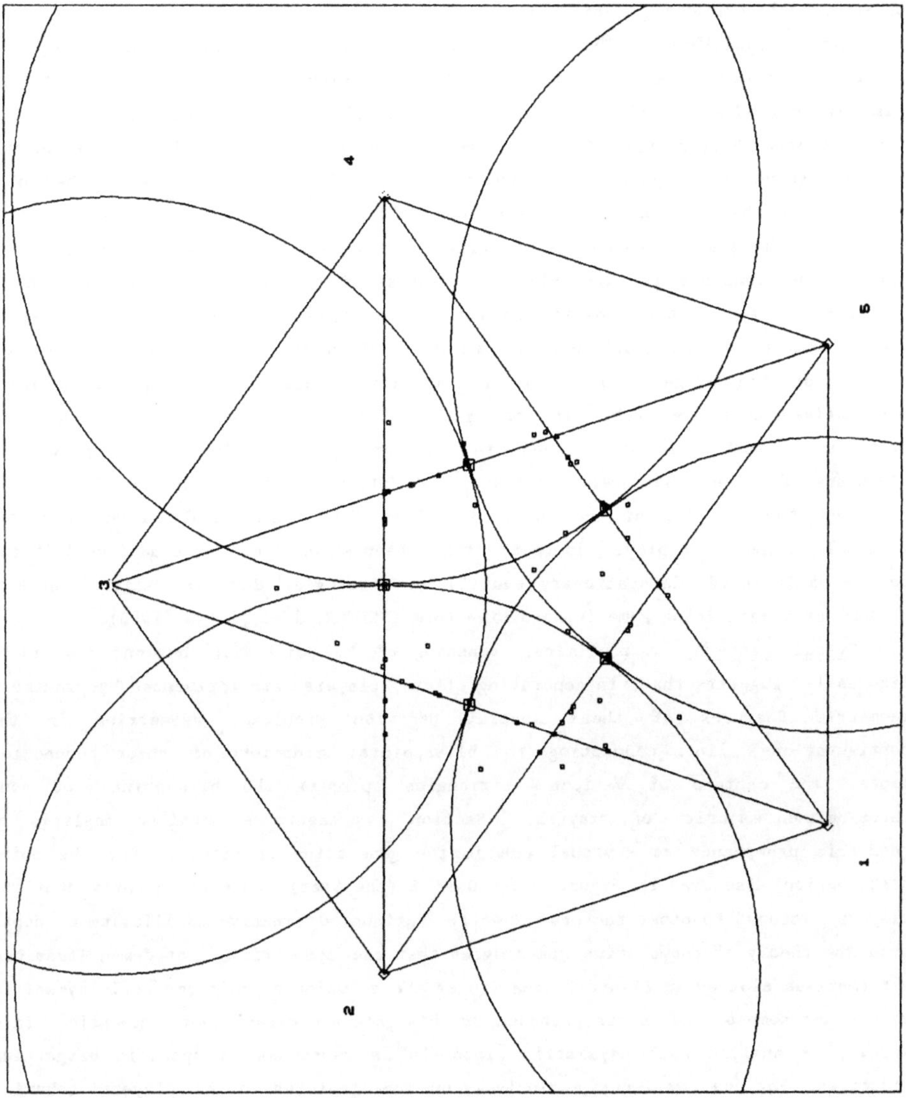

FIGURE 7. GAME 5: TABLE DEMANDS AND ALL(N=73) POINTS PROPO ED.

OLMSTED, 1978; ORDESHOOK, 1986:436).

Demand equilibria. The five circles plotted in each figure correspond to a particular vector of *demands* (or *quotas*) that constitutes a type of solution to the game, the demand equilibrium (ALBERS, 1987). According to this theory, it is possible for a majority S of players to enact a point x in the set X of decision alternatives as the outcome of the game only if the payoff associated with x to each member of S is no worse than his demand. Given the stable vector of demands, any potential outcome of the game in which the majority S of players enacts the alternative x is said to be *feasible* if and only if x satisfies the demand of everyone in S. Accordingly, we say that a decision point x or a majority of players S is *stable* if there is some outcome (x,S) that is feasible, given the stable demand vector. The centers of the large squares in each of the figures constitute the decision alternatives that are stable in that game. The alternative majorities that are feasible in each game are evident in the corresponding figure. To provide a benchmark for the subsequent analyses, we interpret the demand equilibrium to represent the solution of the decision problem that would result under perfectly rational, competitive play. In fact, it has been shown that the demand equilibrium can be implemented via stationary equilibrium strategies that are rational in each subgame of a bargaining game in recursive form (SELTEN, 1981; LAING, 1988).

Overall pattern. A preliminary scanning of the point distributions plotted in Figures 1-7 suggests that, in generating offers, players are influenced by prominent geometric features of these spatial decision problems: symmetries in the configurations, lines connecting the blisspoints, midpoints of these connection lines, and centers of various triangles spanning the blisspoints of some three-person majority of players. Section 4 presents a detailed analysis of geometric prominence as a visual cue in the generation of offers. But the point distribution displayed in Figure 7 for Game S (the Star) indicates clearly that the players respond to other factors as well. Various alternative equilibrium concepts from the theory of cooperative games agree that the symmetric set of demand-feasible alternatives plotted in Figure 7 constitutes the solution of this perfectly symmetric game. Yet some of the points proposed in this game are clearly not symmetric. In a sequel, we analyze such asymmetric proposals as responses to specific bargaining histories. For now, we focus exclusively on some features of the players' behavior that are independent of the previous negotiations. Given the wide dispersion of proposed point evident in the figures, one might suspect that many factors play a role in the formulation of such offers.

2.2 Directness of Bargaining

Shortness of perfectly rational play. Noncooperative game-theoretic analyses of stationary strategies in extensive-form bargaining problems usually (but not always) identify solutions such that, at equilibrium, the player who is selected to initiate the game makes one proposal that is promptly enacted as the final decision outcome, thus ending the game (SELTEN, 1981; LAING, 1988). But equilibrium theory presupposes perfectly rational play.

Duration of play under limited rationality. It is no surprise that the players in our experiments, although highly intelligent and strongly motivated, are not perfectly rational. But their negotiations become less protracted as they gain experience in the laboratory situation. Although the players complete only a few games in Session 1, they require considerably less time per game in later sessions. The first sessions last between 7 1/4 and 12 hours, but produce only 3 (in 5 cases) or 4 (in 1 case) decision outcomes and require, on average, more than 2 hours per game. In contrast, the experienced players in Session 4 average only about 1 hour per game.

Time invested to generate an offer. On the other hand, the time that the initiator occupies the floor before transmitting a proposal increases with experience, averaging roughly 3 1/2 minutes in Session 1, 4 1/2 minutes in Session 2, and nearly 5 1/2 minutes in Sessions 3 and 4. Since the players develop proficiency in the operation of their computers early in Session 1, these results suggest that, as players gain experience in the decision problems they select proposals less spontaneously and with greater care, presumably after drawing conclusions from the bargaining history and attempting to predict what scenario would be triggered in the subsequent play by each alternative proposal under consideration.

Redistributive bargaining. Table 1 presents additional data indicating that the length of the bargaining sequence decreases as players gain proficiency in the situation. The high mean frequency of proposals in Session 1 results in part from the tendency of inexperienced players to engage in protracted redistributive bargaining, in the sense that they make a long series of counterproposals of different grid points for the same coalition. These redistributive negotiations often resemble two-person bargaining, such that two members of the proposed majority monopolize the floor with a series of counterproposals, beginning with an exchange of reasonably extreme claims and continuing through a series of small concessions, in an attempt to strike a joint agreement before presenting it to a third player. In later sessions, such sequences of redistributive bargaining occur less frequently and, when they do occur, are relatively brief.

Shortness of negotiations by experienced players. The results presented in Tabl 1 indicate that negotiations become considerably less protracted as players gain experience. In Session 1, protracted redistributive bargaining is partly responsible for the large number of offers preceeding the first tentative decision. In th results averaged over Games A through W as displayed in column 6 of Table 1, th number of proposals required to reach the first tentative decision in the game decreases sharply with xperi nc : 15.5 in Session 1, 5.2 in Session 2, and 3.6 in Sessions 3 and 4. Similarly, the total numbers of proposals, tentative decisions,

	GAME							
	A	F	H	M	W	AFHMW	B	S
Session 1:								
P	48.3	18.3	36.8	21.5	27.8	29.7	–	–
T	4.8	2.7	9	3.5	3.5	4.9	–	–
P1	28	12	8.8	10.8	17	15.5	–	–
R	45.3	18	26.3	19	26	29	–	–
(n)	(4)	(3)	(4)	(4)	(4)	(19)	(0)	(0)
Session 2:								
P	10.2	11.5	7.4	18.8	19.5	13.3	5.3	–
T	2.5	4	2.1	6.3	5.2	4	2.5	–
P1	5.3	5.3	5.3	4.3	5.7	5.2	3.3	–
R	9	9.8	6.7	12.8	17.3	11.6	5	–
(n)	(6)	(6)	(7)	(6)	(6)	(31)	(4)	(0)
Sessions 3 and 4:								
P	7.3	6.2	5.1	11	8.8	9.1	4.8	2.9
T	2.3	2	2.3	4	2.5	2.7	2.4	2
P1	4.3	3.5	3.3	4.1	3	3.6	2	1.6
R	6.2	5.7	4.7	8.6	7	6.6	3	3
(n)	(6)	(6)	(7)	(7)	(6)	(32)	(8)	(9)

Legend
- P: mean number of proposals, omitting counterproposals to final decisions
- T: mean number of tentative decisions, including final outcomes
- P1: mean number of proposals leading to first tentative decision, including that decision
- R: mean number of rounds to final decision
- n: number of final decisions

TABLE 1. Mean Number of Proposals, Tentative Decisions and Rounds, by Game and Session.

and rounds in the game decrease markedly across the sessions. With experience, players proceed more directly to a final agreement.

Solution difficulty. Table 1 also reveals large variations among these decision probl ms, suggesting that players find some decision problems more difficult than oth rs. Evidently this also is a function of experience. In Session 1, Game A app ars to be the most difficult, whereas Games F and M are relatively easy. Yet these differences among the games attenuate greatly when players encounter them for the second or third time, so that most of the variations disappear in Sessions 3 and 4. We find it interesting that players find Game M easy in Session 1 but appear to discover some if its subtleties in later encounters: in the last sessions Game M r quires more extensive bargaining than do the other decision problems. It comes as no surprise that the perfectly symmetric Game S, the Star, is the easiest of these problems for highly experienced players . On the other hand, it seems remarkable that they so readily solve the Bear, despite the challenge that Game B presents analysts seeking its game-theoretic solution. Indeed, players adopt a simple approach to this game. Ignoring complications in the Bear that trouble analysts, they enact points that, at least from th vi wpoint of cooperativ game theory, seem more appropriate for a thre -person, simpl majority game involving only play rs 1,

2, and 4. In fact, this approach to the Bear se ms reasonable, given the level of inexactness, discussed in the next section, with which players select points to offer in this game.

3 EXACTNESS OF BARGAINING

Technical features of the laboratory design permit us to investigate the precision with which players select points in generating offers.

3.1 Presentation of the Grid

A 1400 (horizontal) x 1200 (vertical) array (XxY) of integer-valued points constitutes the set of decision alternatives, but the resolution of EGA graphics is too coarse for us to display all of these points simultaneously. The number of decision alternatives exceeds the number of pixels in the available space on the computer monitor. To overcome this difficulty, we provide options so that each subject can view a subset of the alternatives by selecting the view's center and one of three levels of zoom (magnification). This choice determines which points can actually be selected from the screen.

Zoom 0. This view (the default) displays a grid of decision alternatives on the entire 1400 x 1200 coordinate system. But this grid is coarse. In Zoom 0 one can select only 4 of every 15 values on the X axis, and every third value on Y. Moreover, blisspoints and points that are proposed by other players do not appear exactly at their true location in Zoom 0. Instead they are plotted at the near st point in the coarse grid. The location of the cursor in the set of d cision alternatives also is converted in the same way to the coarse grid, and the distances from blisspoints to this point are computed in the coarse grid. Thus, Zoom 0 provid s an overall picture that only approximates the true bargaining situation. The coars grid is fixed in Zoom 0, and the view is not affected by the selected center of zoom.

Zoom 1. This option displays a particular 950 x 600 subset of th XxY coordinates that depends on the selected center of zoom. In Zoom 1 one can select 8 of very 15 X-values and 2 of every 3 Y-values. Thus, Zoom 1 employs a somewhat finer grid, and provides a better approximation of the bargaining problem than does Zoom 0. The specific points that can be chosen in Zoom 1 depend on the center that is selected.

Zoom 2. This highest available level of magnification provides an accurat repr sentation of the bargaining problem. It displays a 630 x 400 subset of th XxY coordinate system about the selected center of zoom. Typically, on can pick a point without difficulty in Zoom 2, although a specific point that is spann d by th displayed subset of th coordinates can, in som instances, b s l ct d only by adjusting the center of zoom. Using Zoom 2 with an appropriate view center, one can pick any point in the set of decision alternatives exactly.

3.2 Selection of Zoom

These features of the grid presentation enable us to investigate the extent to which players used the higher levels of magnification in selecting points to propose.

Zoom tradeoffs. Choosing the zoom level involves tradeoffs. Zoom 0 presents the "big picture" that is useful for assessing the overall bargaining situation, but it provides a very coarse grid for picking points. At the other extreme, Zoom 2 enables one to select and plot points exactly, but makes it difficult to evaluate these point in relation to other points or game features that lie outside the selected view.

Zoom frequencies. We can exploit the coarseness of the Zoom 0 grid in order to conduct some quantitative investigations into the precision with which players select points in generating an offer. One can select a new point for a proposal in this situation only by picking this point from the screen. But only very special points — those that are represented in the coarse grid — can be selected under Zoom 0. [Ex ante, these constitute 1/3 x 4/15 = 8.9% of the decision alternatives.] Therefore we can determine which proposals in the data cannot be created in Zoom 0 and thus compute the results displayed in Table 2.

SESSION:	1	2	3	4	Total
PROPOSALS	21.3	10.4	12.9	4.4	14.2
	(138/648)	(59/569)	(40/310)	(9/203)	(246/1730)
TENTATIVE DECISIONS	10.9	13.4	10.4	5.7	11.0
	(10/92)	(18/134)	(7/67)	(3/53)	(38/346)
FINAL DECISIONS	21.1	14.3	17.9	9.5	15.5
	(4/19)	(5/35)	(5/28)	(2/21)	(16/103)

Tabl 2. Offered Points that Can be Selected ONLY under Zoom 1 or 2, as Percent, by Session.

Table 2 indicates that 14% of all proposals and 15.5% of all final decisions can be selected only under Zoom 1 or 2. By assuming that the percentages given in the table also apply to the remaining points which can be selected in all three zoom levels, we can estimate the overall proportions of cases in which players employ Zoom 1 or 2, as giv n in Table 3.

It should be noted that the software in some ways encourages the use of Zoom 0. First, Zoom 0 is the default. Second, each player's view is shifted to Zoom 0 automatically whenever the scr n is r freshed in response to a bargaining communication that mak s or r jects a new proposal during the sel ction phase of the

SESSION:	1	2	3	4	Mean*
PROPOSALS	38.2	19.7	24.1	8.6	22.6
TENTATIVE DECISIONS	20.6	25.1	19.8	11	19.1
FINAL DECISIONS	37.7	26.5	32.5	18.1	28.7

*Unweighted.

Table 3. Estimated Percent of Offered Points Selected under Zoom 1 or 2, by Session.

procedure, converts a proposal to a tentative decision, or cancels a tentative decision. Our impression from observations in the laboratory is that players usually employ Zoom 0, shifting to Zoom 2 only on occasion to compare alternatives in a small neighborhood or to select a point with precision.

Zoom and the proposal's success. The estimates in Table 3 suggest that a tentative decision is more likely to be enacted as the final decision if it is selected under Zoom 1 or 2. This observation is supported by the results in Table 4. Although the use of zoom does not increase the chances that a proposal becomes the tentative decision, the entries in Table 4 for Sessions 3 and 4 indicate that, ceteris paribus, the probability that the tentative decision is enacted as the final result by highly experienced players increases roughly from .4 to .7 if the decision point must be selected under Zoom 1 or 2.

SESSION:	1	2	3	4	Mean*
PERCENT OF PROPOSALS THAT BECOME A TENTATIVE DECISION:					
Only Zoom 1 or 2	7.2	30.5	17.1	33.3	22
Otherwise	14.2	23.6	21.5	26	21.3
PERCENT OF TENTATIVE DECISIONS ENACTED AS THE FINAL RESULT:					
Only Zoom 1 or 2	40	27.8	71.4	66.7	51.5
Otherwise	20.7	26.1	41.8	39.6	32

*Unweighted.

Table 4. Success of Proposals and Zoom Level Required, by Session.

Zooming and experience. According to estimates in Table 3, although Zoom 1 or 2 is used to select points for 38% of the proposals and final decisions in Session 1, it is employed less frequently as players gain appreciable experience in the bargaining situation. In Session 4 we estimate that zoom is employed in creating only 9% of all proposals and 18% of the final results.

Table 5(A) presents the distribution of players across various frequencies with which they use Zoom 1 or 2, controlling for session. Across sessions, these distributions seem to shift towards less frequent use of zoom, but some of this shift could be attributed to the fact that Sessions 3 and 4 involved only players who volunteered to continue beyond Session 2. However, the same basic tendency is present in the distributions for just those ten subjects who participated in all four sessions, as displayed in Table 5(B). Therefore, the use of zoom by highly motivated players evidently decreases with experience.

We believe that as players gain proficiency, they shift their attention from microscopic comparisons of alternative decision points to focus on overall strategic considerations, seeking to fashion a reasonably attractive, tentative agreement that can survive a barrage of counterproposals and thus be enacted as the outcome of the game. Nonetheless, the results for Sessions 3 and 4 in Table 4 indicate that, other things being equal, a tentative decision is much more likely to be enacted as the final result by experienced players if the offered point is selected with precision.

	PERCENT OF PROPOSALS REQUIRING ZOOM 1 OR 2*											
SESSION	0	10	20	30	40	50	60	70	80	90	100	(n)
(A) All subjects:												
1	8	4	5	3	1	3	1	2		2	1	(30)
2	13	2	3	2	5		3	2				(30)
3	4	2	3	1	2	1	1			1		(15)
4	5	4			1							(10)
(B) Four-session subjects only:												(10)
1	2	2	1	1	1	2		1				
2	4	1	1	1	2		1					
3	3	2	2	1	1		1					
4	5	4			1							

*Least upper bound in 10% increments.

Table 5. Number of Players Using Different Proportions of Zoom 1 or 2, by Session.

3.3 Sources of Inexactness

Later we investigate the extent to which players tend to offer special types of points in their proposals, such as points that are equidistant from the blisspoints of two players or that lie on a line in the boundary of the alternatives that are Pareto efficient for the proposed coalition. For various reasons, a player who apparently intends to select such a special point might fail to do so exactly.

Inaccuracies in the representation. Imprecision in the selection or comparison of points can result from misrepresentations in the display of the game under Zoom 0 and 1, as discussed previously. For example, a point can appear in Zoom 0 to lie on the line connecting two blisspoints even though, in Zoom 2, it does not. To select points precisely, one must use an appropriate view center and Zoom 2. Subjects might not realize this fully until they gain experience in the situation.

Trembles. Even in Zoom 2 it takes manual skill to plot the target point exactly. Using the mouse, one must first move the cursor on the display to the precise location desired, then hold the mouse perfectly still while pushing a button to plot the selected point. Experienced play rs with good hand-eye coordination can take half a minute or so before they succeed at this. Less skilled players can require a series of attempts before they are able to plot a point that hits the target or is "close enough."

Intentional deviations. Of course, not all apparent inaccuracies are mistakes. Under the payoff scheme, every distance unit is worth DM 0.20. Even if it is common knowledge that all players in the game are highly skilled, there may be compelling strategic reasons to offer a point that deviates somewhat from a purported "solution." [We investigate this in a sequel.] On the other hand, if it is generally believed that small deviations likely result from trembles or inaccuracies in the representation, then a clever player can try to exploit this. This could be one reason why players, as they gain experience, increasingly offer points that can be selected in Zoom 0.

3.4 Accuracy in Point Selection

This subsection analyzes the extent to which points proposed by the players lie on or near (1) particular lines of division that bound the outcomes that are Pareto efficient for some coalition that is, in a sense to be defined, extreme, or (2) points that are equidistant from two or more players' blisspoints. The results provide additional insights concerning the precision with which players select points for their proposals.

Extreme coalitions and lines of division. We shall analyse the extent to which the proposed points lie on or near lines connecting the blisspoints of pairs of players that are, in the sense defined here, "crucial" in the game. Let $\beta(S)$ denote the set of all points in X that are Pareto optimal for the coalition S. Given that each player's payoff is a linear function of the distance from his blisspoint, then $\beta(S)$ is the smallest closed and convex subset of X that contains the blisspoint of every player in S. We seek to identify certain pairs of players such that the line connecting the blisspoints of any two players in such a pair divides the Pareto set of the grand coalition, $\beta(N)$, into two parts.

Towards this end, we define some terms. Following LAING and OLMSTED (1978:261), we say that a winning coalition $S \subseteq N$ of players is *consolidated* if there is no decision alternative in X that is Pareto efficient both for S and for the set of all players who are not members of S: that is, $\beta(S) \cap \beta(N \setminus S) = \emptyset$. In a sense, a consolidated coalition does not "split the opposition." Define the coalition S to be *extreme* if (1) S is both consolidated and minimum winning (here, a three-person coalition) and (2) $\beta(N) \setminus \beta(S)$ is convex. For example, these are the extreme coalitions in Game A (Figure 1): 123, 345, 125, 145, 234. [For the first two of these coalitions, the Pareto optimal set is a line and thus has an empty interior; the Pareto set for each of th other coalitions in this list is a triangl .] Th pair of distinct players *i* and *j* belonging to coalition S is *crucial* to S if (1) S is an xtreme coalition whose

Pareto set has nonempty int rior, and (2) $\beta(ij) = \beta(S)\cap\beta[(ij)\cup(N\backslash S)]$. In Game A, for example, the following pairs of players are crucial to some coalition: 14, 24, and 25. There is a one-to-one correspondence between the extreme coalitions whose Pareto sets have positive measure and the crucial pairs of players in these games: every such extreme coalition contains exactly one crucial pair, and any pair is crucial for at most one extreme coalition. We refer to any line that connects the blisspoints of a crucial pair of players as a *line of division* in the game.

Accuracy in hitting lines of division. The distributions of proposed points displayed in Figures 1-7 indicate that much of the bargaining in these games occurs near lines of division. We now analyze the observed deviations of proposed points from these lines of division in order to learn more about the accuracy with which players select points.[2]

Table 6 displays for each laboratory session the deviations of points proposed for an extreme coalition from the line of division associated with that coalition. Any such deviation is measured normal to the line in grid distance units and has a negative (positive) sign if the point lies outside (inside) the associated extreme coalition's Pareto set. Table 6 is restricted to points that can be selected by players only in Zoom 1 or 2 and that lie within 16 units of the line.

DEVIATION:	-16...-7	-6	-5	-4	-3	-2	-1	0	1	2	3	4	5	6	7	8	9...16
SESSION																	
1		1				1	14	45	6	2	1		2	1	1	1	
2					2	1	1	25	2	1							
3							3	16	2	1	2						
4							1	5	2								
TOTAL:	0	1	0	0	2	2	19	91	12	4	3	0	2	1	1	1	0

Table 6. Deviations of Points that Can be Selected ONLY in Zoom 1 or 2 from Line of Division, by Session.

It is evident in Table 6 that points lying within 16 units of the relevant line of division deviate little from this line. The largest deviations occur in Session 1, and probably were intentional. Most of the large deviations lie in the interior of the set of alternatives that are Pareto efficient for the relevant extreme coalition, to the benefit of the third member of the coalition who does not belong to its crucial pair.[3] Especially in the later sessions, players attain a high level of accuracy in selecting these points, undoubtedly by using the precise representation of Zoom 2, rather than the coarser grid of Zoom 1, since it is possible in the latter representation to pick only about (8/15 x 2/3) 35.6% of the points spanned by the

---

[2]These analyses are restricted to those games that were played in all four sessions. These are the crucial pairs for each of these games. A: 14, 24, 25. F: none. H: 13, 14, 24, 25, 35. M: 13, 14, 25, 35. W: 13, 14, 25, 35. With thre exceptions (24 in A, and 13 and 35 in M), the line of division for each crucial pair contains the solution point identified by the demand equilibrium for the associated extreme coalition.

selected intervals of the X and Y coordinates. This agrees with our observations during the laboratory sessions: players usually employ Zoom 0, but opt for Zoom 2 rather than Zoom 1 when they seek precision.

On the other hand, in the truncated distribution displayed in Table 7, there are greater deviations from the line of division if the offered points can be (and probably were) selected in the coarse grid provided in Zoom 0. In the Zoom 0 representation, an error of one pixel in picking a point on the line as displayed on the computer monitor can yield a deviation of up to 7 grid units from the actual line. In fact, nearly all the deviations in the portion of the distribution shown in Table 7 for Session 4 fall within this range. Table 7 also indicates that, for those proposed points that can be selected under Zoom 0 that lie within 14 units of the appropriate line of division, few deviate more than 3 units from the extreme coalition's Pareto efficient set. Thus, even in Zoom 0, the players appear to achieve some accuracy in selecting points for their proposals.

<u>Accuracy in hitting midpoints.</u> Whether in search of equity or recognition of a symmetry in the game, players sometimes offer points that are at least approximately equidistant from two or more players' blisspoints. We now investigate the accuracy

<u>Deviations Outside the Extreme Coalition's Pareto Set:</u>

DEVIATION(x-1):	14	13	12	11	10	9	8	7	6	5	4	3	2	1
SESSION														
1			1		1	3			1	6	7	11	10	20
2					2	1		4	2	5	7	15	17	24
3					1	1	2		3	3	7	3	15	24
4							2		1	1	5	6	7	22
TOTAL:	0	0	1	0	4	5	4	4	7	15	26	35	49	90

<u>Deviations Inside the Extreme Coalition's Pareto Set:</u>

DEVIATION:	0	1	2	3	4	5	6	7	8	9	10	11	12	13	14
SESSION															
1	21	16	13	6	4	4	1	1	2	1	1	3	1	1	
2	40	28	32	6	12	5	5	3	1	2		1	2	1	2
3	23	19	25	9	12	4	3	2		1	1				
4	19	22	9	3	3	3	2	1		2				2	
TOTAL:	103	85	79	24	31	16	11	7	3	6	2	4	3	4	2

<u>Table 7.</u> <u>Deviations of Points that Can be Selected in Zoom 0 from Line of Division, by Session</u>

[3] In Session 1, 8 of the 14 points that lie one unit outside the Pareto set were created by one group in Game A. This group considered points near the horizontal line connecting players 2 and 4 that differ only on the X-axis from a previously offer d point that could have been selected in Zoom 0 and became a starting point of the subsequent ast-west bargaining.

with which players offer such midpoints. To do so, we compute for every pair of players the difference in distances to each proposed point, then examine the tail of the distribution of these pair-wise differences in the neighborhood of equidistance.

Tables 8 and 9 display these results. Table 8 is restricted to differences with regard to points that can be selected only by using Zoom 1 or 2, whereas Table 9 pertains to points that can be selected in Zoom 0. Both tables have two parts: (A) presents a tail of the distribution of differences for all ten player-pairs per point, whereas (B) pertains just to that player pair that is most nearly equidistant from the proposed point.

Given the way in which the data shown in Table 8 are skewed, one can assume with confidence that in many cases players employ Zoom 1 or 2 in order to pick a midpoint. In fact, Table 8(B) indicates that, of the 248 points proposed in this study as a whole that require Zoom 1 or 2, fully 130 (52%) lie within one unit of equidistance for some pair of players. (Sometimes it is not possible to select points closer than one unit to equidistance, because the grid coordinates are integers.) Inspection of the entire distributions, not just the tails presented in Table 8, indicates that the global maximum is reached at zero difference and the frequencies generally tend to decrease sharply until a roughly uniform distribution apparently attributable to noise is reached.

Table 9 displays the tails of differences between player pairs for proposed points that can be selected in the coarse grid of Zoom 0. An inspection of the entire distributions, not just the tails shown in Table 9, reveals a pattern similar to that

Table 8. Differences of Distances from Two Players' Blisspoints of Offered Points that Can be Selected ONLY in Zoom 1 or 2, by Session

(A) Tail of the distribution, all ten player pairs per point

	DIFFERENCE											TOTALS		
	0	1	2	3	4	5	6	7	8	9	10	tail	(%)	all
SESSION														
1	119	33	21	7	6	5	1	1	1	2		196	(14.1)	1390
2	46	48	16	5			4		1			120	(20.3)	590
3	33	9	5	4			3		1			55	(13.4)	410
4	5	4	2									11	(12.2)	90
all	203	94	44	16	6	5	8	1	3	2	0	382	(15.4)	2480

(B) Tail of the distribution, player pair nearest equidistance per point

	DIFFERENCE											TOTALS		
	0	1	2	3	4	5	6	7	8	9	10	tail	(%)	all
SESSION														
1	59	12	4	3	5	3	1	1		1		89	(64.0)	139
2	25	10	1	2					1			39	(66.1)	59
3	18	3	2	2			1		1			27	(65.9)	41
4	3		1									4	(44.4)	9
all	105	25	8	7	5	3	2	1	2	1	0	159	(64.1)	248

described above for points that must be selected in Zoom 1 or 2. But, as one might expect, the distributions for points that can be selected in Zoom 0 contain more noise. For this reason, we display in Table 8 the tail of the distribution within 15 rather than 10 units of equidistance. The distributions under Zoom 0 attain a global maximum in the immediate neighborhood of zero difference, and tend to decrease roughly monotonically to the right until a level is reached that can be attributed to noise. But in Zoom 0 the decrease is less abrupt and the noise level is higher than in Zoom 1 or 2. Subjects attain an impressive level of exactness using Zoom 1 or 2, but less accuracy using Zoom 0, in selecting points that are equidistant from two or more players.

These results indicate the extent to which the laboratory design induced players to bargain intensely over precisely selected points. We now consider some basic features of the decision problems that help to shape the proposals that players make in these games.

Table 9. Differences of Distances from Two Players' Blisspoints of Offered Points that Can be Selected in Zoom 0, by Session

(A) Tail of the distribution, all ten player pairs per point

	DIFFERENCE																TOTALS		
	0	1	2	3	4	5	6	7	8	9	10	11	12	13	14	15	tail	(%)	all
SESSION																			
1	134	71	73	38	23	25	23	17	13	17	22	8	9	3	9	10	495	(9.7)	5110
2	77	49	64	31	35	26	17	14	12	11	8	4	13	13	13	9	396	(7.8)	5100
3	19	35	41	14	16	5	9	5	9	5	3	2	7	8		2	180	(6.7)	2700
4	32	17	17	6	8	4	3	2	9	3	3	1	1	2	2		110	(5.6)	1950
all	262	172	195	89	82	60	52	38	43	36	36	15	30	26	24	21	1181	(7.9)	14860

(B) Tail of the distribution, player pair nearest equidistance per point

	DIFFERENCE																TOTALS		
	0	1	2	3	4	5	6	7	8	9	10	11	12	13	14	15	tail	(%)	all
SESSION																			
1	70	43	36	20	14	14	13	8	2	4	7	8	4	1	6	6	256	(50.1)	511
2	33	35	45	21	18	13	12	9	8	6	3	3	12	8	3	8	237	(46.5)	510
3	14	25	20	12	11	3	4	1	8	4	1	1	2	3		1	110	(40.7)	270
4	16	8	8	4	3			2	9	2	2	1	1	2	1		59	(30.3)	195
all	133	111	109	57	46	30	29	20	27	16	13	13	19	14	10	15	662	(44.5)	1486

4 PROMINENCE, COMPETITION, AND LEARNING

This section analyzes the coalitions and points proposed in the laboratory situation to gain insight into the criteria by which these offers are generat d. It focuses on the extent to which the proposals can be explained by prominent features of the game that are not related to its equilibrilium solution, and whether behavior adjusts in the direction of the demand equilibrium as players gain experience in the situation. We analyze as aggregate data the 1553 proposals that players generated in the five decision problems which were addressed in all four laboratory sessions: Games A, F, H, M, and W (see Figures 1-5). A sequel will analyze these proposals as responses to particular bargaining histories. We first investigate the proposed coalitions, then extend the analyses of points that were offered in these games.

4.1 Coalition Selection: Proximity, Stability, and Experience

Various criteria can be used to characterize the alternative majorities in these five games, and some of these criteria are not independent. As these terms are defined in the first subsection of Section 3.4, a coalition is *extreme* only if it is both minimum winning (3-person) and *consolidated*. We describe any coalition that is supported by the demand equilibrium as *stable*. In Games A, H, M, and W, a majority is stable if and only if it is extreme. In Game F, although there is no extreme coalition, every stable majority is consolidated, but some consolidated majorities are not stable. Moreover, all but one (N\3 in Game F) of the stable majorities in these five games are minimum winning. Thus it is difficult to separate any effects of prominence associated with these properties from those due to competition, interpreted in terms of the demand equilibrium. Note that these stable coalitions are consolidated, hence do not "split the opposition", yet players may hold the preconception that center ("middle-of-the-road") alliances are most natural.

Proximity. An important influence on the selection of coalitions in these games is associated with a geometric property that is not related to the stability concept. The *diameter* of a coalition S refers to the maximum distance between blisspoints of any two players in S. Define any coalition S to be *proximal* if it is minimum winning and, across all winning coalitions, it has minimum diameter. A proximal coalition might appear to be a natural alliance, since its members are comparatively near to one another. This agrees with the adage: "Birds of a feather flock together." (Characteristically, folk wisdom subscribes also to an apparent contradiction: "Politics makes strange bedfellows." Perhaps this refers to sophisticated play.) Proximal coalitions might arise also if players expect that any winning coalition would tend to choose a point that minimizes the maximum distance from anyone in the coalition. For whatever reason, close proximity of its members appears to make a coalition prominent in this situation. In these games, three of the seven proximal coalitions are not stable. Thus we can constrast stability versus proximity in the coalitions that are proposed.

Preponderance of stable coalitions. Most proposals in these games are made to coalitions that are supported by the demand equilibrium. In a five-person gam , there are ten simple majorities and six larger majorities. Of these, half of the three-person coalitions and none of the larger coalitions are stabl in Games A, H,

M, and W. In Game F, the stable coalitions consist of one four-person coalition and six bare majorities. Thus, except for Game F, the stable coalitions are bare majorities. In the data, 98.4% of all proposals are made to three-person coalitions.

The results summarized in Table 10 demonstrate that most of the proposed coalitions are stable. The percentage of first proposals that are made to stable coalitions increases from a low of 63% in the first session to 81% in Sessions 3 and 4 (combined). The preponderance of stable coalitions in the combined data for all proposals in each session indicates a strong tendency towards stable coalitions after the first proposal in the game, and across sessions. In Sessions 3 and 4, nearly 95% of all proposals are made to coalitions that are stable. Most notably, all but 2 of the 82 coalitions that enacted final decision outcomes in these five games support the demand equilibrium.

Table 10. Percent of Coalitions Proposed in Five Games that are STABLE

		SESSION			
		1	2	3-4	All
FIRST PROPOSALS:	Stable	63.2	77.4	81.2	75.6
	(n)	(19)	(31)	(32)	(82)
ALL PROPOSALS:	Stable	85.0	91.3	94.7	89.5
	(n)	(648)	(531)	(374)	(1553)
TENTATIVE DECISIONS:	Stable	95.7	92.7	96.5	94.7
	(n)	(92)	(124)	(85)	(301)
FINAL OUTCOMES:	Stable	100.0	96.8	96.9	97.6[a]
	(n)	(19)	(31)	(32)	(82)

[a]In two final outcomes an unstable coalition formed: N\4 in Game M, session 2, and 234 in Game F, session 3.

The same basic pattern is evident in the coalitions proposed within each of the five games, as displayed in Table 11. Most proposals, especially in Sessions 3 and 4 and Games H, M and W, are made to coalitions that support the demand equilibrium. Some of the stable coalitions become popular only after players gain experience, but others are frequent from the beginning.

Proximity and stability in coalition selection. An asterisk in Table 11 flags each coalition that is proximal. Of these, 234 in Games A and H, and 123 and 345 in Game W are both proximal and stable. These four coalitions are proposed frequently, particularly in Session 1, although they become somewhat less prevalent in later sessions as players learn to appreciate the viable alternatives. Two of the proximal coalitions that are not stable — 135 in Games M and W — seldom are proposed after Session 1. The third, 234 in Game F, attracts fully one quarter of the proposals in both Sessions 1 and 2, and appears in 8% of the proposals generated in Sessions 3–4.

Table 12 displays the percent of coalitions proposed that are proximal, giv n that such a coalition is available in the game, controlling for the coalition's stability and the session in which it is proposed. In the first proposals and in all proposals combined, there is a strong interaction effect of proximity and stability on the frequency with which a coalition is proposed, but this attenuates somewhat with experience. Proximity appears to make a coalition prominent to players, and this is reinforced markedly if the coalition also is stable. Although proposals to prominent but unstable coalitions decrease with experience, offers to coalitions that are both proximal and stable remain important even after players become experienced in these games. Of the proximal coalitions that enacted final outcomes, one — 234 in Game F (Session 3) — is unstable. Coalitions that are both proximal and stable enacted 12.5% of the final outcomes in Game A, 33.3% in Game H, and fully 75% in Game W. Combining the results for these three games, 40% of the final decision outcomes are based on coalitions that are proximal and stable, and this rate is fairly constant across sessions. Players seem attracted to a coalition that is both proximal and stable.

Table 11. Coalitions Proposed in Five Games, as Percent of All Offers, by Game, Stability, and Session

Notes: Symmetric coalitions are paired if both are observed.
Coalitions never proposed are omitted.
Proximal coalitions are indicated by *.

			SESSION			
	STABILITY	COALITIONS	1	2	3-4	All
GAME A						
	Stable	123, 345	33.0	55.3	50.0	41.7
		125, 145	7.6	21.2	25.8	14.4
		234*	39.1	15.3	13.6	28.4
		all stable	79.7	91.8	89.4	84.5
	Unstable	124, 245	18.8	3.6	9.1	13.3
		N\3	1.5	4.7	1.5	2.3
		all unstable	20.2	8.3	10.6	15.6
		(n)	(197)	(85)	(66)	(348)
GAME F						
	Stable	123, 345	36.4	54.9	59.7	50.7
		124, 245	33.3	16.5	24.2	23.7
		125	—	1.1	4.8	1.8
		N\3	3.0	2.2	3.2	2.7
		all stable	72.7	74.7	91.9	79.0
	Unstable	134	1.5	—	—	.5
		234*	25.8	25.3	8.1	20.5
		all unstable	27.3	25.3	8.1	21.0
		(n)	(66)	(91)	(62)	(219)

(continued next page)

	STABILITY	COALITIONS	SESSION 1	2	3-4	All
GAME H						
	Stable	123, 345	25.0	32.5	34.4	28.9
		125, 145	28.1	33.8	37.5	31.6
		234*	42.5	33.8	26.6	36.8
		all stable	95.6	100.0	98.4	97.4
	Unstable	124, 245	1.2	—	—	.6
		235	—	—	1.6	.3
		135	.6	—	—	.3
		N\2	1.2	—	—	.7
		N	1.2	—	—	.7
		all unstable	4.4	—	1.6	2.6
		(n)	(160)	(80)	(64)	(304)
GAME M						
	Stable	123, 345	47.9	37.8	31.4	38.7
		125, 145	12.5	28.9	38.1	27.1
		234	19.8	28.1	26.7	25.3
		all stable	80.2	94.8	96.2	91.1
	Unstable	135*	16.7	3.7	2.9	7.1
		235	—	.7	—	.3
		N\2, N\4	2.1	.7	1.0	1.2
		N\3	1.0	—	—	.3
		all unstable	19.8	5.1	3.9	8.9
		(n)	(96)	(135)	(105)	(336)
GAME W						
	Stable	123*, 345*	78.3	62.9	58.4	67.6
		125, 145	11.6	20.7	28.6	19.1
		234	—	10.0	9.1	6.1
		all stable	89.9	93.6	96.1	92.8
	Unstable	135*	8.5	6.4	3.9	6.6
		N\2, N\4	1.6	—	—	.6
		all unstable	10.1	6.4	3.9	7.2
		(n)	(129)	(140)	(77)	(346)

4.2 Point Selection: Prominence, Stability, and Experience

This section investigates the extent to which the initiator of a proposal t nds to offer certain types of points that are prominent in the situation. It first discusses salient features of the situation that might draw the players' att ntion to certain types of points, then examines the frequency with which players g nerate offers to enact such points. Again, we focus on th five gam s played in all sessions to identify interactions among prominence, stability, and xperience.

Table 12. Percent of Coalitions Proposed that are PROXIMAL, by Stability and Session

	STABILITY	GAMES	SESSION 1	2	3-4	All
FIRST PROPOSALS	Stable	AHW	66.7	52.6	52.6	56.9
		(n)	(12)	(19)	(19)	(50)
	Unstable	FMW	36.4	38.9	15.8	29.2
		(n)	(11)	(18)	(19)	(48)
	OVERALL	AFHMW	63.2	54.8	40.6	51.2
		(n)	(19)	(31)	(32)	(82)
ALL PROPOSALS	Stable	AHW	50.6	42.0	34.3	44.6
		(n)	(486)	(305)	(207)	(998)
	Unstable	FMW	15.1	10.1	4.5	10.2
		(n)	(291)	(366)	(244)	(901)
	OVERALL	AFHMW	44.8	31.1	21.9	34.6
		(n)	(648)	(531)	(374)	(1553)
TENTATIVE DECISIONS	Stable	AHW	42.9	39.3	35.6	39.8
		(n)	(70)	(61)	(45)	(176)
	Unstable	FMW	8.1	6.4	3.6	5.9
		(n)	(37)	(94)	(55)	(186)
	OVERALL	AFHMW	35.9	24.2	21.2	26.9
		(n)	(92)	(124)	(85)	(301)
FINAL OUTCOMES	Stable	AHW	41.7	36.8	42.1	40.0
		(n)	(12)	(19)	(19)	(50)
	Unstable	FMW	—	—	5.3[a]	2.1
		(n)	(11)	(18)	(19)	(48)
	OVERALL	AFHMW	26.3	22.6	28.1	25.6
		(n)	(19)	(31)	(32)	(82)

[a]One final outcome was based on an unstable but proximal coalition: 234 in Game F.

Sources of prominence in point selection. Various types of points might be prominent in this situation. A *midpoint* is the halfway point of a line connecting the blisspoints of two players. An *intersection point* is defined for this discussion to be a point that lies in the interior of the convex span of the five players' blisspoints at an intersection of two or more distinct lines, each connecting the blisspoints of two players. (By this definition, the blisspoints of players 2 and 4 and the intersection of line 14 with line 25 constitute the three intersection points of Game F.) An *equidistance point* is a point that is equidistant from the blisspoints of all members of the proposed coalition. Thus, the equidistance point of coalition S, if it exists, is the center of a circle that passes through the blisspoints of everyone in S. An equidistance point exists (respectively, does not exist) for a three-person coalition S if the convex span of blisspoints of everyone in S is a triangle (resp., a line). Some of the larger majorities have equidistance points, but others do not. Note that tendencies toward midpoints or equidistance points might be explained by appealing to some concept of equity (min max distance, or even split). Any apparent proclivities in these data toward two other potentially prominent types of points — the center of gravity of the proposed coalition S (mean of blisspoints of everyone in S) or a point whose distance from one or more blisspoints is a multiple of 100 — can be attributed to one or more of the first three types of prominent points. Therefore, we focus here on midpoints, intersections, and equidistance points.

Prominent points offered in each of five games. Table 13 displays the percent of all proposals in each game that offered a point lying no farther than 5 decision grid units from a prominent point. The table also controls for the laboratory session and whether the prominent point is stable. A prominent point is stable if it is the stable point for the proposed coalition. Otherwise it is unstable.

In Game A, all prominent points are unstable. Midpoint proposals seem to be popular in this decision problem: in Session 1, the point offered in fully 26.9% of all proposals lies within the closed neighborhood (radius=5) of the nearest midpoint (labeled "Mid LIJ"). Nearly half of these propose the midpoint of line 24 to coalition 234. In later sessions, more than 9% of all proposals offer midpoints; roughly a third of these propose the midpoint of line 24 to coalition 234. The other prominent points, the intersection of lines 14 and 25, and the equidistance point of the proposed coalition S (labeled "Mid S") were offered rarely in Game A.

In Game F, midpoint proposals again are popular, but intersections are not. The unstable intersection is the blisspoint of player 2 or 4, whichever is closer to the proposed point. Such a point was proposed only once. The stable intersection of lines 14 and 25 was never proposed. Yet, in one sense, the unstable midpoint of line 24 differs little from the stable intersection point, although they are 22 units apart: the midpoint is 4 units worse than the intersection for players 1 and 5, but 2 units better for players 2 and 4. Neither point can be selected from the screen in Zoom 0.

In Game H, most prominent point proposals are unstable intersections, and they persist as players gain experience. Surprisingly, the stable midpoint of line 24 accounts for a scant 3% of proposals in Session 1, although it is offered a little more frequently in later sessions.

In Game M, an equidistance point is offered in 8% of all proposals in Session 1, but seldom in later sessions. The stable midpoint of line 24, at the blisspoint of player 3, is offered in only 6% of the Session 1 proposals, but attracts a littl more attention as players gain experience.

In Game W, intersections, particularly when they are also midpoints, account for most of the prominent point proposals. Again, the stable midpoint of line 24, th blisspoint of player 3, is seldom proposed. In Game W, as in H and M, players s em reluctant to propose a point so close to player 3.

Table 13. Perc nt of All Points Proposed in Five Games Within 5 Unit of a Prominent Point, by Game, Type, Stability and S ssion

	TYPE OF PROMINENCE	SESSION 1	2	3-4	All
GAME A					
Prominent Points (ALL Unstable)	Mid LIJ	26.9	9.4	9.1	19.3
	L14IntL25	1.0	3.5	1.5	1.7
	Mid S	2.5	2.4	1.5	2.3
	ALL	29.9	15.3	12.1	23.0
	(n)	(197)	(85)	(66)	(348)
GAME F					
Unstable Prominent Points:	Mid LIJ	13.6	14.3	6.5	11.9
	Intrsctn	1.5	—	—	0.5
	Mid S	3.0	2.2	1.6	2.3
	ALL	18.2	16.5	8.1	14.6
Stable Prominent Point:	L14IntL25	—	—	—	—
	(n)	(66)	(91)	(62)	(219)
GAME H					
Unstable Prominent Points:	Mid LIJ	3.8	8.8	3.1	4.9
	Intrsctn	6.9	18.8	14.1	11.5
	Mid S	2.5	6.2	3.1	3.6
	ALL	8.1	21.2	14.1	12.8
Stable Prominent Point:	Mid L24	3.1	8.8	6.2	5.3
ALL Prominent Points:		11.2	30.0	20.3	18.1
	(n)	(160)	(80)	(64)	(304)
GAME M					
Unstable Prominent Points:	Mid LIJ	2.1	0.7	—	0.9
	Intrsctn	3.1	3.0	2.9	3.0
	Mid S	8.3	1.5	1.0	3.3
	ALL	13.5	5.2	3.8	7.1
Stable Prominent Point:	Mid L24	6.2	6.7	9.5	7.4
ALL Prominent Points:		19.8	11.9	13.3	14.6
	(n)	(96)	(135)	(105)	(336)
GAME W					
Unstable Prominent Points:	Mid IJ	6.2	7.9	7.8	7.2
	Intrsctn	7.0	10.0	11.7	9.2
	Mid S	4.7	2.9	2.6	3.5
	ALL	11.6	12.1	14.3	12.4
Stable Prominent Point:	Mid L24	—	5.7	2.6	2.9
ALL Prominent Points:		11.6	17.9	16.9	15.3
	(n)	(129)	(140)	(77)	(346)

Prominent point proposals and xperience. Table 14 presents, for the pooled data from Games A, F, H, M, and W, the percent of proposals that offer a point within 5 units of a prominent point.

Effects of prominence are most appar nt in the *first proposals* offered in each decision problem, as shown in the top section of Table 14. Nearly a quarter of all

Table 14. Percent of Points Proposed in Five Games Within 5 Units of a Prominent Point, by Type, Stability, and Session

	TYPE OF PROMINENCE	SESSION 1	2	3-4	All
FIRST PROPOSALS					
Unstable Prominent Points:	Mid LIJ	10.5	6.5	9.4	8.5
	Intrsctn	—	12.9	6.2	7.3
	Mid S	10.5	12.9	9.4	11.0
	ALL	21.1	29.0	21.9	24.4
Stable Prominent Points:	Mid L24	5.3	16.1	18.8	14.6
	L14IntL25	—	—	—	—
ALL Prominent Points:		26.3	45.2	40.6	39.0
	(n)	(19)	(31)	(32)	(82)
ALL PROPOSALS					
Unstable Prominent Points:	Mid LIJ	12.0	7.5	4.8	8.8
	Intrsctn	4.0	6.8	5.9	5.4
	Mid S	3.9	2.8	1.9	3.0
	ALL	17.3	13.0	9.9	14.0
Stable Prominent Points:	Mid L24	1.7	4.5	4.3	3.3
	L14IntL25	—	—	—	—
ALL Prominent Points:		19.0	17.5	14.2	17.3
	(n)	(648)	(531)	(374)	(1553)
TENTATIVE DECISIONS					
Unstable Prominent Points:	Mid LIJ	8.7	4.8	1.2	5.0
	Intrsctn	9.8	4.8	2.4	5.6
	Mid S	1.1	1.6	—	1.0
	ALL	16.3	8.9	3.5	9.6
Stable Prominent Points:	Mid L24	4.3	6.5	1.2	4.3
	L14IntL25	—	—	—	—
ALL Prominent Points:		20.7	15.3	4.7	14.0
	(n)	(92)	(124)	(85)	(301)
FINAL OUTCOMES					
Unstable Prominent Points:	Mid LIJ	10.5	6.5	—	4.9
	Intrsctn	5.3	—	—	1.2
	Mid S	—	—	—	—
	ALL	10.5	6.5	—	4.9
Stable Prominent Points:	Mid L24	10.5	6.5	3.1	6.1
	L14IntL25	—	—	—	—
ALL Prominent Points:		21.1	12.9	3.1	11.0
	(n)	(19)	(31)	(32)	(82)

first proposals offer unstable prominent points, and this does not change much across sessions. On the other hand, the frequency with which first proposals offer stable prominent points increases noticeably after Session 1.

The contrast of these results with those displayed in the second section of Table 14 for *all* proposals indicates that prominent points are selected much less frequently as the bargaining continues beyond the first round. Moreover, the proportion of prominent points offered in all proposals decreases noticeably across sessions.

The third section of Table 14 indicates that the frequency of unstable prominent point proposals that reach a *tentative decision* is much like that of *all* proposals in Session 1, but declines sharply across sessions. Note that only 4% of the tentative decisions are based on a stable prominent point.

There is considerably less prominence in those tentative agreements that become *final outcomes*, as shown in the bottom section of Table 14. Notice that the frequency of unstable prominent points in the final outcomes decreases with experience. No unstable prominent point was enacted in Sessions 3 and 4. It seems curious that stable prominent point outcomes also decrease across sessions and that a scant 6% of all final outcomes enacted stable prominent points. Moreover, only 5% of all points proposed to (and 14% of the final outcomes enacted by) a stable coalition lie within 5 units of the coalition's stable point. Most proposals to stable coalitions offer points that are neither prominent nor stable.

Prominence, stability, and experience in point selection. To develop further insight into the interactions of prominence, stability and experience, we focus on proposals to stable coalitions. Table 15 displays the percent of proposals to a stable coalition that lie closer to the nearest prominent but unstable point or to the coalition's center of gravity (if this point is unstable) than to the stable point for that coalition. In Session 1, fully two thirds of all proposals to a stable coalition lie closer to an unstable prominent point than to the demand equilibrium. To make allowances for deviations from stability toward the center of the coalition's Pareto set that are not captured by the prominent points, we can also include the coalition's center of gravity, if it is unstable, with the unstable prominent points. The proposals that are closer to the nearest such unstable point decrease across sessions, but they still account for half of the proposals in Sessions 3-4. Of the final outcomes enacted by stable coalitions in these games, virtually half in Session 1, but only 16% in Sessions 3-4, lie closer to one of the unstable prominent points. Table 16, although it is based on *all* prominent points and the proposed coalition's center or gravity, whether stable or unstable, reveals a similar pattern.

The influence of prominence on the selection of points is strong, but attenuates gradually across sessions. As players gain experience, they tend to shift their proposals from the unstable prominent points toward the stable solution. Nonetheless, the final decision outcomes enacted in Sessions 3-4 still deviate on average 59 units from, and only 14% lie within 5 units of, the solution. With experience, players tend to approach, but seldom reach, the demand equilibrium.

Table 15. Percent of Points Proposed to a Stable Coalition that are Closer to Some Specified Unstable Point than to the Stable Point

	SESSION			
	1	2	3-4	All
FIRST PROPOSALS				
Prominent point	58.3	50.0	50.0	51.6
Prominent or CG point	66.7	58.3	50.0	56.5
(n)	(12)	(24)	(26)	(62)
ALL PROPOSALS				
Prominent point	67.7	52.8	48.3	57.6
Prominent or CG point	69.0	56.1	50.0	59.6
(n)	(551)	(485)	(354)	(1390)
TENTATIVE DECISIONS				
Prominent point	50.0	39.1	25.6	38.6
Prominent or CG point	50.0	43.5	29.3	41.4
(n)	(88)	(115)	(82)	(285)
FINAL OUTCOMES				
Prominent point	47.4	40.0	16.1	32.5
Prominent or CG point	47.4	50.0	16.1	36.2
(n)	(19)	(30)	(31)	(80)

Table 16. Average Deviation of Points Proposed to a Stable Coalition from Nearest Prominent Point (Stable or Not) or Coalition's Center of Gravity, and from Coalition's Stable Point

	SESSION			
	1	2	3-4	All
FIRST PROPOSALS				
Prominent point	69	58	53	58
Prominent or CG point	63	46	47	50
Stable point	195	108	86	115
(n)	(12)	(24)	(26)	(62)
ALL PROPOSALS				
Prominent point	70	87	84	80
Prominent or CG point	66	81	80	75
Stable point	127	114	90	113
(n)	(551)	(485)	(354)	(1390)
TENTATIVE DECISIONS				
Prominent point	64	88	104	85
Prominent or CG point	58	81	96	78
Stable point	96	81	75	84
(n)	(88)	(115)	(82)	(285)
FINAL OUTCOMES				
Prominent point	72	72	99	82
Prominent or CG point	63	66	96	77
Stable point	72	60	49	59
(n)	(19)	(30)	(31)	(80)

5 CONCLUSION

This paper analyzes data from an experimental study of processes and outcomes of simple majority decision making in five-person spatial games. The laboratory design provides a rich source of machine-readable details about highly motivated and increasingly sophisticated bargaining processes in spatial games governed by simple majority rule.

The results are consistent with this scenario. Although at the beginning of the first session players are confronting a novel and complex situation, they come equipped with some rough understanding of bargaining problems in general that they have gleaned from their previous experience. They know that alliances form more readily among people who are similar, and that agreements are reached through compromise when there is some conflict of interests. They use such insights heuristically to cope with this new environment. They key on features in the spatial representation that flag their attention, and investigate whether such cues might provide clues to the problem's solution. They initiate the bargaining tentatively, "testing the waters" with prominent proposals. But with experience, through multiple rounds of intense negotiations within a game, and across games and sessions, they begin to appreciate more fully the competitive structure of the situation, and to progress toward the demand equilibrium. They move away from proposals that, while prominent, are far from the stable solution. But they continue to respond to spurious cues that signal proposals reasonably close to the equilibrium. Through competition and learning, players soon discover the stable coalitions, but progress more slowly toward decision points of the equilibrium solution.

This synopsis suggests an important task for subsequent research: to analyze proposals generated in this situation not as aggregates, but as sequences of responses to particular bargaining histories. The objective is to contribute toward the development of rigorous predictive theory about the processes and outcomes of competition and learning in highly motivated, increasingly sophisticated, but only quasirational collective decision making.

APPENDIX A. A BARGAINING MECHANISM

In the following procedure, the symbol "M:" identifies an action by the mechanism, whereas "P:" denotes a choice opportunity for the player who has the floor.

A. Tentative Decision (Selection) Phase

A.0 Initialization (START)

M: Specify the set N of n players, the set X of decision alternatives, the default option $z \epsilon X$, and the outcomes in X that each coalition $S \subseteq N$ can enact (thus defining the admissible proposals).

M: Set the parameters $N' := N$, $t := 0$, t^* and a^*, where t indexes the round, t^* is the maximum number of rounds available for negotiations (default value: t^*=999), and a^* is the minimum number of opportunities that must be given each player in the verification phase to initiate an alternative agreement before the current tentative decision becomes the final decision outcome (default value: a^*=2).

A.1 Chance move

M: Select $i \in N'$ at random and set *initiator*=*i*.

A.2 Initiation of tentative proposal

M: Set $t := t+1$. If $t > t^*$, go to B.4 (default outcome). Otherwise, announce that this is round t and that i has the floor as initiator.

P: The initiator i must choose just one of the following options:

- *pass*[optional: *pass to* $j \in N \backslash i$].

  [M: Announce that i passes[to j, if specified]. If j is specified, then set initiator $i := j$ and go to A.2. Otherwise, set $N' := N \backslash i$ and go to A.1.]

- *propose* x to $R \subseteq N \backslash i$ (ordering R).

  [M: Set $S := i \cup R$.]

A.3 Response to tentative proposal.

M: If $R = \emptyset$, go to B.0. Otherwise, set *responder*, $r :=$ first player in R. and announce that r has the floor as responder to (x,S).

P: The responder, r, must choose just one of the following options:

- *reject* (x,S) [and assume the initiator position].

  [M: Set the initiator, $i := r$. Go to A.2.]

- *accept* (x,S).

  [M: Set $R := R \backslash r$. Go to A.3.]

B. <u>Final Decision (Verification) Phase</u>

B.0 Set parameters

M: Set the tentative decision $(x^*,S^*) := (x,S)$, $N' := N$, and each player's number of opportunities to initiate a counterproposal: $(\forall i \in N) a_i := 0$.

B.1 Chance move

M: If $N' = \emptyset$, go to B.5. Otherwise, select $i \in N'$ at random.

Set *initiator*$:= i$ and $a_i := a_i + 1$.

Set $N' := \{ j \in N \mid a_j < a^* \}$.

Open floor to the initiator.

B.2 Initiation of counterproposal

P: The *initiator* i must choose just one of the following options:

- *pass*.

  [M: Announce that i passes.

  Set $N' := N' \backslash i$ if $N' \backslash i \neq \emptyset$.

  Go to B.1.]

- *propose* x to $R \subseteq N \backslash i$ (ordering R).

  [M: Set $S := i \cup R$.

  If $i \in S^*$, then (1) set $t := t+1$ and,

  (2) if $t \leq t^*$, announce that this is round t and go to A.3, but

  (3) if $t > t^*$, go to B.4 (default outcome).]

B.3 Response to counterproposal

M: If $R = \emptyset$ go to B.1.

Otherwise, set $r :=$ first player in R and open floor to r.

P: The responder, r, must choose just one of the following options:

- *reject* (x,S).

  [M: Go to B.1.]

- *accept* (x,S).

  [M: Set $R := R \backslash r$. If $r \in N \backslash S^*$, go to B.3. Otherwise (1) set $t := t+1$ and (2) if $t \leq t^*$, then announce this is round t and go to A.3.]

B.4 Default outcome

M: Announce that the final decision outcome is the default option, z.

STOP.

B.5 Decision outcome

M: Announce that S^* has enacted x^* as the final decision outcome.

STOP.

APPENDIX B. LABORATORY INSTRUCTIONS

The problem. This is a decision problem in which 5 persons must pick a point in the plane. The 5 persons have different points in the plane as ideal positions. Every person tries to obtain a result that is as close as possible to his ideal position. The position of the resulting point is determined by simple majority rule: that is, at least 3 of the 5 persons have to agree to the point. — The distances are measured in the Euclidean way, i.e. a player evaluates all points on the same circle round his ideal position as being equally good. — The persons are denoted as G (green),) (orange), R (red), V (violet) and B (blue).

Overview of the decision procedure. The decision is done via computers. It happens in two phases, A and B: The aim of phase A is to find a point supported by a majority of the players ("tentative decision" = T). — Frequently it happens that not all players are content with the result of phase A. — Therefore, in a control phase (phase B) the opportunity is given to modify the tentative decision or to replace it by another one. — When a tentative decision survives the control phase, it becomes the final result.

The payoffs of the players are determined by the distances of their ideal positions from the final result. If a final point is not obtained the players get a default outcome (here - 2000 points per player).

Notations.

A proposal is a point in the Euclidean plane, and the specification of two or more players in a given sequence.
— Example: G suggests [(73,107), B,O,V] - here G proposes the point (73,107) first to B, then to O, then to V.

If B,O,V agree to this proposal, then a tentative decision T = [(73,107),G,B,O,V] is made. Its stability will be checked in phase B.

The procedure in detail

Phase A (when no tenative decisions exists)

A.1 A player is determined via computer by chance (every player has the same probability to be selected).

A.2 The selected player chooses between

pass	i.e. the game moves to A.1 (selection of player by chance).
pass to..	i.e. the player selects a specific other player who continues from A.2 (i.e. he has the right of initiation).
send proposal	i.e. the player specifies a proposal.

A.3 When the proposal is made, then the first play r of the proposal is selected.

A.4 The selected player chooses between

not agree i.e. he does not agree to the proposal. — Then the proposal is rejected, and the game moves to A.2, where the rejecting player gets the opportunity to initiate a proposal.

agree i.e. he (preliminary) agrees to the proposal. Then follows

either (if not yet all players where asked if they "agree" or "not agree"): the selection of the next player of the proposal, he will be asked next at A.4.

or (if all players of the proposal agreed): the proposal becomes a tentative decision "T," and the game enters the control phase B (next figure).

Phase B (control phase)

— Phase B is initiated by every new tentative decision.

— Phase B is ended, when every player has had two chances to make a new proposal (at B.1) (or to select "pass" instead).

B.1 A player is selected by chance from those those players who have not yet been in B.1 for two times.[1] He can select between:

pass* i.e. another player gets his chance at B.1

make proposal* then this proposal will be submited to the first player listed in the proposal.

*Only these actions "count" in phase B.

B.2 The player who gets a proposal in phase B can select between:

agree i.e. he agrees to the proposal. There ar two cases:

	<u>Case 1:</u> if thereby the proposal becomes a new tentative decision (because all players agreed) then a new phase B starts (at B.1) <u>Case 2:</u> otherwise the next player of the proposal is asked.
not agree	i.e. he does not agree to the proposal. — then the proposal is rejected, and another player will be selected at B.1.
END	The game is ended, when for the same tentative decision every player has had two opportunities to make a proposal (at B.1) (Count yourself!)

<u>Comment:</u> If you want a tentative decision to become the final result in Phase B, then you must click only "pass" and "not agree."

1) The counting starts over again for each new tentative decision.

<u>The Payoff Method</u>

You will play 8-9 different games with 6 different figures for overall about 12 hours.

For every separate game the distance to your ideal position from the resulting point of the game will be compared with the mean distance to the same ideal position in this game from the points selected by other groups in which you, in whatever position, did not take part.

<u>For every distance-point that you are better (worse) than this mean, you receive (lose) 20 Pfennigs.</u>

Your wins and losses are added over all the games in which you take part. To the total a constant[2] will be added in such a way that the most successful play r (among all participants of these games) will win about DM 300.–, perhaps even DM 400.–.

Overall you are guaranteed to receive a minimum payoff of DM 40.- for the two days.

So adjust yourself to a payoff between DM 40.– and DM 400.–, and try to maximize your payoff by optimal behavior.

If, during the game, you get the impression that will not get more than the minimum payoff because your preceding results are extremely bad, and if you therefore do not feel motivated any more, then contact the person who is conducting the game.

---

2)This constant is already fixed, so you cannot increase your own payoff by reducing the outcomes of others. Your result is only affected by your own distance from the final result.

1. The Mouse

- moves the cross hairs / the arrow
- left button -> sets a point (at cross hairs)
  -> gives command (if arrow)

2. The points.

- the points are characterized by coordinates
- for the point in the crosshairs, the coordinates and the distance to own ideal position are given in the upper white window

3. The screen.

- shows the ideal positions of the players
- shows circles (= "iso-utility lines") of the players through the tentative decision T and through the proposal S you are planning to send.
- the headline gives: your color - the coordinates of your ideal position - your payoff if no final point is obtained - (in case) the coordinates of the current tentative decision T and the distance of T to your ideal position.
- the second line gives the opportunity to open windows (see below)
- the upper white panel gives coordinates and distance to your ideal position from the point in the cross hairs.
- the lower white panel gives a complete protocol of the bargaining process. If the protocol is long, you can move the view of the protocol: click the up and down arrows.

4. The windows.

- are opened by "click" in the second (light blue) line
- commands within the windows are also activated by "click."

Remark:

- if white windows appear in the middle of the screen, you have to answer them first by clicking "continue", "OK", or "cancel", before further activities are possible.
- the other windows can be removed by clicking "cancel" in the second line (light blue).

The windows

The following windows are available.

A "distances"

— enables the command to show the distances to the ideal position of all players in the upper white window

B "zoom"

— enables the command to zoom in any of 3 steps: normal / fine / very fine

— it is possible to select the center of zoom (to change the view of the zoomed picture). The center of zoom is clicked at the cross hairs.

C "plot"

— puts points and circles on the screen.

— the following options are available:

"point selected from screen": select an arbitrary point from the screen with the cross hairs and draw it by "click."

"circles through point selected from screen": draws the iso value-line of all players through a given point (the point must be drawn on the screen first)

"circles through point selected from list": analogously (the list contains all points that appeared during the game)

"lines connecting players": draws all (or selected) connection lines between the ideal points of the players

D "unplot"

— analogous "plot," but for removing points, circles or lines

— the points "T" (and, in case, "D") and corresponding circles cannot be removed

— the point to be removed must be centered in the cross hairs quite precisely (if necessary, click several times)

E | "send" | (= transmit message)

— all messages are sent to all other players

— the following messages are possible:

"pass" — the computer then selects one of the other players by chance (each of them has the same probability of being selected)

"pass to..." — selects another player, who may send the next message.

"create proposal"

— does not need to be clicked

— a proposal offers a point to other players. (Point, players and the order of the players must be selected.)

— a proposal is created as follows:

"select point from screen": select the point at the center of the cross hairs by clicking the mouse (this point must first be drawn on the screen).

Attention: check if the point that the system recorded is the the point you intended. The selected point is the selected point is labeled on the screen by "S."

"select point from list": here the point can be selected from a list — again the selected point is marked by "S."

"select players": enables the command to select players in a given order, to whom the proposal will be sent.

"send proposal": when a proposal (point + players) is ready, it can be sent by clicking "send proposal" (otherwise it does not leave your terminal). — to avoid errors, the computer shows the whole message again on your screen and asks, if it should really "send", or "cancel".

Remark: "agree" and "not agree" are only permitted as answers to proposals of others. They are the only permitted answers on proposals of others. - If you choose "not agree" then you can give your own message.

Point Labels:

T = current tentative decision

P = last proposal that was sent

S = the point in your "send" window

D = default point

REFERENCES

Albers, W. (1987). Aspirations and aspiration adjustment in location games. Working Paper 159, Institute for Mathematical Economics Research, Univ. of Bielefeld.

Albers, W., and Laing, J.D. (1988). A noncooperative mechanism for negotiations in two classes of cooperative games. Working note, Dept. of Decision Sciences, the Wharton School, Univ. of Pennsylvania.

Kormendi, R.C., and Plott, C.R. (1980). Committee decisions under procedural rules: an experimental study applying a new nonmonetary method of preference inducements. Social Sciences Working Paper 345. California Institute of Technology.

Laing, J.D. (1988). A noncooperative model of bargaining in simple spatial games. Working Paper 88-11-03, Dept. of Decision Science, the Wharton School, Univ. of Pennsylvania.

Laing, J. D., and Olmsted, S. M. (1978). An experimental and game-theoretic study of committees. In P. C. Ordeshook (ed.), *Game Theory and Political Science*: 215-281. New York NYU Press.

Ordeshook, P.C. (1986). *Game Theory and Political Theory: An Introduction*. Cambridge Univ. Press.

Selten, R. (1981). A noncooperative model of characteristic-function bargaining. In V. Boehm and H. Nachthanp, eds., *Essays in Game Theory and Mathematical Economics in Honor of Oskar Norgenstern*. Wissenschaftsverlag Bibliographisches Institut Mannheim, Wien – Zürich, 131-151.

ORIGINAL OR FAKE -
A BARGAINING GAME WITH INCOMPLETE INFORMATION

by

Reinhard Selten and Werner Güth

Summary: We consider a special class of noncooperative bargaining games with incomplete information and two agents who bargain about the price of a given object. The object can be either of high value or of low value. Whereas the seller knows the real value, the buyer is not completely informed in that respect. With probability w he expects that the object is of low value and with the probability $1-w$ that it is of high value. The parameter w is common knowledge. One can therefore distinguish two types of the seller, namely the one who tries to sell an object of low value and the one who offers the more valuable object. To have something specific in mind, the commodity is assumed to be a work of art which can be either a fake or an original. The two types of the seller are accordingly called fake seller and honest seller. The buyer is of a unique type. Bargaining is supposed to proceed in form of a unanimity game, i.e., the players simultaneously determine their price offers. A contract results only if a seller and the buyer have chosen the same prices. As it is typical for bargaining games, the game has many equilibrium points. We apply the equilibrium selection theory of Harsanyi and Selten to determine a unique solution point for each of the bargaining games.

The solution as a function of parameters exhibits a surprising discontinuity which one can loosely describe as follows: Whereas in a certain subset of the parameter space the solution price can be considered as a compromise within the set of equilibrium prices, the solution price outside this subset is always the worst equilibrium price for the buyer. Knowing the unique solution of each game one can ask who has to suffer more from the risk of art forgery, the buyer or the honest art dealer. To answer this question we compare the payoffs of both players for $w = 0$ with those for $w > 0$. It is an interesting result that it is always the buyer who suffers more from the risk of art forgery. An intuitive argument is given which tries to explain this result.

0. Introduction

In his pathbreaking paper on games with incomplete information John C. Harsanyi (1968) has developed a Bayesian model of game situations in which the players are not fully informed about payoffs or strategic probabilities of their opponents. In this paper, a special bargaining situation will be modeled as a game with incomplete information. We consider a seller and a buyer bargaining on the price of a work of art. The seller knows whether the work of art is an original or a fake, but the buyer is uninformed in this respect. In the language of the theory of games with incomplete information we may say that there are two types of sellers, an honest seller and a fake seller. There is only one type of buyer. His subjective probabilities for both types of sellers are common knowledge.

The bargaining situation is modeled in the simplest possible way as a unanimity game. Both players simultaneously name a price. If both name the same price, the object is sold at this price. If the prices are different, bargaining ends in conflict and the object remains in the possession of the seller. This simple bargaining model has many equilibrium points. The theory of equilibrium selection developed by Harsanyi and Selten (1988) is applied in order to select one of them as the solution of the game.

It is an important feature of the situation modeled in this paper that an honest seller cannot prove that the work of art is an original and not a fake. Otherwise, the situation of incomplete information could not prevail since an honest seller would be able to reveal his type and a fake seller could be recognized by the fact that he cannot prove originality.

Obviously the situation considered here is quite common and not restricted to the art market. It should be pointed out, however, that we are not dealing with a bargaining problem embedded into a larger competitive market like Akerlof's 'lemons' (Akerlof, 1970). The situation considered here is strictly bilateral.

The simple framework of the unanimity game does not leave any room for signaling. Games of incomplete information where signaling is important have been analysed in the literature (Spence, 1973 and 1974; Selten, 1975a). In the situation considered here, the signaling moves before the beginning of bargaining would not add anything essential to the strategic possibilities of the players as long as the signals are costless and the same set of signals is available to both types of sellers. Any signal sent by an honest seller could be imitated by a fake seller. It is an inherent feature of the situation that a fake seller always has an interest to appear to be an honest seller. This is the reason why signaling is unimportant here.

In the literature, bargaining under incomplete information has been modeled by multistage-bargaining models (Harsanyi and Selten, 1972). Such models implicitly provide possibilities for signaling. In our case a one stage-model where both players make just one decision simultaneously is a sufficient approximation of the bargaining situation since signaling would be ineffective.

In the solution derived by equilibrium selection theory the risk that the work of art is a fake is not shared equally by buyer and seller. In a sense to be specified in the paper, the buyer has to bear the greater part of this risk. This is true for all parameter constellations which permit trading in equilibrium.

In Section 1, we describe the bargaining situation underlying the bargaining game which will be defined in Section 2. In Sections 3 and 4, we determine the strong equilibrium points of the bargaining game and thereby its first candidate set.

Dominance comparisons are defined in Section 5 and actually performed for the game at hand in Sections 6, 7, 8, 9, and 10. An overview over all dominance comparisons is given in Section 11 with the help of dominance diagrams. Section 12 explains how dominance relations can be used to eliminate solution candidates and why a substitution step has to be performed if no candidate can be eliminated in this way. Three different principles which determine the solution in three subregions of the parameter space are discussed in Sections 15, 16, and 17. The final Section 17 summarizes the results and discusses how the solution depends on the parameters of the bargaining situation.

1. The Bargaining Situation

It is our intention to investigate a bargaining situation in which a seller offers an object of art, say, a painting, to a buyer. The seller knows whether the painting is an original or a fake. The buyer, however, is in a situation of incomplete information. He does not know whether the painting is an original or a fake. He knows the probability w for the event that the painting is a fake. This parameter w is assumed to be common knowledge of both bargainers.

We assume that both bargainers are risk neutral. This means that we do not have to distinguish between expected utilities and expected monetary payoffs. If the painting is an original, it has a value of 1 for the seller and a value of h for the buyer. It is assumed that we have $h > 1$. If the painting is a fake, it has 0-value both for the seller and the buyer. These assumptions on the value of the painting for buyer and seller do not entail any loss of generality since the zero points of the utility scales and the money unit can be chosen arbitrarily.

The bargaining situation is modeled in the simplest possible form as a unanimity game: Buyer and seller simultaneously and independently make price proposals. Each of them names one price. If both prices agree, the painting is sold at this price. If both prices disagree, the painting is not sold. The buyer also has the possibility not to make any proposal at all which means that he does not want to engage in bargaining.

The game will be analysed with the help of the equilibrium selection theory developed by Harsanyi and Selten (1988). This theory selects one of the equilibrium points as the solution of the game. In our case, the solution uniquely determines a price of the painting if mutually profitable bargaining is possible at all.

2. The Bargaining Game

The bargaining game is a game with incomplete information in the sense of Harsanyi (1968). In such games it is useful to replace the original game by another one which is called the type game. Types are possible descriptions of actors. Incompleteness of information means that it is not known with certainty which of the possible descriptions of an actor is the correct one.The type game models each type of an actor as a different player.

In our case, the seller has two types which will be named player 1 and player 2. Player 1 is the seller type who owns a fake and player 2 is the seller type who owns an original. The buyer has only one type, named player 3.

It is assumed that proposed prices must be integer multiples of a smallest money unit g. Moreover, only such price proposals will be admissible which enable the seller and the buyer of an original to make positive profits. Proposed prices must be integer multiples kg of g with

$$1 < kg < h. \tag{1}$$

For players 1 and 2 these price proposals are their pure strategies. Player 3 also has these pure strategies but in addition he has a pure strategy ϕ which is interpreted as the decision not to make any price proposal. We refer to ϕ as player 3's non-bargaining strategy.

Figure 1 shows the extensive form for the case of only two possible price proposals $r = 1+g$ and $s = 1+2g$.

Equilibrium selection is not applied to the extensive form but to the "standard form" which in our case agrees with the normal form of the type game.

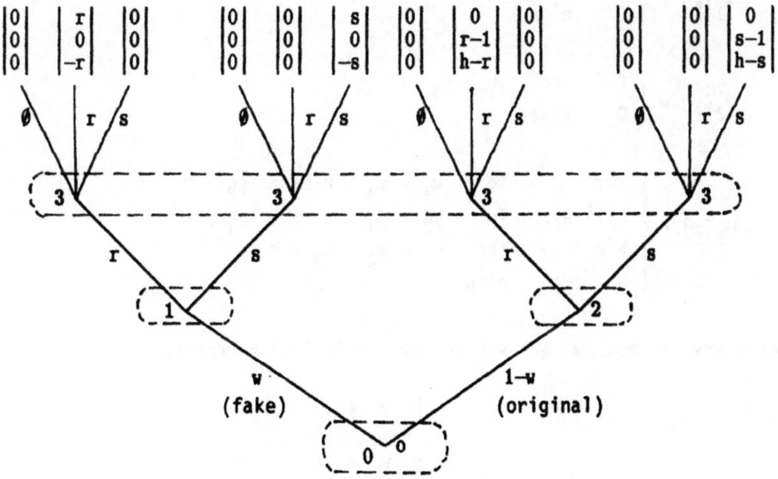

Figure 1: The bargaining game for the case of only two price proposals r and s. Dotted lines indicate information sets. Payoffs of players 1, 2, and 3 are shown in this order from above to below above the endpoints.

An n-person game in normal form $G = (\Phi_1,\ldots,\Phi_n;H)$ consists of n finite sets $\Phi_1,\ldots,\Phi_n$, the sets of pure strategies for players $1,\ldots,n$, respectively, and a payoff function H. For every strategy combination

$$\varphi = (\varphi_1,\ldots,\varphi_n) \text{ with } \varphi_i \in \Phi_i \text{ for } i = 1,\ldots,n \tag{2}$$

the payoff function H assigns a payoff vector

$$H(\varphi) = (H_1(\varphi),\ldots,H_n(\varphi)). \tag{3}$$

In our case $\Phi_1 = \Phi_2$ is the set of all admissible price proposals (integer multiples of g satisfying (1)). Player 3's pure strategy set contains the same pure strategies and an additional pure strategy Φ.

It is convenient to base the analysis on a game which is equivalent to the type game but slightly different. Instead of the expected payoff derived from the extensive form, players 1 and 2 receive the conditionally expected payoffs for the case that the seller is of the concerning type. This means that we omit a constant factor w or $1-w$, respectively. Define

$$b = (1-w)h . \tag{4}$$

With the help of (4) the payoffs for a strategy combination $\varphi = (\varphi_1,\varphi_2,\varphi_3)$ can be described as follows:

$$H_1(\varphi) = \begin{cases} r & \text{for } \varphi_1 = \varphi_3 = r \\ 0 & \text{else} \end{cases} \tag{5}$$

$$H_2(\varphi) = \begin{cases} r-1 & \text{for } \varphi_2 = \varphi_3 = r \\ 0 & \text{else} \end{cases} \tag{6}$$

$$H_3(\varphi) = \begin{cases} -wr & \text{for } \varphi_1 = \varphi_3 = r \text{ and } \varphi_2 \neq r \\ b-(1-w)r & \text{for } \varphi_2 = \varphi_3 = r \text{ and } \varphi_1 \neq r \\ b-r & \text{for } \varphi_1 = \varphi_2 = \varphi_3 = r \\ 0 & \text{else.} \end{cases} \tag{7}$$

The parameters h and w are subject to the following constraints:

$$h > 1 \tag{8}$$

$$0 < w < 1. \tag{9}$$

As we shall see later on only those parameter constellations are of interest for which in addition to this the following condition is satisfied:

$$b > l \,. \tag{10}$$

If (10) is not satisfied, then no equilibrium point exists which results in a sales contract.

We shall assume that the parameters l, h, w, and b are integer multiples of $4g$. This simplifying assumption is not really necessary but it considerably reduces the complexity of the analysis. A paper on a different bargaining situation shows the problems which would arise without such a divisibility assumption and how these problems can be solved (Selten and Leopold, 1982).

The divisibility assumption has the consequence that there are at least three admissible prices in the interval (l,h) given by (1), namely, $l+g$, $l+2g$, $l+3g$, since in view of (8) the smallest possible value of h is $l+4g$.

3. Strong Equilibrium Points of the Bargaining Game

An equilibrium point can be described as a strategy combination with the property that no player can improve his payoff by a deviation if all other players use their equilibrium strategy. An equilibrium point is called strong if every deviation of a player entails a loss if the others play their equilibrium strategies. A strong equilibrium point must be an equilibrium point in pure strategies.

Let (π_1, π_2, π_3) be a strong equilibrium point of the bargaining game. The equilibrium strategy of the buyer cannot be his non-bargaining strategy ϕ since he can also obtain his equilibrium payoff by a price proposal which differs from π_1 and π_2 (at least three prices are admissible). If player 3 proposes an admissible price $\pi_3 = r$, then $\pi_1 = r$ and $\pi_2 = r$ are the only best replies for both types of sellers. Therefore a strong equilibrium point must have the form $\pi = (r,r,r)$. A strategy combination of this form cannot be a strong equilibrium point unless it yields a positive payoff for player 3. This follows by the fact that the player can enforce 0 by his non-bargaining strategy ϕ. With the help of (7) it can be seen immediately that player 3's payoff for (r,r,r) is positive if and only if $b > r$ holds.

We have seen that the strong equilibrium points of the bargaining game are the pure strategy combinations of the form (r,r,r) with

$$l < r < b. \tag{11}$$

It is now clear why parameter combinations with $b \leq l$ are not interesting. There is no bargaining range unless $b > l$ holds. In the following we shall always assume $b > l$.

In general, the bargaining game has many strong equilibrium points. With the help of equilibrium selection theory one of these strong equilibrium points will be singled out as the solution of the game.

Equilibrium selection theory is not applied to the game itself but to its uniformly perturbed games. In these games the choice of strategy is restricted to strategies which choose every pure strategy with a small minimum probability ε. Apart from this, the uniformly perturbed games agree with the original game. One determines a solution for every perturbed game and then obtains a limit solution for the original game as the limit of the perturbed game solution for $\varepsilon \to 0$.

The indirect determination of the solution as the limit of perturbed game solutions secures perfectness of the selected equilibrium point (for the definition of perfectness see Selten, 1975b).

We do not want to burden the analysis with tedious technical details. Therefore, in this paper equilibrium selection theory will be applied directly to the unperturbed bargaining game. In cases where the solution is a strong equilibrium point this shortened procedure almost always yields the same result as the application to the uniformly perturbed games. Especially for the case at hand this is true. The reasons shall not be discussed here (see Harsanyi and Selten, 1988).

It is not possible in the framework of this paper to explain equilibrium selection theory in full detail. We shall give precise descriptions of only those parts of the theory which are relevant for our problem. The explanations will be given where they are needed for the continuation of the analysis.

The type game has only one information set for each player. Such games will be called games with normal form structure. Some of the concepts of equilibrium selection theory can be explained more easily with the help of special definitions which are valid only for games with normal form structure. Wherever this will be convenient we shall rely on such special definitions. The more general definitions can be found in the book by Harsanyi and Selten (1988).

4. The First Candidate Set

The first step in the application of equilibrium selection theory is a process of reduction and decomposition which replaces the task of solving the game by the task of solving one or several smaller games. In the case at hand this process is without significance since it does not change the bargaining game. Therefore, we shall not explain the process of reduction and decomposition.

Before the next steps in the application of equilibrium selection theory can be discussed, some definitions and notations have to be introduced.

A mixed strategy q_i of player i is a probability distribution over the set Φ_i of his pure strategies; $q_i(\varphi_i)$ denotes the probability assigned to the pure strategy φ_i by q_i. A combination of mixed strategies $q = (q_1,q_2,q_3)$ contains one mixed strategy for each of the three players.

An i-incomplete strategy combination contains one strategy for each player except player i. For such combinations we use the notation $q_{-1} = (-,q_2,q_3)$, $q_{-2} = (q_1,-,q_3)$, $q_{-3} = (q_1,q_2,-)$. The pure i-incomplete combinations whose components are pure strategies are of special importance. Φ_{-i} denotes the set of pure i-incomplete strategy combinations φ_{-i}.

A joint mixture $q_{.i}$ is a probability distribution over Φ_{-i}; here $q_{.i}(\varphi_{-i})$ is the probability assigned to φ_{-i} by $q_{.i}$. A joint mixture $q_{.i}$ is interpreted as a possible subjective probability distribution of player i concerning the behavior of the other players. Even if the other players are known to act independently, it is possible that player i forms an expectation which requires a description by a joint mixture. Thus player 3 can expect that with probability z both other players choose price r and with probability 1-z both other players choose price s. For this joint mixture we also use the following notation:

$$q^z_{.3} = z(r,r,-) + (1-z)(s,s,-). \tag{12}$$

As we shall see expectations of this form are important for equilibrium selection theory.

A hybrid combination $(q_i,q_{.i})$ consists of a mixed strategy q_i of player i and a joint mixture $q_{.i}$ for the other players. $H_i(q_i,q_{.i})$ is the expected payoff of player i if he has the beliefs described by $q_{.i}$ and if he plays q_i. In this way we extend the definition of H_i to hybrid combinations. i-incomplete mixed strategy combinations can be looked upon as special joint mixtures. In this sense mixed strategy combinations are special hybrid combinations.

A strategy $\tilde{q}_i$ is a best reply to q_{-i} or $q_{.i}$, respectively, if it maximizes the payoff $H_i(q_i,q_{-i})$ or $H_i(q_i,q_{\cdot i})$, respectively.

Substructures, called formations, have to be investigated in games which cannot be further reduced or decomposed. A formation F of a game $G = (\Phi_1,\ldots,\Phi_n;H)$ in normal form is a game $F = (\Psi_1,\ldots,\Psi_n;H')$ in normal form with the following properties:

a) The pure strategy sets Ψ_i are subsets of the corresponding strategy sets Φ_i.

b) H' is the restriction of H to the strategy combinations of F.

c) F is closed with respect to best replies in G in the following sense: Let $q_{.i}$ be a joint mixture in G which assigns positive probability only to those i-incomplete strategy combinations φ_{-i} which also occur in F. A joint mixture in G of this kind is called F-admissible. Closedness against best replies means that each pure best reply φ_i to an F-admissible joint mixture $q_{.i}$ in G always belongs to the pure strategy set Ψ_i of the concerning player in F.

A formation is called primitive if it does not have any formations different from itself.

Let $\pi = (\pi_1,\ldots,\pi_n)$ be a strong equilibrium point of G. Then $F = (\Psi_1,\ldots,\Psi_n;H)$ with $\Psi_i = \{\pi_i\}$ for $i = 1,\ldots,n$ and H' as in b) is a primitive formation of G. This formation F is called the formation generated by π. It is, of course, also possible that primitive formations have strategy sets with more than one element.

In equilibrium selection theory, the solutions of primitive formations are regarded as natural candidates for the solution of the game. The set of all solutions of primitive formations is called the first candidate set and is denoted by Ω_1.

This kind of preselection is motivated by the idea that as far as possible the solution should have properties similar to those of a strong equilibrium point. Of course, one cannot restrict one's attention to strong equilibrium points alone since some games do not have strong equilibrium points.

We shall now show that the bargaining game has no other primitive formations than those generated by its strong equilibrium points. Let F be a primitive formation and let r be a price proposal with $r \in \Psi_3$. Then r must also be contained in Ψ_1 and Ψ_2 since for both players r is the only best reply to player 3's strategy r. If $r < b$ holds, F must be the primitive formation generated by (r,r,r). If $r \geq b$ holds, then ϕ is a best reply of player 3 to (r,r,-). Each strategy s with $s \leq b$ is a best reply of players 1 and 2 to player 3's strategy ϕ. This shows that s must be in Ψ_3 and that F cannot be primitive. The same reasoning also shows that F cannot be primitive if $\phi \in \Psi_3$ holds. We can conclude that the primitive formations of the bargaining game are exactly those generated by its strong equilibrium points. Consequently, the first candidate set Ω_1 is the set of all strong equilibrium points or, in other words, the set of pure strategy combinations (r,r,r) with $1 < r < b$.

5. Dominance Comparisons

Equilibrium selection theory makes use of two dominance relationships between equilibrium points, payoff dominance and risk dominance, in order to narrow down the first candidate set Ω_1. Dominance comparisons between two candidates U and V are not made in the original game but in a restricted game. If all players use different strategies in U and V, then the restricted game is nothing else than the smallest formation containing U and V. This is always the case in the bargaining game considered here.

U payoff dominates V if $H_i(U) > H_i(V)$ for all players i of the restricted game. In our case the inequality must be satisfied for $i = 1,2,3$. It can be seen immediately that in the bargaining game it does not happen that one strong equilibrium point payoff dominates the other. This is due to the fact that a higher payoff for the buyer is always connected to a lower payoff for the seller. Equilibrium selection theory combines payoff dominance and risk dominance to one dominance relationship which gives preference to payoff dominance and which is based on risk dominance alone where no payoff dominance is present. Since payoff dominance relationships do not occur in our bargaining game, dominance can be identified with risk dominance for the purpose of this paper.

The definition of risk dominance requires some preliminary remarks. If all players use different strategies in U and V, then apart from exceptional cases it is not important whether risk dominance is examined in the restricted game or in the original game. The exceptional cases are those in which the logarithmic version of the tracing procedure has to be applied (the tracing procedure will be explained later). Such exceptional cases will not occur in the analysis of the bargaining game. Therefore, for the sake of simplicity, we shall define risk dominance in a way which neglects the difference between the restricted game and the original game.

The concept of risk dominance makes use of the tracing procedure developed by John C. Harsanyi (1975). This procedure models a thinking process which starting from a prior strategy combination $p = (p_1,...,p_n)$ leads to an equilibrium point by a process of continuous adaptation. The prior strategy combination can be interpreted as a 'naive' theory of behavior which serves as the starting point of further deliberations. The result of the application of the tracing procedure to a prior strategy combination p of a game G is a uniquely determined equilibrium point $T(G,p)$ of G.

The definition of risk dominance between two candidates U and V is based on a special prior strategy combination p which is called the bicentric prior strategy combination for U and V. If for this p the result $T(G,p)$ of the tracing procedure is U, then U risk dominates V. Analogously, V risk dominates U for $T(G,p) = V$. It is also possible that $T(G,p)$ is neither U nor V. If this happens, none of both equilibrium points risk dominates the other.

In the next two sections the bicentric prior strategy combination and the tracing procedure will be explained in more detail.

6. The Bicentric Prior Combination

The bicentric prior combination $p = (p_1,\dots,p_n)$ for the comparison of two equilibrium points $U = (U_1,\dots,U_n)$ and $V = (V_1,\dots,V_n)$ can be interpreted as a naive behavioral theory based on the following assumptions:

(i) Player i believes that either all other players use their equilibrium strategy in U or all other players use their equilibrium strategies in V $(i=1,\dots,n)$,

(ii) Player i has a subjective probability z for U_{-i} ; accordingly, his probability for V_{-i} is $1-z$.

(iii) Player i chooses a pure best reply to $z\,U_{-i} + (1-z)V_{-i}$. If there are several pure best replies, all of them are chosen with the same probability.

(iv) The subjective probability z of player i for U_{-i} is a random variable which is evenly distributed over $[0,1]$.

(v) The probability $p_i(\varphi_i)$ for $\varphi_i \in \Phi_i$ in the bicentric prior strategy p_i of player i is his probability of choosing φ_i under assumptions (i) to (iv).

Assumption (i) is based on the idea that player i thinks that U and V are the only possible candidates for the solution of the game and that he is the only player not knowing which of the two equilibrium points is the solution.

Now, we can investigate which bicentric prior combination is obtained in our bargaining game for the comparison of the two candidates $U = (r,r,r)$ and $V = (s,s,s)$. For this purpose we have to determine the best replies to $z\,U_{-i} + (1-z)V_{-i}$. r is the best reply of player 1 to $z\,U_{-1} + (1-z)V_{-1}$ if we have

$$z\,H_1(r,r,r) + (1-z)H_1(r,s,s) \geq zH_1(s,r,r) + (1-z)H_1(s,s,s). \tag{13}$$

This is equivalent to

$$r\,z \geq s(1-z). \tag{14}$$

Since z is evenly distributed over $[0,1]$, the probability that (14) is satisfied is as follows:

$$p_1(r) = \frac{r}{r+s}\ . \tag{15}$$

This yields

$$p_1(s) = \frac{s}{r+s}\ . \tag{16}$$

Analogous considerations for players 2 and 3 lead to the following results:

$$p_2(r) = \frac{r-1}{r+s-2} \tag{17}$$

$$p_2(s) = \frac{s-1}{r+s-2} \tag{18}$$

$$p_3(r) = \frac{b-r}{2b-r-s} \tag{19}$$

$$p_3(s) = \frac{b-s}{2b-r-s} \ . \tag{20}$$

7. The Tracing Procedure

Let $G = (\Phi_1,\ldots,\Phi_n;H)$ be an n-person game in normal form and let $p = (p_1,\ldots,p_n)$ be a mixed strategy combination for G. In order to determine the result $T(G,p)$ of the tracing procedure, one looks at the family of auxiliary games $G^t = (\Phi_1,\ldots,\Phi_n;H^t)$ whose payoff functions H_i^t are related to H_i as follows:

$$H_i^t(\varphi) = (1-t)H_i(\varphi_i,p_{-i}) + t\,H_i(\varphi) \tag{21}$$

for $i = 1,\ldots,n$ and $0 \leq t \leq 1$. For $t = 0$ the payoff H_i^t only depends on player i's strategy and the prior strategies of the others. This corresponds to a situation where player i forms his beliefs exclusively on the basis of the prior strategy combination p. For $t = 1$ one obtains the game G. The greater the parameter t is, the less trust the players put on the naive theory expressed by the prior strategy combination.

For every t with $0 \leq t \leq 1$ let E^t be the set of all pairs (q,t) with the property that q is an equilibrium point of G^t. The union of all E^t with $0 \leq t \leq 1$ is denoted by E. We call E the equilibrium graph. Obviously, E can be looked upon as a subset of the Euclidean space containing all pairs of the form (q,t) where q is a mixed strategy combination for G.

Apart from exceptional cases there is only one best reply to the prior strategy combination p. Let $\pi = (\pi_1,\ldots,\pi_n)$ be this best reply. Obviously, π is an equilibrium point of G^0. As Harsanyi (1975) has shown, there is almost always a unique continuous path in the equilibrium graph E which connects $(\pi,0)$ with a pair of the form $(q^*,1)$. The strategy combination q^* is an equilibrium point of G which is uniquely determined by the application of the tracing procedure to p. This q^* is denoted by $T(G,p)$.

The tracing procedure, as it just has been explained, is called the linear tracing procedure. In border cases, in which there is more than one path from $t = 0$ to $t = 1$ in E, the linear tracing procedure does not yield a unique result. In such cases it is necessary to apply a modified version of the tracing procedure, the so-called logarithmic tracing procedure, to determine T(G,p). However, we shall not investigate such border cases here. Therefore, we omit the description of the logarithmic tracing procedure.

It is tempting to interpret the parameter t as a time variable. However, this interpretation is not quite correct since the path from $(\pi,0)$ to $(q^*,1)$ may sometimes move back in the sense that t decreases. Nevertheless, we shall use convenient ways of speaking based on the time interpretation of t.

In simple cases like those which we shall investigate here the path determined by the tracing procedure has a typical shape which can be described as follows: The path begins with $(\pi,0)$ and continues with combinations of the form (π,t) up to a point of time t'. In the game $G^{t'}$ there is exactly one player i for whom π_i is no longer the only best reply to π_{-i}. In $G^{t'}$ this player i has a second pure best reply ψ_i to π. We call this player i the first defector and t' the first reversal time.

Let q_i be a mixed strategy of player i which assigns positive probability only to π_i and ψ_i. In many cases all combinations (q_i,π_{-i}) with a q_i of this kind are equilibrium points of $G^{t'}$. This is the case whenever (ψ_i,π_{-i}) is an equilibrium point of $G^{t'}$. If the path is uniquely determined by the linear tracing procedure, then after going through these equilibrium points of $G^{t'}$ it is continued by combinations of the form $((\psi_i,\pi_{-i}),t)$ up to a second reversal time t''. In the game $G^{t''}$ the situation is the same as in the game $G^{t'}$: A player j (the case $i = j$ is not excluded) is the second defector who becomes indifferent between two pure strategies at $t = t''$.

In this way, the tracing procedure can reach reversal points several times until finally an equilibrium point of the original game is reached. However, it is also possible that for some reversal time, say t', a situation arises where (ψ_i,π_{-i}) fails to be an equilibrium point of $G^{t'}$ and that therefore the path cannot be continued as described above. In such situations it frequently happens that the path must be continued backward, i.e., starting with t' the path goes through decreasing values of t. However, in this bargaining game we shall not have to look at such cases.

It may happen that the uniquely determined best reply π to p is already an equilibrium point of G. The path of the tracing procedure then stays at this strategy combination π which finally becomes the result T(G,p) of the tracing procedure.

If the best reply π to the bicentric prior p is no equilibrium point of G, the first task in the computation of the trace is the determination of the first defector. Therefore, one determines for each player i a potential reversal time t_i with the property that in G^{t_i}, but not earlier, another pure strategy φ_i besides π_i becomes a best reply to π_{-i}. Of course, there may be players who, in the interval $0 \leq t \leq 1$, never face this situation. In order to determine the first defector, one has to find out which potential reversal time t_i is the smallest one. The computation of the second reversal time is analogous.

8. Best Replies to the Bicentric Prior Combination

As has been said before, $U = (r,r,r)$ risk dominates $V = (s,s,s)$ if for the corresponding bicentric prior combination p we have $T(G,p) = U$. The examination of risk dominance between U and V requires the application of the tracing procedure to p. The first task is to investigate which pure strategy combinations are best replies to p for which parameter combination.

In the following we shall always assume $r < s$. This does not entail any loss of generality. First we examine the conditions under which r and s are best replies to p_{-1}. We have:

$$H_1(r,p_{-1}) = p_3(r)r \tag{21}$$

$$H_1(s,p_{-1}) = p_3(s)s. \tag{22}$$

Since both payoffs are positive and since for all other strategies 0-payoffs are obtained, it is clear that only r and s can be best replies to p_{-1}.

r is the only best reply to p_{-1} if and only if the payoff in (21) is greater than that in (22). Equations (19) and (20) show that this is true if and only if we have

$$\frac{r(b-r)}{2b-r-s} > \frac{s(b-s)}{2b-r-s} . \tag{23}$$

Since r and s are smaller than b, (23) is equivalent to the following condition:

$$(s-r)(r+s-b) > 0 . \tag{24}$$

If the reversed inequality holds, s is the only best reply of player 1. Since $s > r$ holds, the only best reply π_1 to p_{-1} is as follows:

$$\pi_1 = \begin{cases} r & \text{for } b < r+s \\ s & \text{for } b > r+s . \end{cases} \tag{25}$$

In the same way we examine the best reply π_2 of player 2 to p_{-2}. We have

$$H_2(r,p_{-2}) = \frac{(r-1)(b-r)}{2b-r-s} \tag{26}$$

$$H_2(s,p_{-2}) = \frac{(s-1)(b-s)}{2b-r-s} \ . \tag{27}$$

Since both payoffs are positive, only r and s can be best replies. The condition for r being the only best reply is as follows:

$$b + 1 < r + s. \tag{28}$$

Therefore we have

$$\pi_2 = \begin{cases} r & \text{for } b+1 < r+s \\ s & \text{for } b+1 > r+s \ . \end{cases} \tag{29}$$

The payoffs of player 3, obtained by r and s against p_{-3}, are as follows:

$$H_3(r,p_{-3}) = -w\,\frac{r^2}{r+s} + (1-w)\,\frac{(h-r)(r-1)}{r+s-2} \tag{30}$$

$$H_3(s,p_{-3}) = -w\,\frac{s^2}{r+s} + (1-w)\,\frac{(h-s)(s-1)}{r+s-2} \ . \tag{31}$$

In order to show that the payoff in (31) is always positive, we transform the expression on the right-hand side:

$$H_3(s,p_{-3}) = \frac{(b-s)(s-1)(r+s)+ws(s-r)}{(r+s)(r+s-2)} \ . \tag{32}$$

Since r and s are between 1 and b and $s > r$ holds, the right-hand side of (32) is positive. Therefore, only r and s can be best replies to p_{-3} ; all other pure strategies yield the payoff 0. The inequality

$$b + 1 + w < r + s \tag{33}$$

is the necessary and sufficient condition for r being the only best reply to p_{-3}. If the reversed inequality holds, s is the only best reply. This yields

$$\pi_3 = \begin{cases} r & \text{for } b+1+w < r+s \\ s & \text{for } b+1+w > r+s \ . \end{cases} \tag{34}$$

Equations (25), (29), and (34) show that only four combinations $\pi = (\pi_1, \pi_2, \pi_3)$ can result as combinations of only best replies to p:

$$\pi = \begin{cases} (s,s,s) & \text{for } r+s<b \\ (r,s,s) & \text{for } b<r+s<b+1 \\ (r,r,s) & \text{for } b+1<r+s<b+1+w \\ (r,r,r) & \text{for } b+1+w<r+s. \end{cases} \tag{35}$$

Apart from border cases, (35) gives a complete overview over the best replies to the bicentric prior combination. In the two cases in which the equilibrium points $U = (r,r,r)$ and $V = (s,s,s)$ are found as only best replies to p the question of risk dominance is already decided. U risk dominates V for $b+1+w < r+s$ and V risk dominates U for $r+s < b$.

9. The Case $\pi = (r,s,s)$

Two cases remain open in (35) in which we have to examine risk dominance with the help of the tracing procedure. We first consider the case $b < r+s < b+1$ for which $\pi = (r,s,s)$ holds. Player 3 cannot be the first defector since $H_3^t(r,s,s)$ is increasing with t. The probability for the unfavorable event of getting a contract with player 1 decreases with increasing t. Moreover, player 3's payoff for (r,s,s) is positive. His payoff for (r,s,r), however, is decreasing in t. Without detailed computation these considerations show that player 3 cannot be the first defector.

In the same way it can be seen easily that player 2 cannot be the first defector. His payoff $H_2^t(r,s,s)$ is positive and increasing in t since the probability of a contract increases with t. The payoff $H_2^t(r,r,s)$, however, decreases with t.

We can conclude that player 1 is the first defector. As soon as he shifts to (s,s,s), a combination is reached which is not left by the trace until $t = 1$. The fake seller's shift from r to s decreases the buyer's payoff. However, player 3's payoff remains positive since (s,s,s) is a strong equilibrium point and since his payoff for (s,p_{-3}) is also positive and greater than that for (r,p_{-3}). This shows that for $b < r+s < b+1$, too, the equilibrium point $V = (s,s,s)$ is the result T(G,p) of the tracing procedure. Here, too, V risk dominates U.

In the special case $r+s = b$, player 1 has two best replies to p_{-1}, namely r and s. Nevertheless, the linear tracing procedure determines a unique path in this case, too, since for very small t player 1's strategy s is his only best reply to π_{-1} in G^t. In this case, one could speak of a first reversal time $t' = 0$.

We have obtained the result that $V = (s,s,s)$ risk dominates $U = (r,r,r)$ for $r+s < b+1$.

10. The Case $\pi = (r,r,s)$

We now consider the case $b+1 < r+s < b+1+w$ in which $\pi = (r,r,s)$ is the only best reply to the bicentric prior combination π. Unlike in the preceding section, it is not possible to exclude one of the players as the first defector by simple qualitative arguments. For all three players we have to determine the potential reversal time t_i and to compare them with each other. Only by choosing r or s player 1 can obtain positive payoffs in all games G^t if player 3 chooses s. Therefore, a reversal must involve a shift form r to s. The potential reversal time t_1 is the solution of the following equation:

$$H_1^{t_1}(r,r,s) = H_1^{t_1}(s,r,s) \tag{36}$$

or, equivalently

$$(1-t_1)H_1(r,p_{-1}) = (1-t_1)H_1(s,p_{-1}) + t_1 s \tag{37}$$

$$(1-t_1)(H_1(r,p_{-1}) - H_1(s,p_{-1})) = t_1 s. \tag{38}$$

An easy computation shows:

$$H_1(r,p_{-1}) - H_1(s,p_{-1}) = (s-r)\,\frac{r+s-b}{2b-r-s}\,. \tag{39}$$

Equations (38) and (39) yield:

$$\frac{t_1}{1-t_1} = \frac{s-r}{s} \cdot \frac{r+s-b}{2b-r-s}\,. \tag{40}$$

It is not necessary to solve this equation for t_1 since instead of the potential reversal times t_i the reversal quotients

$$\tau_i = \frac{t_i}{1-t_i} \tag{41}$$

can be compared with each other.

We now turn our attention to the potential reversal time t_2. Like player 1, player 2 obtains a positive payoff in the games G^t only for r and s if player 3 chooses s. Player 2, too, can shift to s only. Instead of (38) the equation analogous to (36), yields

$$(1-t_2)(H_2(r,p_{-2}) - H_2(s,p_{-2})) = t_2(s-1). \tag{42}$$

An easy computation shows

$$H_2(r,p_{-2}) - H_2(s,p_{-2}) = (s-r)\,\frac{r+s-b-1}{2b-r-s} \;. \tag{43}$$

This yields

$$\tau_2 = \frac{s-r}{s-1} \cdot \frac{r+s-b-1}{2b-r-s} \;. \tag{44}$$

We now consider the situation of player 3. Since $H_3(s,p_{-3})$ as well as $H_3(s,s,s)$ are positive, player 3 can shift only to r at a reversal point. With increasing t player 3 will finally find (r,r,r) more attractive than (r,r,s) in spite of the risk of being cheated since the probability to reach a contract at the price s becomes smaller with increasing t. Instead of of (38) the equation, analogous to (36), yields

$$(1-t_3)(H_3(s,p_{-3}) - H_3(r,p_{-3})) = t_3(b-r). \tag{45}$$

We have

$$H_3(r,p_{-3}) - H_3(s,p_{-3}) = (s-r)\,\frac{r+s-b-1-w}{r+s-2} \;. \tag{46}$$

This yields

$$\tau_3 = \frac{s-r}{b-r} \cdot \frac{1+b+w-r-s}{r+s-2} \;. \tag{47}$$

It will now be shown that the following inequality always holds:

$$\tau_1 > \tau_2 \;. \tag{48}$$

Inequality (48) has the consequence that player 1 cannot be the first defector. In view of $s > r$ (48) is equivalent to

$$(s-1)(r+s-b) > s(r+s-1-b) \;. \tag{49}$$

It can easily be seen that this inequality is always satisfied for $b > r$. Since

$b > r$ must be true for strong equilibrium points (r,r,r), player 1 cannot be the first defector. In order to find out who is the first defector, we have to compare τ_2 and τ_3.

Before comparing τ_2 and τ_3 we want to look more closely at the consequences of $\tau_2 < \tau_3$ and $\tau_2 > \tau_3$, respectively. For $\tau_2 < \tau_3$ player 2 is the first defector. Then in G^{t_2} not only (r,r,s) but also (r,s,s) is an equilibrium point since player 2's shift from r to s does not influence player 1's payoff and strengthens player 3's incentive to choose s. Therefore, in G^{t_2} the trace moves from (r,r,s) to (r,s,s). This leads to a situation which is essentially the same as in the case where π = (r,s,s) is the best reply to the bicentric prior. Only player 1 can be the second defector and a shift to (s,s,s) takes place. This has the consequence that for $\tau_2 < \tau_3$ the candidate V = (s,s,s) risk dominates U = (r,r,r).

For $\tau_2 > \tau_3$ player 3 is the first defector and the trace leads to (r,r,r). This has the consequence that for $\tau_2 > \tau_3$ the candidate U = (r,r,r) risk dominates the candidate V = (s,s,s).

In order to determine which of both equilibrium points U = (r,r,r) and V = (s,s,s) risk dominates the other for π = (r,r,s), one has to look at the condition for $\tau_2 < \tau_3$:

$$\frac{1}{s-1} \cdot \frac{r+s-1-b}{2b-r-s} < \frac{1}{b-r} \cdot \frac{1+b+w-r-s}{r+s-2} . \tag{50}$$

The reversed inequality holds for $\tau_2 > \tau_3$. The candidate V = (s,s,s) risk dominates U = (r,r,r) in the case π = (r,r,s) if (50) holds. If the reversed inequality holds in the case π = (r,r,s), then U = (r,r,r) risk dominates V = (s,s,s).

We shall not look at risk dominance in the border case where both sides of (50) are equal. However, we want to examine the border cases $r+s = 1+b+w$ and $r+s = 1+b$. In the case $r+s = 1+b+w$ player 3 has the two pure best replies r and s to p_{-3}. For sufficiently small positive t only r is player 3's best reply in G^t to (r,r,-). Therefore, the path of the linear tracing procedure is uniquely determined in this case. It leads to (r,r,r). Therefore, U = (r,r,r) risk dominates V = (s,s,s) for $r+s = 1+b+w$.

In the border case $r+s = 1+b$ player 2 has the two pure best replies r and s to p_{-2}. For sufficiently small t, however, only s is a best reply of player 2 to (r,-,s) in G^t. Therefore, in this border case (s,s,s) is the result of the tracing procedure and V = (s,s,s) risk dominates U = (r,r,r).

With the exception of the border cases, where both sides of (50) are equal, we have obtained a complete overview of the risk dominance relationships between two equilibrium points in the first candidate set. With the possible exception of such border cases it is always true that one of both equilibrium points risk dominates the other.

11. The Risk Dominance Diagram

In the following we shall introduce a risk dominance diagram which shows the parameter combinations for which $U = (r,r,r)$ and $V = (s,s,s)$ are risk dominant in pairwise comparisons. It will be convenient to replace the original parameters r and s by the normed deviations α and β from the midpoint of the interval [1,b]:

$$\alpha = \frac{r - \frac{b+1}{2}}{b-1} \tag{51}$$

$$\beta = \frac{s - \frac{b+1}{2}}{b-1} \,. \tag{52}$$

The values of α and β for equilibrium points in the first candidate set are between -1/2 and 1/2. We shall make use of the following notation:

$$\gamma = \alpha + \beta \tag{53}$$

$$\eta = \beta - \alpha \tag{54}$$

$$\omega = \frac{w}{b-1} \,. \tag{55}$$

Equations (51) - (55) yield the following relationships:

$$r = \alpha(b-1) + \frac{b+1}{2} \tag{56}$$

$$s = \beta(b-1) + \frac{b+1}{2} \tag{57}$$

$$r+s = \gamma(b-1) + b+1 \tag{58}$$

$$s-1 = (\beta + \tfrac{1}{2})(b-1) \tag{59}$$

$$b-r = (\tfrac{1}{2} - \alpha)(b-1) \tag{60}$$

$$r+s-1-b = \gamma(b-1) \tag{61}$$

$$1+b+w-r-s = (b-1)(\omega-\gamma) \tag{62}$$

$$2b-r-s = (b-1)(1-\gamma) \tag{63}$$

$$r+s-2 = (b-1)(\gamma+1). \tag{64}$$

Figure 2 shows the combinations of best replies to the bicentric prior as a function of the normed deviations α and β. With the help of (35) together with (51) to (64) it can be shown that the borders or regions of best reply vectors π are as described by Figure 2. Only in case of the region $\pi = (r,r,s)$ we shall look at this question in more detail. Since we assume $s > r$ only that part of Figure 1 lying above the main diagonal $\eta = 0$ is of interest here. For the sake of completeness we have added the part below the main diagonal to the diagram.

The case $\pi = (r,r,s)$ in which (50) answers the question of risk dominance is characterized by the inequality

$$b+1 < r+s < b+1+w . \tag{65}$$

The equations above show that (65) is equivalent to

$$0 < \gamma < \omega . \tag{66}$$

Since we continue to assume $r < s$ we also have

$$\eta > 0 . \tag{67}$$

With the help of Equations (56) - (67) inequality (50) can be rewritten as follows:

$$\frac{1}{\beta + \frac{1}{2}} \cdot \frac{\gamma}{1-\gamma} < \frac{1}{\frac{1}{2} - \alpha} \cdot \frac{\omega-\gamma}{1+\gamma} . \tag{68}$$

(68) is equivalent to

$$\omega-\gamma > \frac{1-2\alpha}{1+2\beta} \cdot \frac{1+\gamma}{1-\gamma}\gamma . \tag{69}$$

(53) and (54) yield

$$2\beta = \eta + \gamma \tag{70}$$

$$-2\alpha = \eta - \gamma . \tag{71}$$

If in (69) -2α and 2β are replaced by the right-hand side of (71) and (70), respectively, one obtains the following inequality after some rearrangements:

$$\frac{\omega}{2} > \gamma \frac{\eta+1-\gamma^2}{\eta(1-\gamma)+1-\gamma^2} . \tag{72}$$

This is equivalent to

$$\frac{\omega}{2} > \gamma\left[1 + \frac{\gamma\eta}{\eta(1-\gamma)+1-\gamma^2}\right] \tag{73}$$

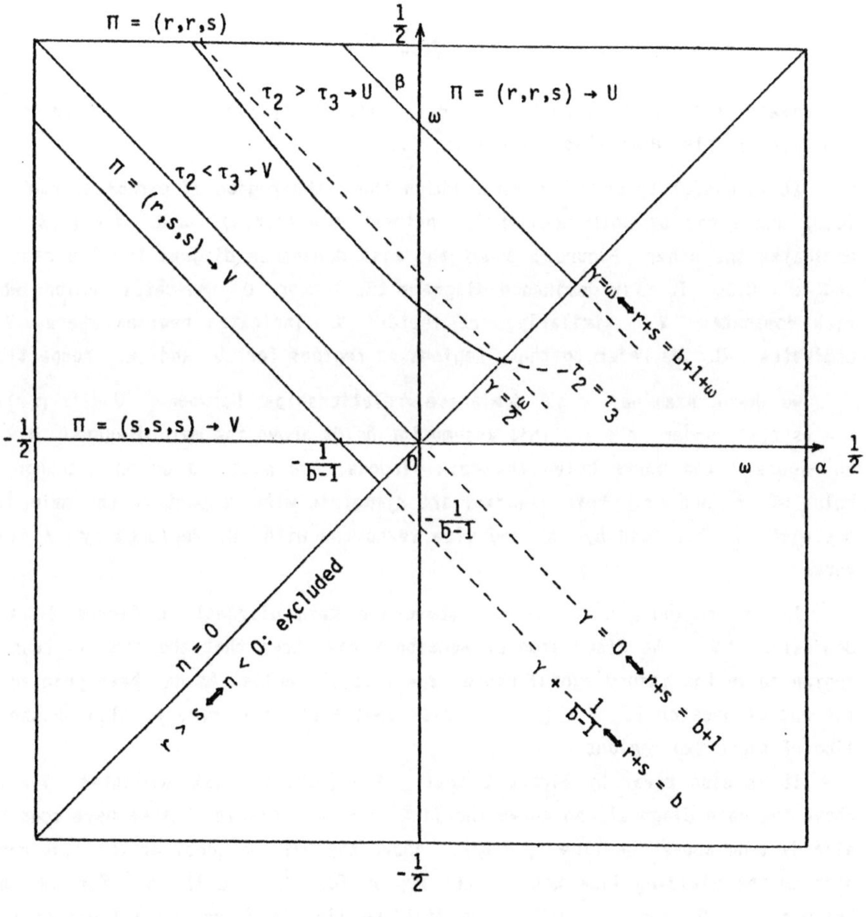

Figure 2: Regions for best replies to the bicentric prior and the outcome of the tracing procedure in the (α,β)-plane.

as well as

$$\eta\left[\frac{\omega(1-\gamma)}{2\gamma} - 1\right] > (1-\gamma^2)(1 - \frac{\omega}{2\gamma}). \tag{74}$$

With (74) we have obtained a transformation of inequality (50) which permits an important conclusion. In view of $\eta > 0$ and $\gamma < 1$ inequality (74) cannot hold unless we have

$$\gamma < \frac{\omega}{2} . \tag{75}$$

If (75) is violated, the left-hand side of (74) is negative and the right-hand side is nonnegative. Therefore, we have $\tau_2 > \tau_3$ for

$$\gamma \geq \frac{\omega}{2} . \tag{76}$$

This means that in the case $\pi = (r,r,s)$ inequality (76) has the consequence that $U = (r,r,r)$ risk dominates $V = (s,s,s)$.

It is useful to draw a diagram which shows the region of parameter combinations (α,β) where one of both equilibrium points $U = (r,r,r)$ and $V = (s,s,s)$ risk dominates the other. Figure 3 shows the risk dominance diagram for the case $h = 2$ and $w = 0.3$. In risk dominance diagrams the letter U indicates regions where U risk dominates V; similarly, the letter V indicates regions where V risk dominates U. We refer to these regions as regions for U and V, respectively.

We have examined risk dominance relationships between $U = (r,r,r)$ and $V = (s,s,s)$ under $r < s$. This assumption holds above the main diagonal $\alpha = \beta$. As in Figure 2 the parts below the main diagonal are obtained by an exchange of the roles of r and s. Both diagrams are symmetric with respect to the main diagonal but with r replaced by s and vice versa and with U replaced by V and vice versa.

In the region $\pi = (s,s,s)$ above the main diagonal in Figure 2 V risk dominates U. The discussion of Section 9 has shown that the same is true in the region above the main diagonal where $\pi = (r,s,s)$ holds. As has been pointed out at the end of Section 9, $V = (s,s,s)$ risk dominates $U = (r,r,r)$ also on the border line of these two regions.

It is also clear by Figure 2 that $U = (r,r,r)$ risk dominates $V = (s,s,s)$ above the main diagonal and above the line $\gamma = \omega$. Moreover, as we have seen this is already true above the line $\gamma = \omega/2$. Above the main diagonal of the risk dominance diagram the dividing line between the region for U and the one for V must be between $\gamma = 0$ and $\gamma = \omega/2$. The dividing line is formed by all points (α,β) in this area for which both sides of (74) are equal.

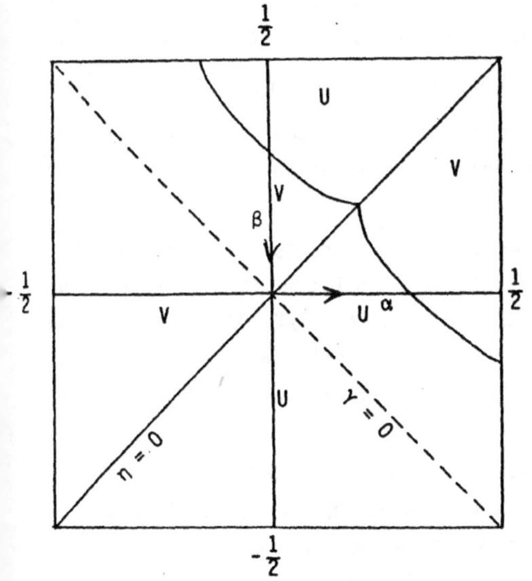

Figure 3: The risk dominance diagram for $h = 2$ and $w = 0.30$.

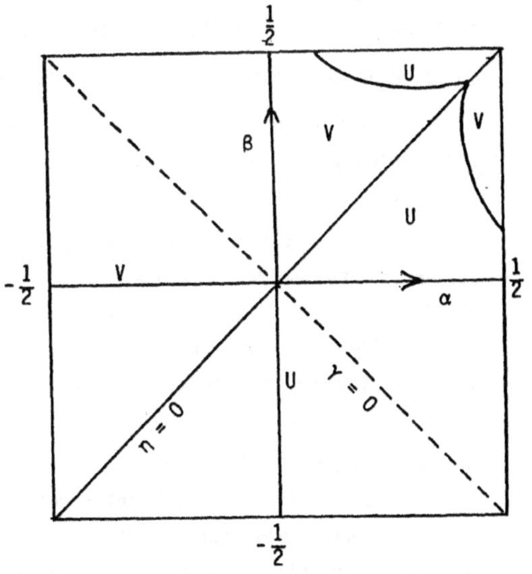

Figure 4: The risk dominance diagram for $h = 2$ and $w = 0.38$.

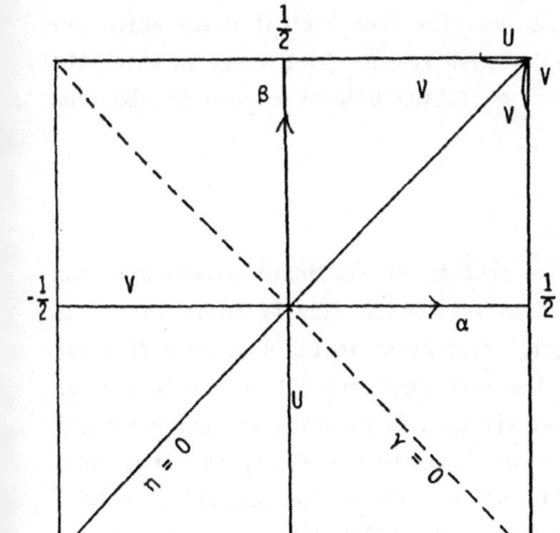

Figure 5: The risk dominance diagram for $h = 2$ and $w = 0.42$.

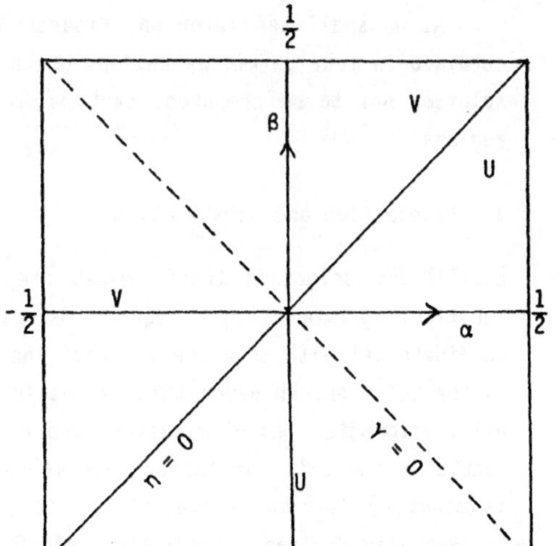

Figure 6: The risk dominance diagram for $h = 2$ and $w = 0.45$.

Figure 4 shows a case in which the shape of the dividing line is different from that in Figure 3. It is also possible that the dividing line does not meet the 45^0-line in the interior of the diagram but at the right upper corner. A case of this kind is shown in Figure 5. It is also possible that above the 45^0-line the equilibrium point V is risk dominant everywhere. A case of this kind is shown in Figure 6.

In Figures 3 and 4 the dividing line meets the 45^0-line. As we shall see, this happens at the point $\alpha = \beta = \omega/4$. In (74) the left-hand side is 0 in view of $\eta = 0$. Since γ is smaller than 1, γ must be equal to $\omega/2$ if the right-hand side of (74) is equal to 0. Therefore, $\alpha = \beta = \omega/4$ holds at the intersection point of the curve generated by the equality of both sides in (74) and the main diagonal.

With the help of (73) it can be seen that the curve generated by the equality of both sides in (74) does not intersect the line $\gamma = 0$. We can conclude that the dividing line between the regions for U and V above the main diagonal is entirely composed of points where both sides of (74) are equal as long as the risk dominance diagram contains any points above the main diagonal at which U risk dominates V.

It is clear from what has been said that, independently of the size of γ, V is risk dominant above the main diagonal if inequality (74) is satisfied, and that U is risk dominant if this inequality is satisfied with "<" instead of ">".

As we shall see later on, Figures 3 - 6 describe four typical cases which are obtained in four parameter regions which differ with respect to the way in which the solution has to be computed. Each parameter combination belongs to one of the four regions.

12. Elimination and Substitution

Equilibrium selection theory makes use of a process of candidate elimination and substitution generating a sequence of candidate sets which finally leads to a last candidate set with only one element. The equilibrium point in the last candidate set is the solution. Whenever this is possible, the next candidate set is obtained by an elimination step. An elimination step removes all candidates which are not maximally stable in the set. For the purpose of this paper it is not necessary to define what is meant by "maximally stable". It is sufficient to look at the special case of a "universally dominant" candidate. Let Ω be a set of equilibrium points of the game G. An equilibrium point $U \in \Omega$ is called universally dominant in Ω if U dominates all other equilibrium points in Ω.

If the first candidate set Ω_1 contains a universally dominant equilibrium point, then an elimination step removes all other equilibrium points and the universally dominant equilibrium point in Ω_1 becomes the solution.

Whenever in the bargaining model investigated here the first candidate set has no universally dominant element, we encounter a situation where each equilibrium point in Ω_1 is dominated by another equilibrium point in Ω_1. In principle, such situations still permit the existence of less than maximally stable elements in Ω_1 since the definition of maximal stability gives more weight to comparisons between equilibrium points which, in a certain sense, are more similar to each other. The degree of similarity is measured by a number called strategic net distance. In our case, the strategic net distance between any two equilibrium points in the first candidate set happens to be the same one. Therefore, no distinctions between risk dominance comparisons on the basis of strategic net distances can be made and no equilibrium point can be eliminated as less stable than another equilibrium point in such cases. This means that all members of the first candidate set Ω_1 are maximally stable in Ω_1 ; an elimination step cannot be performed.

If an elimination step fails to reduce the first candidate set, then a substitution step has to be performed. A substitution step determines a "substitute" for the candidate set. The substitution is computed by the application of the tracing procedure to a special prior strategy combination, called the centroid of the candidate set. If the first candidate set is replaced by a substitute the second candidate set contains only one equilibrium point, namely the substitute which then becomes the solution. The centroid of a candidate set will be defined in Section 15.

13. The Case of an Internal, Universally Dominant Equilibrium

It can easily be seen that in the case of Figure 3 a universally dominant equilibrium point is present. Let (ξ,ξ) be the intersection point of the 45^0-line $\alpha = \beta$ with the dividing curve between the region for U and the one for V. As we have seen in Section 11, the value of ξ is as follows:

$$\xi = \frac{\omega}{4} . \tag{77}$$

This value corresponds to an equilibrium point X = (x,x,x) with

$$x = \frac{1+b}{2} + \frac{w}{4} . \tag{78}$$

The assumption that b+1 and w are integer multiples of 4g guarantees that x is an admissible strategy. Figure 3 shows that U = (x,x,x) dominates all equilibrium points V = (s,s,s) with $r < s$ and $r > s$. In order to see this, it is sufficient to observe that the line $\alpha = \xi$ with the exception of the point (ξ,ξ) entirely belongs to the region for U.

Figure 4 shows the case in which $\alpha = \xi$ intersects the region where V is risk dominant. If the money unit g is sufficiently small one finds points (ξ,β) in the intersection of the region for V and the vertical line through (ξ,ξ) such that β is an admissible strategy. Consequently, for sufficiently small g the equilibrium point X fails to be universally dominant. Moreover, one can see that in the case of Figure 4 and for sufficiently small g every equilibrium point in Ω_1 is dominated by another one.

A universally dominant equilibrium point may also be present in a situation which corresponds to Figure 6. Here Y = (b-g,b-g,b-g) is universally dominant. In the risk dominance diagram the value

$$\mu = \frac{1}{2} - \frac{g}{b-1} \tag{79}$$

corresponds to the price b-g. All points (μ,β) corresponding to equilibrium points V = (s,s,s) with $s \neq b-g$ are in the region for U below the 45^0-line.

If X = (x,x,x) is universally dominant, we call X an internal, universally dominant equilibrium point. For the sake of simplicity, we will apply the word "internal" also to the special case where the intersection point (ξ,ξ) corresponds to the price x = b-g. If Y = (b-g,b-g,b-g) is universally dominant and b-g does not correspond to the intersection point coordinate ξ we call Y an extreme, universally dominant equilibrium point. In this section, we want to investigate the case of an internal, universally dominant equilibrium point.

It is necessary to find out under which conditions one obtains situations which correspond to Figures 3, 4, 5, and 6, respectively. Situations described by Figures 3 and 4 require that the intersection point coordinate ξ is smaller than 1/2. This is the case if and only if we have

$$b-1 > \frac{w}{2} . \tag{80}$$

In the remainder of this section we shall assume that (80) holds. We want to look at the question which conditions must be satisfied in addition to (80) to secure universal dominance of X for arbitrarily small g.

It is clear that X = (x,x,x) always risk dominates all equilibrium points V = (s,s,s) with $s > x$ since in such cases we have $r+s > 2x$ and therefore $\gamma > \omega/2$. As we have seen, U is always risk dominant under condition (76).

We still have to examine the question under which conditions X = (x,x,x) dominates all equilibrium points of the form U = (r,r,r) with $r < x$. Let

$$\alpha = \xi - \delta \tag{81}$$

be the value which corresponds to the price r in the risk dominance diagram.

$V = (x,x,x)$ risk dominates $U = (r,r,r)$ if (69) is satisfied. If α in (69) is replaced by $\xi-\delta$ and β by ξ, one obtains in view of (77):

$$\frac{\omega}{2} + \delta > (\frac{\omega}{2} - \delta)\,\frac{1 - \frac{\omega}{2} + 2\delta}{1 + \frac{\omega}{2}} \cdot \frac{1 + \frac{\omega}{2} - \delta}{1 - \frac{\omega}{2} + \delta} \quad . \tag{82}$$

Easy computations yield that this is equivalent to:

$$(1 + \frac{\omega}{2})(\frac{\omega}{2} + \delta)(1 - \frac{\omega}{2} + \delta) > (1 + \frac{\omega}{2} - \delta)(\frac{\omega}{2} - \delta)(1 - \frac{\omega}{2} + 2\delta). \tag{83}$$

Further transformations show that this inequality is equivalent to the following one:

$$2-\omega^2 > 2\delta^2 - \delta(2 + 3\omega). \tag{84}$$

The derivative of the right-hand side with respect to δ, namely $4\delta-2-3\omega$, is always negative for $0 \leq \delta \leq 1/2$. Therefore, (84) is satisfied in the whole interval $0 < \delta < 1/2$ if and only if the following condition holds:

$$\omega^2 \leq 2 . \tag{85}$$

In view of (55) inequality (85) is equivalent to the following condition:

$$w \leq \sqrt{2}(b-1) . \tag{86}$$

If (86) is violated and if g is sufficiently small the equilibrium point (r,r,r) with $r = x-g$ dominates the equilibrium point (x,x,x). The δ corresponding to r is then sufficiently small to reverse the direction of inequality (84).

From what has been shown up to now it is clear that (x,x,x) is universally dominant if and only if the conditions (80) and (86) are satisfied. It can easily be seen that (80) is implied by (86).

For sufficiently small g we have obtained the following result: The equilibrium point $X = (x,x,x)$ is an internal, universally dominant equilibrium point if and only if (86) is satisfied. The equilibrium price x is determined by (78). If (86) holds $X = (x,x,x)$ is the solution of the game.

14. The Case of an Extreme, Universally Dominant Equilibrium Point

In this section, the conditions shall be investigated under which $Y = (b-g,b-g,b-g)$ is universally dominant for $x \neq b-g$. This is the case of an extreme, universally dominant equilibrium point. As we have seen, this case can occur only if (80) is violated. Otherwise, $X = (x,x,x)$ risk dominates Y. Therefore, in the following we shall assume

$$\omega \geq 2 . \tag{87}$$

It is necessary to examine under which conditions $V = Y$ risk dominates all $U = (r,r,r)$ with $r < b-g$.

The following value β corresponds to Y in the risk dominance diagram:

$$\beta = \frac{1}{2} - \varepsilon \tag{88}$$

where

$$\varepsilon = \frac{g}{b-1} . \tag{89}$$

Let

$$\alpha = \frac{1}{2} - \delta \tag{90}$$

be the value of α which corresponds to the equilibrium point $U = (r,r,r)$ in the risk dominance diagram. Inequality (69) indicates under which conditions U is dominated by V. With the notation introduced above this inequality assumes the following form:

$$\omega + \varepsilon + \delta - 1 > \frac{2\delta}{2-2\varepsilon} \cdot \frac{2-\varepsilon-\delta}{\varepsilon+\delta}(1-\varepsilon-\delta). \tag{91}$$

A transformation of the right-hand side yields the following condition which is equivalent to (91):

$$\omega + \varepsilon + \delta - 1 > 2\delta \frac{(2-\delta)(1-\delta)-\varepsilon(3-2\varepsilon-2\delta)}{2\delta+\varepsilon(2-2\varepsilon-2\delta)} . \tag{92}$$

If ε is neglected in this inequality, the left-hand side is decreased and the right-hand side is increased. In this way, one obtains the following necessary condition for (92):

$$\omega + \delta - 1 > (2-\delta)(1-\delta) . \tag{93}$$

(93) is equivalent to

$$\omega - 3 > -\delta(4-\delta). \tag{94}$$

Inequality (94) shows that (91) always holds if we have

$$\omega \geq 3 . \tag{95}$$

Consequently, Y dominates every other number of the first candidate set for $\omega \geq 3$. Y is universally dominant for $\omega \geq 3$.

In the following it will be shown that for $\omega < 3$ the equilibrium point Y fails to be universally dominant if g is sufficiently small. Inequality

$$\omega - 3 < -\delta(4-\delta) \tag{96}$$

is implied by $\omega < 3$ for sufficiently small δ. If g is sufficiently small, then $\delta > \varepsilon$ can be found such that (96) holds. (96) implies (93) with "<" instead of ">". In view of the continuity of the expressions on both sides of (91) it follows that for sufficiently small ε a condition like (91) with "<" instead of ">" is satisfied.

If neither $X = (x,x,x)$ nor $Y = (b-g,b-g,b-g)$ is universally dominant no universally dominant equilibrium point is present. An equilibrium point $U = (r,r,r)$ cannot be universally dominant unless the value ρ corresponding to r in the risk dominance diagram satisfies the condition that the vertical line $\alpha = \rho$ does not intersect the region for V. For $r \neq b-g$ this is possible only if $\alpha = \rho$ meets the point where the dividing curve and the 45^0-line intersect.

For sufficiently small g we have obtained the following results:

(i) The equilibrium point $X = (x,x,x)$ is universally dominant for $0 < \omega \leq \sqrt{2}$ (see Section 13).

(ii) The first candidate set does not contain a universally dominant equilibrium point for $\sqrt{2} < \omega < 3$.

(iii) The equilibrium point $Y = (b-g,b-g,b-g)$ is universally dominant for $\omega \geq 3$.

15. Candidate Substitution in the Case Without a Universally Dominant Equilibrium Point

We shall now discuss the case without a universally dominant equilibrium point. As we have seen above, in this case every equilibrium point is dominated by some other equilibrium point. All equilibrium points are maximally stable (see Section 12). The whole candidate set Ω_1 has to be replaced by a substitute which becomes the final solution of the game.

The substitute is obtained by the application of the tracing procedure to a special mixed strategy combination, called the centroid of the first candidate set. We shall now give a definition of this centroid. Let $q^1,\ldots,q^K$ with $q^k = (q_1^k,\ldots,q_n^k)$ be the equilibrium points in the first candidate set. Here the q_i^k are mixed strategies. Pure strategies are interpreted as special mixed strategies.

The centroid

$$c = (c_1,\dots,c_n) \tag{97}$$

of the first candidate set is the mixed strategy combination whose components c_i are defined as follows:

$$c_i(\varphi_i) = \frac{1}{K}\sum_{k=1}^{K} q_i^k(\varphi_i) \tag{98}$$

for all $\varphi_i \in \Phi_i$ and for $i = 1,\dots,n$. The result $T(g,c)$ of the tracing procedure is the substitute for the first candidate set. In the case without a universally dominant equilibrium point the substitute of the first candidate set is the solution of the game.

Even if it is not possible to justify all ideas of equilibrium selection theory in this paper, it is perhaps appropriate to add a short remark on the replacement of the first candidate set by its substitute. If all elements of the first candidate set are maximally stable, the dominance relations have failed to be useful as selection criteria. It is then reasonable not to pay any further attention to dominance relationships and to base the decision among the equilibrium points in the first candidate set on another principle. For this purpose, the centroid of the first candidate set suggests itself as a natural prior strategy combination to which the tracing procedure can be applied. It is possible that a solution $T(G,c)$ is obtained which does not belong to the first candidate set. As we shall see, this is not the case in the bargaining game examined here.

16. The Case Without a Universally Dominant Equilibrium Point

As we have seen in Section 14, the case without a universally dominant equilibrium point arises if the following condition is satisfied:

$$\sqrt{2} < \omega < 3. \tag{99}$$

In this section, we shall always assume that (99) holds. Let K be the number of equilibrium points in the first candidate set. These equilibrium points have the form (r,r,r) with $r = 1+kg$ and $k = 1,\dots,K$. The centroid of the first candidate set $c = (c_1,c_2,c_3)$ has the following components c_i for $i = 1,2,3$:

$$c_i(1+kg) = \frac{1}{K} \quad \text{for} \quad k = 1,\dots,K\,. \tag{100}$$

The non-bargaining strategy ϕ being available to player 3 in addition to the price proposals $1+kg$ receives probability $c_3(\phi) = 0$.

Consider the payoff obtained by player i if he chooses $r = 1+kg$ whereas the other players use the i-incomplete centroid combination c_{-i}:

$$H_1(r,c_{-1}) = \frac{1}{K}\,r \tag{101}$$

$$H_2(r,c_{-2}) = \frac{1}{K}\,(r-1) \tag{102}$$

$$H_3(r,c_{-3}) = \frac{1}{K}\,(b-r) \quad . \tag{103}$$

Moreover, we have

$$H_3(\phi,c_{-3}) = 0 \ . \tag{104}$$

These equations show that the best reply σ_i of player $i = 1,2,3$ to c_{-i} is as follows:

$$\sigma_1 = 1+Kg = b-g \tag{105}$$

$$\sigma_2 = 1+Kg = b-g \tag{106}$$

$$\sigma_3 = 1+g \ . \tag{107}$$

We now determine the payoffs obtained against the i-incomplete combinations σ_{-i}, contained in $\sigma = (\sigma_1,\sigma_2,\sigma_3)$, for the auxiliary games G^t arising in the application of the tracing procedure to c.

$$H_1^t(r,\sigma_{-1}) = \begin{cases} (1-t)\frac{1}{K}\,r & \text{for } r \neq 1+g \\ [(1-t)\frac{1}{K} + t](1+g) & \text{for } r = 1+g \end{cases} \tag{108}$$

$$H_2^t(r,\sigma_{-2}) = \begin{cases} (1-t)\frac{1}{K}\,(r-1) & \text{for } r \neq 1+g \\ [(1-t)\frac{1}{K} + t]g & \text{for } r = 1+g \end{cases} \tag{109}$$

$$H_3^t(r,\sigma_{-3}) = \begin{cases} (1-t)\frac{1}{K}\,(b-r) & \text{for } r \neq b-g \\ [(1-t)\frac{1}{K} + t]g & \text{for } r = b-g \ . \end{cases} \tag{110}$$

It is clear that players 1 and 2 can shift only to $r = 1+g$ if one of them is the first defector since this strategy is the only one whose payoff increases with increasing t. Similarly, player 3 can shift only to $r = b-g$. As in Section 10, it is sufficient to compute the potential reversal quotients $\bar{\tau}_i = \bar{t}_i/(1-\bar{t}_i)$ for the three players $i = 1,2,3$ instead of the corresponding reversal times $\bar{t}_1$, $\bar{t}_2$, $\bar{t}_3$:

$$\bar{\tau}_1 = \frac{1}{K}\frac{b-2g-1}{1+g} \tag{111}$$

$$\bar{\tau}_2 = \frac{1}{K}\frac{b-2g-1}{g} \tag{112}$$

$$\bar{\tau}_3 = \frac{1}{K}\frac{b-2g-1}{g} \quad . \tag{113}$$

Since all numerators agree, it is obvious that $\bar{\tau}_1$ is the smallest potential reversal quotient. Player 1 is the first defector.

It is now necessary to investigate how the path of the tracing procedure continues after the time $\bar{t}_1$ has been reached. If (1+g,b-g,1+g) were an equilibrium point of $G^{\bar{t}_1}$, the path could be continued in $E^{\bar{t}_1}$ to this equilibrium point and then stay there for a time.

At this point we have to make a case distinction. The continuation of the path depends on whether we have

$$w > \frac{1}{1+b} \tag{114}$$

or

$$w < \frac{1}{1+b} \; . \tag{115}$$

We do not look at the border case where w is equal to $1/(1+b)$. In the case (114) we shall see that (1+g,b-g,1+g) is no equilibrium point of $G^{\bar{t}_1}$. It will turn out that the path continues in $E^{\bar{t}_1}$ up to a point $(q_1,b-g,1+g)$; from there the path goes on with increasing t and mixed strategies for player 1 and player 3; player 2 continues to play his pure strategy b-g. The probabilities with which players 1 and 3 use their strategy b-g are increasing. Eventually, at a time $\bar{t}$ the equilibrium point (b-g,b-g,b-g) is reached.

In the case (115) the strategy combination (1+g,b-g,1+g) is an equilibrium point of $G^{\bar{t}_1}$. It will turn out that the path continues to stay there until player 3 becomes indifferent between 1+g and b-g. From there on the path continues with mixed strategies for player 1 and player 3 until the equilibrium (b-g,b-g,b-g) is reached which then becomes the final solution. The result is the same as in the case (114) but the path is a different one.

We now begin to examine the case (114). In order to show that the path of the tracing procedure has the shape described above, it is necessary to start with the investigation of the game $G^{\bar{t}_1}$. Figure 7 shows the payoffs of players 1 and 3 in G^t under the condition that player 2 uses this strategy b-g.

Figure 8 repeats Figure 7 but with all payoffs multiplied by $K/(1-t)$. With the help of (111) it can be seen that in the special case of $\tau = \bar{\tau}_1$ the bimatrix of Figure 8 has the form shown in Figure 9. Arrows indicate the direction of incentives in the case (114). Thus, the arrow next to the second column indicates by its downward direction that in this column player 1 has a higher payoff in the second

1\3	1+g	b-g
1+g	$((1-t)\frac{1}{K}+t)(1+g)$ $(1-t)\frac{1}{K}(b-1-g)-t\omega(1+g)$	$(1-t)\frac{1}{K}(1+g)$ $(1-t)\frac{1}{K}\cdot g + t(1-\omega)(h-b+g)$
b-g	$(1-t)\frac{1}{K}(b-g)$ $(1-t)\frac{1}{K}(b-1-g)$	$((1-t)\frac{1}{K} + t)(b-g)$ $((1-t)\frac{1}{K} + t)g$

<u>Figure 7</u>: Payoffs for player 1 (above) and player 3 (below) in G^t if player 2 chooses b-g; player 1 is the row player and player 3 the column player.

1\3	1+g	b-g
1+g	$(1+\tau K)(1+g)$ $b-1-g-\tau K\omega(1+g)$	$1+g$ $g+\tau K(1-\omega)(h-b+g)$
b-g	$b-g$ $b-1-g$	$(1+\tau K)(b-g)$ $(1+\tau K)g$

<u>Figure 8</u>: The payoffs of Figure 7 multiplied by $K/(1-t)$; the expression $t/(1-t)$ is replaced by τ.

row. The fact that both payoffs of player 1 in the first column are equal is expressed by the double arrow to the left of the first column. The horizontal arrows concern incentives for player 3.

$1\backslash^3$	1+gb-g	b-g
1+g	b-g $[(1-\omega)(b-1)+g(2\omega-1)]$	$g + \frac{(h-b+g)(b-2g-1)}{1+g} \cdot (1-\omega)$
b-g	b-g b-1-g	$\frac{b-g}{1+g}$ (b-g) $\frac{b-g}{1+g}$ g

Figure 9: The bimatrix of Figure 8 for the special case $\tau = \bar{\tau}_1$; arrows indicate the direction of incentives.

The bimatrix in Figure 9 can be interpreted as a 2-person game. The equilibrium points of $G^{\bar{t}_1}$ in which player 2 chooses b-g must have components for players 1 and 3 which correspond to an equilibrium point of the game in Figure 9. In Figure 9, player 1's strategy b-g is weakly dominant and player 3's strategy 1+g is the only best reply to b-g. Therefore, all equilibrium points of the game in Figure 9 must be of the form $(q_1,1+g)$ with $q_1(1+g) < 1$. In view of $\bar{\tau}_2 > \bar{\tau}_1$, player 2's only best reply to player 3's strategy 1+g in $G^{\bar{t}_1}$ is b-g. Therefore, every equilibrium point $(q_1,1+g)$ of the game in Figure 9 corresponds to an equilibrium point $(q_1,b-g,1+g)$ of $G^{\bar{t}_1}$. Consequently, the part of the path of the tracing procedure in $G^{\bar{t}_1}$ contains only equilibrium points of the form $(q_1,b-g,1+g)$. It is not yet clear, however, at which of these equilibrium points the path leaves $E^{\bar{t}_1}$ and whether the path continues from there on with increasing or decreasing t.

We first exclude the case of a continuation with decreasing t. A combination with decreasing t is impossible since for $t < \bar{t}_1$ players 1 and 2 have the only best reply b-g to the strategy 1+g of player 3.

After having seen that the path is continued with increasing t it is now necessary to investigate the equilibrium points of the games G^t with $t > \bar{t}_1$. For this purpose, we shall first analyse the game in Figure 8 for $\tau > \bar{\tau}_1$. Later it will turn out that in this way one can find out what one needs to know about the continuation of the path. As we shall see, player 2 sticks to his strategy b-g in the remainder of the path.

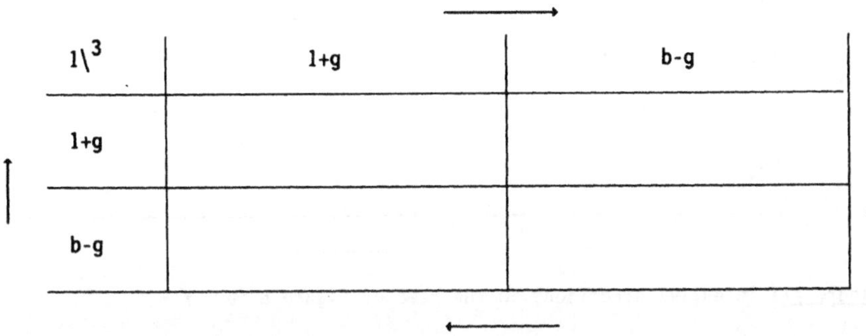

Figure 10: Incentive directions in the game of Figure 8 for $\bar{\tau}_1 < \tau < \bar{\tau}_3$.

We first examine the question how the incentive directions indicated by the arrows in Figure 9 change with increasing τ. The arrows in Figure 9 and the way in which payoffs in Figure 8 depend on τ permit the conclusion that in an interval $\bar{\tau}_1 < \tau < \hat{\tau}$ the incentive directions are as indicated in Figure 10. It is clear that the upper border $\hat{\tau}$ of the interval is nothing else than the reversal quotient $\bar{\tau}_3$ since this reversal quotient has been determined by the condition that player 3 is indifferent between 1+g and b-g if players 1 and 2 both choose b-g.

It is known that 2×2-games with incentive directions like in Figure 10 have exactly one equilibrium point in mixed strategies. Moreover, this equilibrium point is completely mixed in the sense that both players choose both strategies with positive probabilities. For every t with $\bar{\tau}_1 < t/(1-t) < \bar{\tau}_3$ let (q_1^t, q_3^t) be this equilibrium point of the game in Figure 8. Player 2's best reply to (q_1^t, q_3^t) is always b-g for the following reasons: In view of (112) and (113) we have $\bar{\tau}_2 = \bar{\tau}_3$. Player 2's best reply depends only on player 3's strategy. For $\tau < \bar{\tau}_3$

player 2's strategy b-g is his only best reply to player 3's strategy 1+g. Player 2's incentive to choose b-g can only be increased if player 3 changes over from 1+g to a mixture of 1+g and b-g. This shows that for $\bar{\tau}_1 < \tau < \bar{\tau}_3$ the strategy combination $(q_1^t, \text{b-g}, q_3^t)$ is an equilibrium point of G^t. In view of the fact that the game in Figure 8 has no other equilibrium points besides (q_1^t, q_3^t) in the concerning interval it is clear that the path of the tracing procedure must continue through the equilibrium points $(q_1^t, \text{b-g}, q_3^t)$.

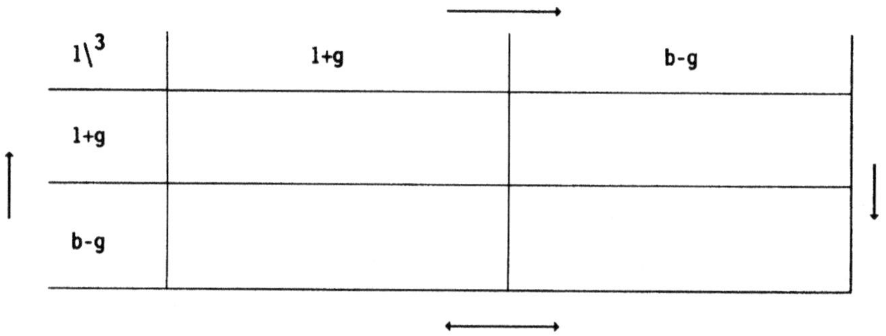

<u>Figure 11</u>: Incentive directions in the game of Figure 8 for $\tau = \bar{\tau}_3$.

The incentive directions in $G^{\bar{t}_3}$ are shown in Figure 11. All equilibrium points of a game of this kind must be of the form $(\text{b-g}, \text{b-g}, q_3)$ where the admissible values of $q_3(\text{b-g})$ fill an interval of the form $\kappa \leq q_3(\text{b-g}) \leq 1$ with $\kappa > 0$. After having reached $\bar{t}_3$, the path of the tracing procedure cannot continue with decreasing t since in the concerning games only the equilibrium points of the form $(q_1^t, \text{b-g}, q_3^t)$ are available for a continuation. Therefore, the path continues with increasing t.

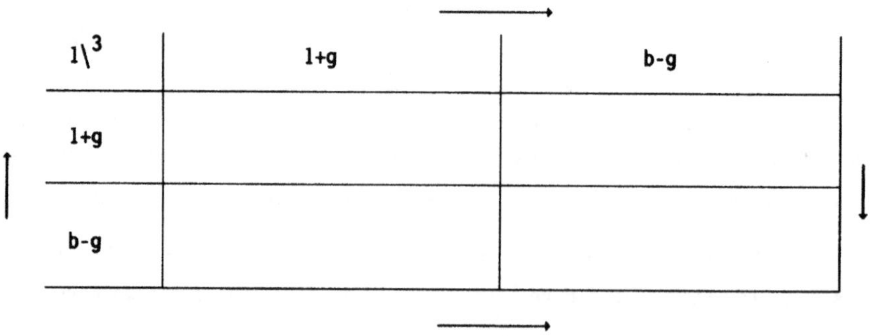

<u>Figure 12</u>: Incentive directions in the game of Figure 8 for $\tau > \bar{\tau}_3$.

The games G^t with $t > \bar{t}_3$ have the incentive directions for players 1 and 3 shown in Figure 12. It can be seen immediately that for all $t > \bar{t}_3$ the path stays at (b-g,b-g,b-g).

We now examine the case (115). Here the upper arrow in Figure 9 is reversed to the direction from right to left. This has the consequence that for $\tau = \bar{\tau}_1$ all strategy combinations of the form $(q_1,b\text{-}g,1\text{+}g)$, where q_1 is a mixture of 1+g and b-g, are equilibrium points. For $\tau > \bar{\tau}_1$, sufficiently near to $\bar{\tau}_1$, the arrows are as in Figure 10 with the exception of the upper arrow which continues to have the direction from right to left. (1+g,b-g,1+g) becomes the only equilibrium point. Therefore, the path moves to this equilibrium point within $\tau = \tau_1$. Then the path of the tracing procedure continues to stay at this equilibrium point until one of the players becomes indifferent between his two strategies 1+g and b-g. It can be seen immediately that this cannot be player 1 whose incentive to stick to 1+g is increasing in τ. We have to determine whether player 2 or player 3 is the next defector.

Player 2's reversal quotient remains $\bar{\tau}_2$ defined by (112) since player 2's incentives depend on the strategy of player 3. Player 3's reversal quotient is

$$\bar{\tau}_3 = \frac{1}{K}\,\frac{b-2g-1}{(1-w)(h-b)+w+g}\ . \tag{116}$$

It can be seen immediately that $\bar{\tau}_3$ is smaller than $\bar{\tau}_2$. It follows that player 3 is the next defector.

For $\tau > \bar{\tau}_3$, sufficiently near to $\bar{\tau}_3$, the situation is the same as in Figure 10. The path of the tracing procedure continues with mixed strategies for players 1 and 3. The remainder of the path is analogous to what happened in the case (114). In the same way as there it can be seen that the path ends with the final result (b-g,b-g,b-g).

We have shown that (b-g,b-g,b-g) is the result T(G,c) of the tracing procedure applied to the centroid c of the first candidate set. Consequently, for sufficiently small g the equilibrium point (b-g,b-g,b-g) is the solution of the game in the parameter region $\sqrt{2} < \omega < 3$.

The condition that the money unit is sufficiently small is not needed for the determination of T(G,c) but the delimitation of the parameter region $\sqrt{2} < \omega < 3$ is based on this assumption.

17. The Solution as a Function of the Parameters

In the following the results obtained up to now will be summarized. It will always be assumed that the money unit is sufficiently small.

The first candidate set contains the equilibrium points of the form (r,r,r) with $1 < r < b$ and $b = (1-w)h$.

The model has two parameters, namely w and h. It has turned out that the solution is a function of only one parameter ω which depends on w and h as follows:

$$\omega = \frac{w}{(1-w)h-1} \quad . \tag{117}$$

Every strong equilibrium point of the form (r,r,r) is characterized by a parameter α with $-1/2 < \alpha < 1/2$. The relationship of α and r is as follows:

$$r = \alpha(b-1) + \frac{b+1}{2} \quad . \tag{118}$$

Figure 13 shows how the parameter $\tilde{\alpha}$ corresponding to the solution $(\tilde{r},\tilde{r},\tilde{r})$ depends on ω. We have:

$$\tilde{\alpha} = \begin{cases} \frac{\omega}{4} & \text{for} \quad 0 < \omega \le \sqrt{2} \\ \frac{1}{2} - \frac{g}{b-1} & \text{for} \quad \omega > \sqrt{2} \quad . \end{cases} \tag{119}$$

It is interesting that $\tilde{\alpha}$ does not depend continuously on ω ; $\tilde{\alpha}$ jumps from $\omega/4$ to $\frac{1}{2} - \frac{g}{b-1}$ at $\omega = \sqrt{2}$. This jump is connected to the lack of a universally dominant equilibrium point in the interval $\sqrt{2} < \omega < 3$. In this interval, risk dominance comparisons do not contribute anything to the choice among equilibrium points of the first candidate set. The solution is determined by the application of the tracing procedure to the centroid of the first candidate set. The jump reflects the transition from one principle of equilibrium selection to another one.

At $\omega = 3$, a similar change occurs again in the opposite direction. For $\omega \geq 3$ the solution is a universally dominant equilibrium point. However, immediately before and after $\omega = 3$ the solution is the same even if it has been selected by different principles. Therefore, at $\omega = 3$ no discontinuity occurs.

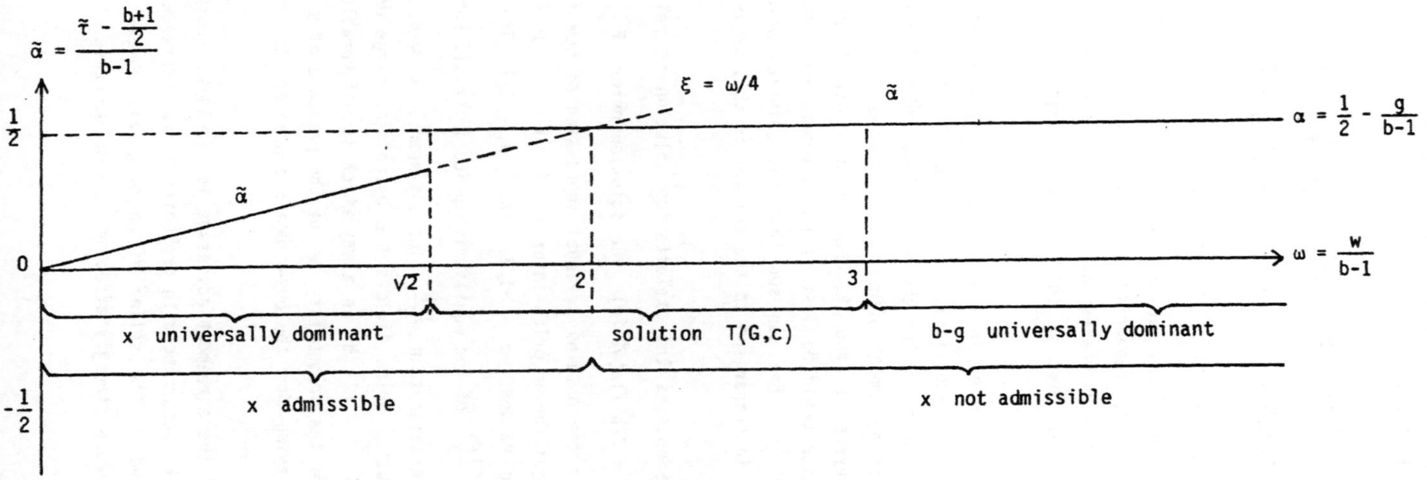

<u>Figure 13</u>: The relationship between the solution parameter $\tilde{\alpha}$ and ω.

The solution price $\tilde{r}$ which belongs to the solution $(\tilde{r},\tilde{r},\tilde{r})$ depends on the parameters w and h as follows:

$$\tilde{r} = \begin{cases} \frac{1+(1-w)h}{2} + \frac{w}{4} & \text{for } 0 < \frac{w}{(1-w)h-1} \leq \sqrt{2} \\ (1-w)h-g & \text{for } \frac{w}{(1-w)h-1} > \sqrt{2} \end{cases} \quad . \tag{120}$$

Figure 14 illustrates the dependence of the solution on the parameters h and w. Above the curve $w = 1 - \frac{1}{h}$ one finds parameter combinations for which the bargaining game has no equilibrium points connected with an agreement. We did not really determine a solution for these uninteresting cases. But since no other equilibrium points are available we know that in this region no agreement is reached with probability 1.

The region between both curves in the diagram contains parameter combinations for which the solution is the equilibrium point $(b-g,b-g,b-g)$. This is the most unfavorable strong equilibrium point for the buyer. Compared with the alternative of no agreement, this solution agreement offers only a minimal advantage of one money unit to the buyer. The players 1 and 2, however, have payoffs of $1+(K-1)g$ and $(K-1)g$, respectively, which would be lost if no agreement were reached. Since K increases with decreasing g, one can see that the solution price is extremely unfavorable for the buyer in comparison to the seller; in this sense we speak of an *exploitation solution*.

The region where one obtains the exploitation solution is relatively narrow. Below the order curve $w = 2(h-1)/(\sqrt{2}+2h)$ the solution price $\tilde{r}$ is not extreme. With the help of Figure 14 one obtains a visual impression of how far $\tilde{r}$ is away from the borders of the equilibrium price interval $1 < r < b$. In Figure 13, these borders correspond to the values $\alpha = -1/2$ and $\alpha = 1/2$. The value $\alpha = 0$ corresponds to the middle $\frac{1+b}{2}$ of the equilibrium price interval. For $\omega < \sqrt{2}$, too, the solution price $\tilde{r}$ is greater than $\frac{1+b}{2}$. The difference is $w/4$. In this sense, one can say that the solution price divides the bargaining range described by the equilibrium price interval $1 < r < b$ in a way which is unfavorable for the buyer and the more so the greater the probability w of the presence of a fake seller is. This may be expressed by saying that the buyer bears a greater part of the fake risk than the seller.

In the case $w = 0$, the agreement advantage is $(h-1)/2$, both for the buyer and the seller. A fake risk, occurring with probability w, decreases the agreement advantage of the buyer and of the seller of an original. However, the buyer is relatively and absolutely more strongly affected by this decrease. Let G_B be the

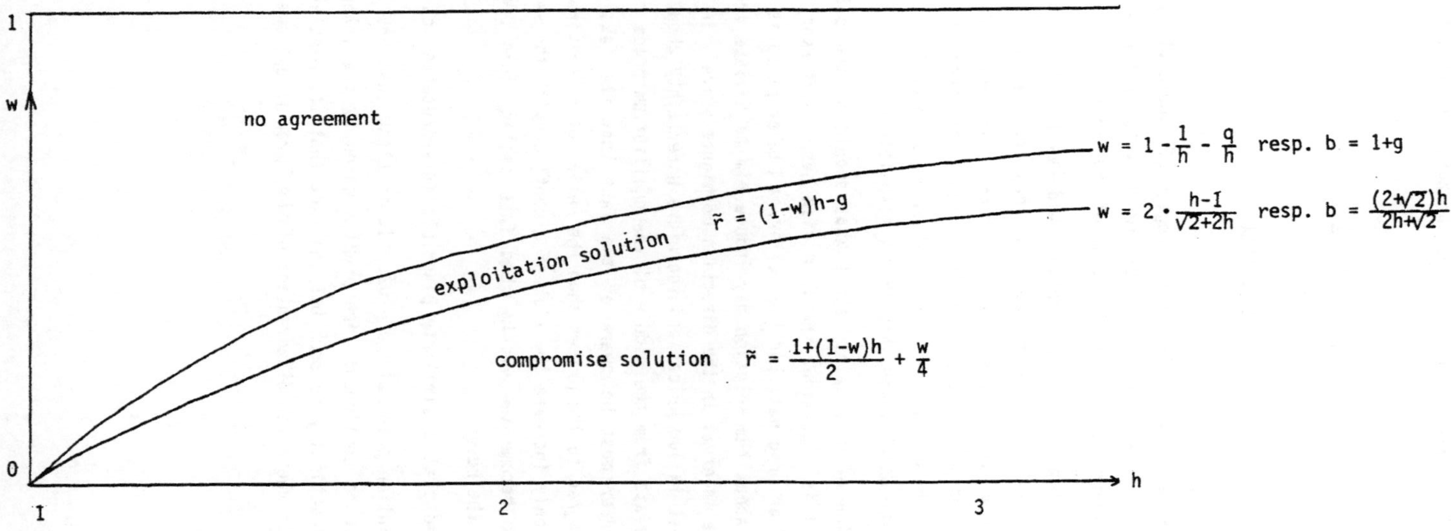

Figure 14: The solution as a function of the parameters h and w .

agreement advantage of the buyer and let G_S be the agreement advantage of a seller of an original. We have

$$G_B = \frac{h-1-w(h+1/2)}{2} \tag{121}$$

$$G_S = \frac{h-1-w(h-1/2)}{2} \quad . \tag{122}$$

Of course, these formulas are valid only for the region of Figure 14 indicated by the word *compromise solution*. As we have seen, the exploitation solution shows a much greater asymmetry of agreement advantages in favor of the seller of an original.

A naive analysis of the bargaining problem could easily come to the conclusion that the solution should be an agreement which divides the fake risk equally among the buyer and the seller of an original. As we have seen, the solution determined by equilibrium selection theory deviates from this 'fair' division in favor of the seller of an original. The way in which the solution is determined by the theory suggests an interpretation which makes this result plausible.

Consider a situation of uncertainty about what should be the solution of the bargaining problem. In this situation, the fake risk has a different impact on the buyer and the seller of an original. For the seller, a higher price is advantageous. Therefore, uncertainty about the solution may induce him to deviate from the middle of the equilibrium price interval in the direction of higher prices. In the same way, the buyer has an interest in low prices and therefore uncertainty about the solution could induce him to deviate from the middle of the equilibrium price range to lower prices. However, the buyer must be aware of the fact that the fake seller has a stronger motivation to agree to low prices than the seller of an original. Therefore, a lower price does not only increase the risk of a conflict with the seller but also increases the danger to become the victim of a fake seller. This additional risk weakens the position of the buyer.

It must be admitted that a similarly plausible interpretation cannot be given for the jump of the solution price at $\omega = \sqrt{2}$. It is clear that one cannot expect equilibrium selection to be continuous. The problem posed by the interpretation of the jump is not the discontinuity as such but the fact that it immediately leads to the exploitation solution where the disadvantage of the buyer is extreme.

References

Akerlof, G.E. (1970). The Market for "Lemons": Quality Uncertainty and the Market Mechanism. Quart. J. Econ., **84**: 488-500.

Harsanyi, J.C. (1968). Games with Incomplete Information Played by "Bayesian" Players. Manag. Sci., **14**: 159-182, 320-334, 486-502.

Harsanyi, J.C. (1975). The Tracing Procedure: A Bayesian approach to defining a solution for n-person non-cooperative games. Int. J. Game The., **4**: 61-95.

Harsanyi, J.C. and R. Selten (1972). A Generalized Nash-Solution for 2-Person Bargaining Games with Incomplete Information. Manag. Sci., **18**: 80-106.

Harsanyi, J.C. and R. Selten (1988). A General Theory of Equilibrium Selection in Games. Cambridge: MIT-Press.

Selten, R. (1975a). Bargaining under Incomplete Information - A Numerical Example - . In: O. Becker and R. Richter (eds.), Dynamische Wirtschaftsanalyse, Tübingen, 203-232.

Selten, R. (1975b). Reexamination of the Perfectness Concept for Equilibrium Points in Extensive Games. Int. J. Game Th., **4**: 25-55.

Selten, R. and U. Leopold (1982a). Equilibrium Point Selection in a Bargaining Situation with Opportunity Costs. Economie Appliquée, **x**: 611-148

Selten, R. and U. Leopold (1982b). Gleichgewichtsauswahl in einer Gehaltsverhandlungssituation mit unvollständiger Information. In: Schriften des Vereins für Socialpolitik, **126**: 155-162.

Spence, A.M. (1973). Job Market Signaling, Quart. J. Econ., **87**: 355-379.

Spence, A.M. (1974). Market Signaling: Informational Transfer in Hiring and Related Screening Processes. Cambridge: Harvard University Press.

WAGE BARGAINING AS A STRATEGIC GAME

Hans Haller

Abstract: Wage bargaining between "capital" and "labor" is analyzed as an alternating offer game, where disagreement leads to surplus sharing according to the contractual status quo or to a strike.

1. Introduction

Recognition of bargaining as an important aspect of economics goes back to Edgeworth (1887). Since then many models of bargaining have been suggested in the literature. A representative collection of articles is provided by Roth (1985) and Binmore and Dasgupta (1987). Roth (1979) summarizes much of cooperative bargaining theory.

Given that theories of wage determination constitute a substantial part of economic thought, an elaborate theory of wage bargaining is desirable. Indeed, the early contributions of Zeuthen (1932) and Hicks (1932) focus on wage bargaining. Some more recent papers also refer to wage bargaining: Selten and Güth (1982), Osborne (1984), Shaked and Sutton (1984), Herrero (1985), Jun (1988).

Two particular features of wage bargaining ought to be emphasized. (1) In most situations, the old contract remains effective until a new contract is signed. Insofar, delaying an agreement is the basic threat with the old contract determining the disagreement outcome. But each party has more drastic, in a sense more physical threats: strike and lock-out, respectively. Some implications of this enriched strategy space are studied in the present article.

(2) Much of wage bargaining is collective bargaining, that is workers and employers delegate the bargaining task to unions and employers' associations, respectively. This situation has been studied in the literature: Jun (1988) has a model of wage bargaining where union formation is endogenized. Fershtman, Judd and Kalai (1987) analyze the outcomes of general non-cooperative games when players can strategically use agents to play on their behalf.

The present article disregards delegation problems. Collective bargaining appears only in a rudimentary form: negotiations take place between two players; the first player is labelled "capital" and stands for a group of employers (or a single firm) while the second player is labelled "labor" and represents the workers employed by "capital".

Capital and labor bargain about a new surplus sharing contract. There is an old contract, the *contractual status quo*, which prevails as long as no new contract is concluded and no disruption of the employment relation occurs. This paper studies the impact of potential *strikes* on the bargaining outcome; i.e. labor has the opportunity to disrupt the employment relation. An opening discussion is devoted to the difficulties to incorporate both a contractual status quo and a different threat point into cooperative bargaining theory. The main analysis, following the Nash program, considers bargaining outcomes as equilibrium points of a strategic (= non-cooperative) game. With regard to the wage bargaining problem, the strategic approach serves three related purposes:

a) provide self-enforcing outcomes;
b) exhibit some interactive and dynamic aspects of the bargaining procedure;
c) explain the difficulties of cooperative game theory to suggest a unique outcome.

More specifically, two subgame-perfect equilibria are derived and discussed in an extension of the alternating offer model of Rubinstein (1982). No effort is made to select a unique equilibrium point. Nor is stationarity of equilibrium points required. Rather the view is taken that in isomorphic situations different self-enforcing conventions can emerge. If for instance labor never intends to strike -- and never has gone on strike, then capital plays a best response such that it is in labor's best interest not to strike. If on the contrary, labor threatens to strike under certain circumstances -- and has gone on strike under these circumstances in the past, then capital's best response is such that labor should strike whenever it has planned to; otherwise it loses its credibility and ends up in the no strike equilibrium, a less favorable regime for labor.

Concluding remarks consider a folk theorem, the possibility of a negotiation impasse along the equilibrium path, and renegotiation-proofness.

2. The Bargaining Problem

Capital and labor bargain about a new surplus sharing contract. There is an old contract, the *contractual status quo*, which prevails as long as no new contract is concluded and if no disruption of the employment relation occurs. Since subtleties of collective bargaining are disregarded here, one would expect the old contract to be renewed in a stationary economic environment. But some economic data may have changed after the conclusion of the old contract. For simplicity, linear utility is assumed and that the old contract was designed to share a surplus of size $S > 0$. Under the old contract, the surplus S was divided in proportions $R > 0$ for capital and $1-R > 0$ for labor so that capital received RS and labor received $(1-R)S$ which, without loss of generality, are also taken as their respective utilities. Bargaining is about the new surplus of size 1. For simplicity, assume $(1-R)S < 1$ which is always satisfied in case $1 > S$. Then define $r > 0$ by $1-r = (1-R)S$.

In the standard interpretation of an old contract, there are no provisions to adjust the share of labor, as long as there is enough surplus to share, i.e. as long as the assumption $(1-R)S < 1$ holds. The continuation of the old contract means that labor continues to receive $1-r$, while capital receives the residual r. Thus $(r,1-r)$ is the status quo division where labor neither benefits from an increase of surplus, $1 > S$, nor suffers from a small decrease of surplus. The status quo is an efficient outcome and, in the absence of disruptive measures, also the conflict outcome. In that latter case, any cooperative bargaining solution divides the surplus 1 into pieces r and $1-r$. This can be achieved in two ways, namely by signing a new contract or by continuing under the old contract. Also, in the absence of disruptive measures, each party can secure its payoff in the division $(r,1-r)$ by not signing a new contract; hence the division $(r,1-r)$ results as Nash equilibrium outcome in any non-cooperative model.

The bargaining problem becomes much more interesting with the possibility of a disruptive measure on the part of labor. During a strike period, the (non-discounted) payoffs are a for capital and b for labor, $a \leq 0$, $b \leq 0$. Notice that a strike means more than merely a delay or, at the limit, non-conclusion of a new contract.

3. Discussion of Cooperative Solutions

For the bargaining problems described in the previous section, cooperative bargaining theory suggests an unambiguous solution, once it is clear which is the conflict outcome. Yet there are at least two potential conflict outcomes, the contractual status quo and the strike outcome, so that the theory can no longer make a definite prediction or suggestion.

The following discussion mainly refers to the *Nash Bargaining Solution* (NBS), as introduced by Nash (1950). Define a two-person bargaining problem as a pair (B,d) where B is a nonempty, compact, and convex subset of $\mathbb{R}^2$ and $d \in B$. B represents the set of feasible utility allocations of the two bargainers and $d = (d_1,d_2)$ represents the pair of conflict utilities (conflict payoffs). For simplicity, any $u = (u_1,u_2) \in B$ is called a bargaining outcome; u_1 is the utility achieved by capital, u_2 is the utility achieved by labor. The NBS, axiomatized by Nash and others, suggests (predicts) the unique outcome $u^* = (u_1^*,u_2^*)$ given as

$$(u^*) = \underset{\substack{(x_1,x_2)\in B \\ (x_1,x_2)\geq d}}{\arg\max} \; (x_1-d_1)\,(x_2-d_2).$$

Without a strike possibility, the natural choices for the wage bargaining problem would be

$$B = \{(x_1,x_2) \in \mathbb{R}^2_+ \mid x_1 + x_2 = 1\},\ d = C = (r,1-r).$$

Then the NBS suggests $u^* = C$.

Next, consider the possibility of a strike with payoff pair $D = (a,b) \leq (0,0)$, while the contractual status quo payoff pair is $C = (r,1-r)$. The set B of feasible utility allocations, allowing joint randomization, is the convex hull of the three points (1,0), (0,1), and D.

First of all, each efficient outcome (α^*,β^*), i.e. $\alpha^* + \beta^* = 1$, which satisfies $(\alpha^*,\beta^*) \geq D$ is a Nash equilibrium of the "Nash demand game" given by the following rules:

Capital demands α, labor demands β.

In case $\alpha + \beta \leq 1$, they "split the difference" so that

capital gets $\alpha + \frac{1-(\alpha+\beta)}{2} = \frac{\alpha+1-\beta}{2}$ and

labor gets $\beta + \frac{1-(\alpha+\beta)}{2} = \frac{\beta+1-\alpha}{2}$.

In case $\alpha + \beta > 1$, the outcome is D.

Secondly, the NBS picks a unique outcome, if a conflict outcome is fixed. When the strike outcome is taken as the conflict outcome, that is $d = D$, then the NBS favors labor, if $|a| > |b|$ and favors capital, if $|a| < |b|$. Now labor should not suffer from the fact that it may or may not use its threat, while capital has no threat. Namely, at some cost, labor can force capital down to the payoff $a \leq 0$, while capital can only sabotage negotiations, thus forcing labor down to $1-r$. With the fictitious conflict outcome $d = E = (a,1-r)$ and B modified such that $E \in B$, the NBS reflects the power of the threat. But then again, this perspective neglects completely that carrying out its threat is costly to labor. After going around in a circle, one has two ways out:

(a) Find a more appropriate cooperative bargaining model.

(b) Study non-cooperative bargaining over time.

Route (a) has been taken by Brito, Buoncristiano, and Intriligator (1977) who investigate bargaining problems with a *status quo* C and a *threat point* D. They derive properties of *allocation functions* $\Phi(C,D)$, given a set of axioms. But their axioms do not determine a unique allocation function Φ; therefore, without further selection criteria, multiple solutions of the wage bargaining problem would be suggested.

Gupta and Livne (1988) consider two-person cooperative bargaining problems, represented by triplets (B,C,D) with $B \subset \mathbb{R}^2$ a convex, compact and comprehensive *bargaining domain*, $C \in B$ a *reference outcome* or *status quo* and $D \in B$ the traditional *conflict outcome* or *threat point*. They assume $D \leq C$. In contrast to the premises of the present paper and of Brito, Buoncristiano and Intriligator, they consider variable bargaining domains and assume C to be inefficient. Gupta and Livne propose a solution function $\Phi(B,C,D)$ which they characterize axiomatically. In case $D = C$, the Gupta-Livne solution coincides with the solution of Kalai and Smorodinsky (1975).

The present article goes into direction (b) and takes a strategic approach.

4. Analysis of an Alternating Offer Model

The strategic approach to the wage bargaining problem yields multiple solutions, like the cooperative approach. But the strategic approach has some advantages:

a) self-enforcing outcomes;
b) consideration of some interactive and dynamic aspects of the bargaining procedure;
c) explanation of multiple solutions.

Point c) is related to points a) and b). Namely, since under the strategic approach a solution is defined as a self-enforcing (= equilibrium) outcome of the strategic game and since equilibrium strategies require a complete description of actions for any history, certain solutions can be associated with specific social conventions. For instance, in the strategic wage bargaining game of this section, there exists a "non-strike equilibrium" where labor never considers a strike. Occurrence of such an equilibrium is most plausible in a society where a strike is taboo. But the taboo is only important as a focal point, inducing the players to select the particular equilibrium. The punishment for violating the taboo may well be null. In a different society, labor's conventional behavior may require a strike after capital has acted improperly from labor's viewpoint. This would explain a "strike equilibrium".

The subsequent analysis exhibits the two types of equilibria just mentioned: a non-strike equilibrium and a strike equilibrium where both equilibria considered are subgame-perfect Nash equilibria in pure strategies. The willingness to strike influences labor's equilibrium payoff in a monotone way: among the two equilibria, labor's payoff is highest in the strike equilibrium, lowest in the non-strike equilibrium.

The formal model is a modification of Rubinstein's (1982) alternating offer model with fixed common discount factor δ, $0 < \delta < 1$. The game has a discrete time structure with infinitely many periods t = 0,1,2,... If a player obtains payoffs π_t

in periods t = 0,1,2,..., then the player's total payoff in the bargaining game is

$$\sum_{t=0}^{\infty} \delta^t \pi_t .$$

A simplified game tree looks as follows, with the subgame starting in period 2 isomorphic to the entire game.

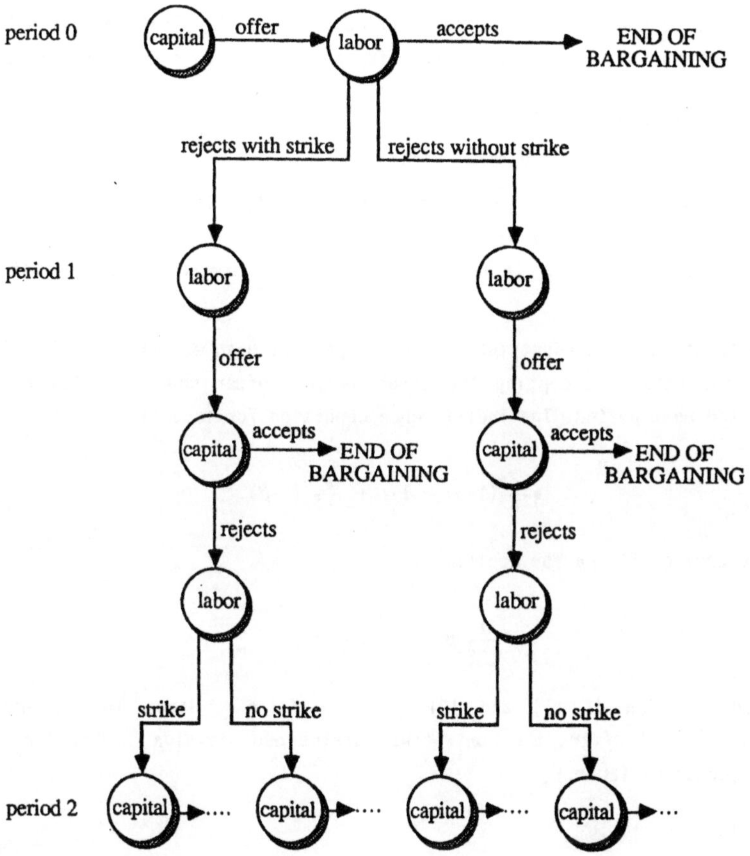

Capital makes the first offer in period 0. An offer has the form $(\alpha,1-\alpha)$, $(A,1-A)$, etc. If in a particular period the offer $(\alpha,1-\alpha)$ is accepted, then the bargaining ends. For that period and all subsequent periods, the non-discounted period payoffs are as specified by the new contract: payoff α for capital and payoff $1-\alpha$ for labor. If in a particular period bargaining is not yet terminated, i.e. all offers until then including the period's offer have been rejected, then the old contract applies to that period -- unless there is a strike. Under the old contract, the non-discounted period payoffs are $r > 0$ for capital and $1-r > 0$ for labor. If in some period a player rejects the offer made by the other player, then the rejecting player is one to make the offer in the next period. Labor has the possibility to strike in any period. In a period where capital makes the offer, labor can reject capital's offer and go on strike. (Alternatively, labor can accept the offer or reject the offer without going on strike.) In a period where labor makes the offer, labor can decide to strike, if its offer is rejected by capital. The strike decision is the third move in such a period, after labor made an offer and capital rejected it. A strike takes place during the same period in which the strike decision is made; a strike yields non-discounted period payoffs a for capital and b for labor.

4.1 The Non-Strike Equilibrium

Let us try two simple threshold strategies:

Labor never strikes. It accepts an offer $(A,1-A)$, if $A \leq \alpha$; otherwise it rejects. It always offers $(\beta,1-\beta)$.

Capital accepts an offer $(A,1-A)$, if $A \geq \beta$; otherwise it rejects. It always offers $(\alpha,1-\alpha)$.

These two strategies constitute a subgame-perfect Nash equilibrium if each party is indifferent between accepting the other party's offer now and getting its own offer accepted next period. The indifference condition for labor is

$$\frac{1}{1-\delta}(1-\alpha) = 1-r + \frac{\delta}{1-\delta}(1-\beta). \tag{1}$$

The indifference condition for capital is

$$\frac{1}{1-\delta}\beta = r + \frac{\delta}{1-\delta}\alpha . \tag{2}$$

This unique solution of (1) and (2) is $\alpha = \beta = r$. Hence in the nonstrike equilibrium, capital offers the contractual status quo division in the first round which is accepted by labor.

4.2 The Strike Equilibrium

Let us try again threshold strategies for acceptance/rejection together with a strike rule.

The strike rule requires labor to strike during a period where it made an offer $(A,1-A)$ with $A \geq \beta^*$ which got rejected by capital. Labor never goes on strike in any other circumstances.

In case labor adhered to the strike rule in the past, the threshold strategies require:

Labor follows the strike rule. Labor accepts an offer (A,1-A), if $A \leq \alpha^*$; otherwise it rejects -- and goes on strike. Labor always offers $(\beta^*, 1-\beta^*)$.

Capital accepts an offer (A,1-A), if $A \geq \beta^*$; otherwise it rejects. It always offers $(\alpha^*, 1-\alpha^*)$.

In case labor violated the strike rule in the past, both capital and labor play the threshold strategies of the non-strike equilibrium, with $\alpha = \beta = r$.

The basic idea is that labor will strike whenever the strike rule prescribes it. A violation causes play of the non-strike equilibrium in the following subgame. It has to be verified that this constitutes a punishment.

Suppose now that labor follows the strike rule. Then in order that the above threshold strategies define a subgame-perfect Nash equilibrium, two indifference conditions for labor and capital, respectively, have to be satisfied.

$$\frac{1}{1-\delta}(1-\alpha^*) = 1-r + \frac{\delta}{1-\delta}(1-\beta^*); \tag{1*}$$

$$\frac{1}{1-\delta}\beta^* = a + \frac{\delta}{1-\delta}\alpha^* . \tag{2*}$$

The unique solution of (1*) and (2*) is

$$\alpha^* = \frac{1}{1+\delta} r + \frac{\delta}{1+\delta} a,$$

$$\beta^* = \delta\alpha^* + (1-\delta)a = \frac{1}{1+\delta} a + \frac{\delta}{1+\delta} r.$$

With $a \leq 0 < r$ and $\alpha^* > \beta^*$ as determined by (1*) and (2*), capital has no incentive to provoke labor to strike. The conditions (1*) and (2*) are sufficient for the above strategy pair to be a subgame-perfect Nash equilibrium, if in addition labor has an incentive to strike whenever required by the strike rule. Suppose that capital has rejected an offer (A,1-A) with $A \geq \beta^*$. If labor strikes, it can achieve as total payoff from the particular period onwards, discounted to that period:

$$b + \frac{\delta}{1-\delta}(1-\alpha^*).$$

If labor does not strike, it can achieve

$$1-r + \frac{\delta}{1-\delta}(1-r) = \frac{1}{1-\delta}(1-r).$$

Hence the incentive constraint for this case is

$$b + \frac{\delta}{1-\delta}(1-\alpha^*) \geq \frac{1}{1-\delta}(1-r). \tag{3}$$

Notice that with the calculated value of α^*, condition (3) is equivalent to

$$r \geq (1-\delta^2)(1-b) + \delta^2 a. \tag{3*}$$

Therefore

<u>Proposition</u>. If condition (3*) holds, then the above strategy pair with α^* and β^* solving (1*) and (2*) is a subgame-perfect Nash equilibrium. ■

Observe that (3*) together with $r < 1$ requires $\delta^2 > |b|/(1+|a|+|b|)$. (3*) is automatically satisfied, if $\delta^2 > (1+|b|)/(1+|a|+|b|)$. So one has a strike equilibrium for sufficiently large discount factors. For a fixed discount factor the strike equilibrium exists, if capital has a high status quo payoff r and high strike costs $|a|$ and labor has a low status quo payoff $1-r$ and low strike costs $|b|$. In the strike equilibrium, capital offers the division $(\alpha^*,1-\alpha^*)$ in the first round which gets accepted. The capital share $\alpha^* = \frac{1}{1+\delta} r + \frac{\delta}{1+\delta} a$ is a convex combination of r and a where $a \leq 0 < r$ and the weight $\frac{1}{1+\delta}$ is decreasing in δ. Labor's strike costs do not enter the strike equilibrium payoffs; but in order to make a strike threat credible, they must be small relative to capital's strike costs. For $\delta \to 1$, α^* converges to $\frac{1}{2} r + \frac{1}{2} a$. So in the limit, one obtains the same division as the Nash Bargaining Solution with conflict outcome $E = (a,1-r)$ in Section 3. In this sense, E is a plausible conflict outcome, provided labor can afford a strike, i.e. labor's strike costs are small relative to capital's strike costs, and provided labor is determined to strike.

5. Concluding Remarks

5.1 <u>Folk Theorem</u>

In the previous section, two equilibria have been studied with the following divisions of a cake of size 1 agreed upon in the first period.

<u>Non-strike equilibrium</u>: $(\alpha,1-\alpha) = (r,1-r)$;

<u>strike equilibrium</u>: $(\alpha^*,1-\alpha^*)$ with $\alpha^* = \frac{1}{1+\delta} r + \frac{\delta}{1+\delta} a$.

Is there room for a *folk theorem* type of result, in particular, is any first period division $(A,1-A)$ with $\alpha^* \leq A \leq r$ a subgame-perfect Nash equilibrium outcome?

Holden (1989) gives an affirmative answer. My conjectured answer was rather no since the bargaining game is not a supergame; actions taken in some period can be payoff relevant for later periods. Rubinstein's (1982) model with fixed discount factors δ_1 and δ_2 has a unique subgame-perfect Nash equilibrium division, $A = (1-\delta_2)/(1-\delta_1\delta_2)$. With three or more bargaining parties, multiplicity of subgame-perfect Nash equilibrium divisions was known before, see Herrero (1985), Haller (1986).

5.2 Efficient Outcomes

With complete information and rational players, bargaining is *efficient*; that is the players settle in the first period. Other outcomes, like delayed agreements or strikes are observed in practice. They can be explained by mistake, bounded rationality, or incomplete information. For the latter case, see Rubinstein (1985) and Herrero (1986).

5.3 Renegotiation-Proofness

In recent years, starting with Farrell's (1983) formulation, the concept of ***renegotiation-proofness*** has been forwarded. This concept refines subgame-perfectness. Not only is it required that the equilibrium strategies induce a Nash equilibrium play in each subgame, but, moreover, the induced (subgame-perfect) equilibrium of any subgame should be undominated by other (subgame-perfect) equilibria of the subgame; however, the comparison should be restricted to those other equilibria which are also immune against renegotiation. The equilibria studied in this paper seem to be renegotiation-proof. For the strike equilibrium, the key argument goes as follows. A crucial subgame starts after capital has rejected an offer (A,1-A) with $A \geq \beta^*$; then labor is supposed to strike. If it strikes, the payoffs are $a + \frac{\delta}{1-\delta}\alpha^*$ for capital, $b + \frac{1}{1-\delta}(1-\alpha^*)$ for labor. If labor does not strike, the payoffs are $\frac{1}{1-\delta} r$ for capital and $\frac{1}{1-\delta}(1-r)$ for labor. Given capital's equilibrium strategy and the strict version of (3*), $r > (1-\delta)^2 (1-b) + \delta^2 a$, labor will definitely strike. Now capital could suggest: "Forget about strike. You will not be punished. We are free to change our plans for the following subgame and can play the strike equilibrium instead of the non-strike equilibrium." For the current subgame, capital suggests a subgame-perfect equilibrium which, indeed, dominates the equilibrium prescribed by the original equilibrium strategies for the entire game. The suggested play pays (r,1-r) for the current period while a strike would only yield (a,b); future payoffs are equal for both alternatives. Yet capital's suggestion, attractive at first glance, makes empty promises. Namely, the suggested play is not renegotiation-proof, unless the original equilibrium is. For if capital can talk labor out of a strike now, it can use the same argument again and again -- thus forcing labor down to its non-strike equilibrium payoff. Similar reasoning applies to other possible suggestions by capital.

5.4 Future Research

The present contribution by-passes some important aspects of bargaining: incomplete information, timing of actions, delegation problems. Bargaining with a sequence of contracts deserves further attention; see related work by Okada (1988). Future work should also consider non-stationary economic environments: surplus may depend on exogenous variables, modelled as a stochastic process, and on endogenous variables, say investment decisions. The latter case was already examined by Selten and Güth (1982) who relied on the Nash Bargaining Solution; a purely strategic analysis is still missing. Finally, bargaining models à la Rubinstein should be contrasted with earlier more descriptive models as presented in Tietz (1983).

Acknowledgement

Part of this research was done while the author participated in the Research Project "Game Theory in the Behavioral Sciences" at the Center for interdisciplinary Research (ZiF) of the University of Bielefeld. A first draft of the paper was circulated as Working Paper E-88-09-02, Department of Economics, Virginia Polytechnic Institute and State University. Independently, Holden (1989) has studied a very similar model. Steinar Holden and myself envisage consolidating our work in a short summary to be published elsewhere.

References

Binmore, K. and P. Dasgupta (1987). Economics of Bargaining. Basil Blackwell.

Brito, Buoncristiano, Intriligator (1977). A New Approach to the Nash Bargaining Problem. Econometrica **45**: 1163-1172.

Edgeworth, F.Y. (1887). In: Paul Kegan, (Ed.), Mathematical Psychics. Also: L.S.E. Series of Reprints of Scarce Tracts in Economics and Political Sciences, No. **10**, 1932.

Farrell, J. (1983). Credible Repeated Game Equilibrium. mimeo.

Fershtman, Ch., K.L. Judd and E. Kalai (1987). Cooperation through Delegation. mimeo.

Gupta, S. and Z. Livne (1988). Resolving a Conflict Situation with a Reference Outcome: an Axiomatic Model. Management Science **34**: 1303-1314.

Haller, H. (1986). Non-Cooperative Bargaining of $N \geq 3$ Players. Economics Lett. **22**: 11-13.

Herrero, J.M. (1985). Bargaining and Involuntary Unemployment. mimeo: London School of Economics.

Herrero, J.M. (1986). Delays in Good Faith Bargaining. mimeo, GSIA, Carnegie-Mellon University.

Hicks, J.R. (1932) The Theory of Wages. Macmillan.

Holden, S. (1989). Non-Cooperative Wage Bargaining. mimeo, University of Oslo.

Jun, B. (1988). Noncooperative Bargaining and Union Formation. mimeo. Forthcoming in Review of Economic Studies.

Kalai, E. and M. Smorodinsky (1975). Other Solutions to Nash's Bargaining Problem. Econometrica **43**: 513-518.

Nash, J.F. (1950). The Bargaining Problem. Econometrica **18**: 155-162.

Okada, A. (1988). A Two-Person Repeated Bargaining Game with Long-Term Contracts. This Volume.

Osborne, M.J. (1984). Capitalist - Worker Conflict and Involuntary Unemployment. Rev. of Economic Studies **LI**: 111-127.

Roth, A.E. (1979). Axiomatic Models of Bargaining. Lecture Notes in Economics and Mathematical Systems, No. **170**, Springer Verlag.

Roth, A.E. (Ed.) (1985). Game-Theoretic Models of Bargaining. Cambridge University Press.

Rubinstein, A. (1982). Perfect Equilibrium in a Bargaining Model. Econometrica **50**: 97-110.

Rubinstein, A. (1985). A Bargaining Model with Incomplete Information about Time Preferences. Econometrica **53**: 1151-1172.

Selten, R. and W. Güth (1982). Game Theoretical Analysis of Wage Bargaining in a Simple Business Cycle Model. Journ. of Mathematical Economics **10**: 177-195.

Shaked, A. and J. Sutton (1984). Involuntary Unemployment as a Perfect Equilibrium in a Bargaining Model. Econometrica **52**: 1351-1364.

Tietz, R. (Ed.) (1983). Aspiration Levels in Bargaining and Economic Decision Making. Lecture Notes in Economics and Mathematical Systems No. **213**, Springer Verlag.

Zeuthen, F. (1932). Problems of Monopoly and Economic Warfare. Routledge & Kegan Paul.

A GAME EQUILIBRIUM MODEL OF THIN MARKETS

by

Reinhard Selten and Myrna H. Wooders[1]

Abstract: We consider games of group, or coalition, formation occuring over infinite, discrete time, with new participants becoming active in the game in each period, and with participants that have successfully formed groups leaving the game each period. Markets may be "thin", in the sense that the number of participants active in the game in any time period is finite and may be small. We construct a subgame perfect equilibrium for an example and show some additional properties of the equilibrium. One property is that, even though markets are thin, the "first mover" within a time period has an advantage (and realizes more than a competitive payoff) only in special circumstances, and, along the equilibrium path, he is the only mover who can have such an advantage. Also, we discuss the limit behavior of the model as costs of waiting (time costs) become small; specifically, the equilibrium payoffs converge to core payoffs of a game with a continuum of players and finite coalitions (f-core payoffs). The static continuum game provides an idealization of the limit of the dynamic games for small waiting costs. Thus our research initiates providing a noncooperative foundation for the core as a solution concept for such games.

1. INTRODUCTION

Our motivation comes from "thin markets"--ones with possibly only a few participants in the market at any one time. It has been demonstrated that economies and games with many players and with the property that "small" groups of players can realize all gains to group formation are competitive, in the sense that cooperative outcomes coincide with price-taking equilibria.[2] However, although "effective" groups may be small, such as buyer-seller pairs, these models require a large total player set. Our study is motivated by the observation that, at any point in time, there may be only a few participants in the market. Also, for many economic interactions, participants may not be confined to buy, sell, or make a deal immediately; they might be able to postpone reaching an agreement. For example, one might rent a temporary residence and postpone buying a house until next month, or next year, or longer. Even casual perusal of the "Homes for Sale" or "Business" sections of the newspapers indicates that the expected economic situation "next period" is a major determinant of current economic variables.

---

[1]The authors are indebted to the Centre for Interdisciplinary Research, Bielefeld, for hospitality and support. We are also indebted to Shouyong Shi, Sage Traviza, and John Wooders for comments on an earlier draft. One author expresses her gratitude for the previous (and ongoing) collaboration of Mamoru Kaneko; this collaboration has had a major influence on the current research. Wooders also gratefully acknowledges the research support of the Social Sciences and Humanities Research Council of Canada, which has greatly faciliated this collaborative research.

[2](See Mas-Colell (1979), Kaneko-Wooders (1986, 1989), Hammond-Kaneko-Wooders (1989), and Wooders (1988)).

In this paper we investigate a highly styalized model of thin markets. The markets are described by a dynamic noncooperative model of group formation and payoff distribution. There is a countable number of time periods and, at most, a finite number of participants within each time period. A finite number of new participants become active in the game each period, and also, each period, there is the possibility of participants completing their market activities and leaving the game. Remaining in the market more than 1 period is costly. We show existence and provide a description of a subgame perfect equilibrium. One notable feature of the equilibrium is that only in special circumstances concerning the exact numbers of participants active in the market is there an advantage to being selected as the first mover, or decision-maker, within a time period. Moreover, we show that as time costs become small, the subgame perfect equilibrium payoffs converge to cooperative/competitive outcomes in the sense that the equilibrium payoffs are in the core of a related static game (a game with a continuum of players and finite coalitions.)

We develop one version of the model in detail, and provide a sketch of another. We sometimes refer to these as "examples", since we have in mind, and are currently working at a model which generalizes both versions in this paper. The first example in this paper has the advantage that it is sufficiently simple to enable us to construct an equilibrium point.

The games we study are ones of group formation and payoff distribution and are related to characteristic function bargaining games. For purposes of exposition, we will call any group that can realize a positive payoff a "firm" and use other terms consistent with this analogy, such as "workers", and "wages". However, it is not to be construed that the groups are necessarily "firms"; they could be "bridge clubs", for example, or "teams" or buyer-seller pairs. The activity of the game is described as occurring in a "market place". Since the total number of active participants in a period is finite, the games are "thin", in analogy with thin markets. The activity of the market-place is firm formation, and payoff distribution via wages and net revenues.

For our first example there are two permissible firm sizes, 1-member firms and n-member firms, where n is an exogenous parameter.[3] At the beginning of a time period, a new participant arrives at the market-place. There is then some finite number of participants in the market place throughout the current period. These participants are randomly ordered, and the jth participant in the ordering is called the jth decision-maker. The decision makers, each in accordance with his position in the ordering, successively choose alternatives from a set containing alternatives of four kinds. The kinds of alternatives are:

---

[3]We will indicate the relaxation of this restriction.

(1) forming a firm, which means that the decision-maker announc s a wage (for n-1 vacancies);

(2) accepting a wage offer from an established firm, if any have unfilled positions;

(3) choosing self-employment (i.e., forming a firm consisting of oneself alone); and

(4) passing (i.e., simply waiting until the next period).

The alternative chosen by a decision maker is required to satisfy several feasibility conditions: (a) a wage offer must be "payable", i.e. it must be a multiple of the smallest money unit (which we assume exists); (b) a wage offer made by the jth decision maker can only be accepted by subsequent decision makers in the ordering, (and not by preceeding decision makers); (c) a wage offer can be accepted by at most n-1 participants. When each decision maker has chosen a feasible alternative from the set available to him, we reach the end of the current period. Those participants who are members of "full firms" (ones with n-1 employees) leave the activity of the game; the workers receive the wages they accepted and the firm-formers, the net revenues of their firms (the exogenously given total revenue of an n-member firm, normalized to equal n, minus the wages paid). Those participants who chose self-employment leave the activity of the game with zero payoff. Partially-filled firms dissolve and their members pay a "waiting cost" of c plus a penalty, for making a mistake, of γ, where both c and γ are payable amounts (multiples of the smallest money unit). Those participants who chose to pass pay the waiting cost of c. A new period begins. The participants in the new period are those who did not leave the game at the end of the preceeding period (who chose pass, or were members of unfilled firms) plus a new participant (a new entrant into the market-place). The activity of the game begins anew with the random ordering of participants.

Before describing our results in more detail, there are several aspects of our model that we wish to emphasize:

(1) We do not keep track of individual players but only of the population, the number of participants, in each period. The game in extensive form played out during a period can be viewed as a "population game" (Selten (1980)), a heuristic term indicating a game which abstracts from identities of particular players and focuses instead on populations of types of players.

(2) At the end of a period, those firms with unfilled vacancies dissolve. At the beginning of the next period, those participants who formed such firms, and those participants who accepted their wage offers are indistinguishable from other participants. It would be desirable to relax this feature of the model, but it appears that to do so would make the model much less tractable.

(3) We do not restrict attention to "steady stat s", but do restrict attention to symmetric stationary equilibria--ones where all identical participants use identical strategies, and strategies depend only on situations[4] and not on choices in previous periods except insofar as these are reflected in numbers of participants in situations.

(4) The expected payoff to a participant is not necessarily the same as his terminal payoff (the payoff received when the participant leaves the game as self-employed or as a member of a full firm) unless the participant is leaving with certainty in the current period. Otherwise his expected payoff this period partially depends on his expected payoff next period, for every period; the game has a recursive structure (see Everett (1954) and Selten (1982)). We describe expected payoffs by defining expected payoffs for truncated games, and then taking limits as the length of the truncated games become infinite.

For our first example, we show that an equilibrium exists. Also, we show that all stationary equilibria with a "shortness" property are "efficient". Moreover, we describe the only stationary equilibrium with, additionally, another shortness property. As the waiting costs in our model go to zero (or the number of participants entering at the beginning of each period increases) the equilibrium expected payoff to a participant in any situation converges to his cooperative payoff in a static game with many players (his f-core payoff in a game with a continuum of players and finite coalitions) which provides an idealized limit model.[5] We can view this cooperative payoff as a price-taking competitive equilibrium payoff[6], and, with this in mind, we can compare equilibrium payoffs in the dynamic model with positive waiting costs to the competitive payoffs. We note again that except in special circumstances, in equilibrium participants can expect no more than their competitive payoffs, even with positive waiting costs.

We prove existence of an equilibrium by describing a particular strategy and then showing that this strategy is a stationary equilibrium point. To enable us to do this, however, we first show that if a global strategy, which prescribes a choice at every decision situation, has the property that at each situation it maximizes the decision makers' expected utility in the subsequent situation then it is a global optimum -- there are no profitable deviations with any other global strategy. In other words, if a strategy specifies a locally optimal choice at every situation,

---

[4]A "situation" is a description of the market place at a point within a period and consists of a number, the number of participants in the period, and a list of the alternatives chosen up to that point by participants who have taken their turns as decision makers.

[5]This sort of game was introduced in Kaneko-Wooders (1986). See also Hammond-Kaneko-Wooders (1989) and Kaneko-Wooders (1985, 1989).

[6]See Wooders (1988).

then it is a global stationary equilibrium. We remark that both an optimal choice and a deviation from a global stationary strategy are defined with respect to a given strategy, and under the assumption that all decision makers in all subsequent situations will employ this strategy.

An equilibrium play of the game for the equilibrium point we construct proceeds as follows: Suppose there are k participants in the current period. If $k < n$, the number required to form a full firm, each participant simply passes (we assume 1, the per-capita revenue of an n-member firm, is greater than $(n-1)c$ so passing is better than self-employment in this situation). If $k = n$, the 1st participant forms a firm with the offer of a wage equal to the smallest payable amount greater than or equal to the expected payoff to a participant at the beginning of a period with $n + 1$ members. (This wage is less than 1, the competitive wage). In equilibrium the remaining participants in the period accept the offerred wage. If $k > n$, $k \neq rn$ for any integer r, the 1st decision maker cannot benefit from being first. In this situation he must offer a wage of 1; if he offered less the next decision maker could offer a wage of 1, his firm would fill, then another decision maker could offer a wage of 1, his firm would fill, and this process of forming and filling firms, all paying wages 1, would continue until there are too few remaining participants to fill the firm of the 1st decision maker. Finally, if $k = rn$ for some positive integer r the 1st decision maker can again offer a wage just sufficient to keep the last $n - 1$ decision makers from refusing, paying waiting costs, and taking the expectation next period. The other $(r-1)n$ decision-makers, between the 1st and the last $n - 1$, each realize a payoff of 1 by either forming a firm offering a wage of 1 or accepting wage a offer of 1. In all cases, when each decision maker has, in his turn, chosen an alternative, we reach a terminal situation. Given $k = rn + \ell$ for some nonnegative integers r and ℓ with $\ell < n$, the equilibrium number of participants who leave the game at the terminal situation as members of full firms is rn, and the number of participants in the next period is $\ell + 1$.

Off the equilibrium path specifying our equilibrium strategy is much more difficult. There may be a number of outstanding wage offers. The optimal alternative for a decision maker depends on the number of participants remaining after all sufficiently attractive vacancies outstanding are filled. But what is a sufficiently attractive wage offer depends on the number of participants in the next period, which itself depends on the number of outstanding vacancies that will be filled. This becomes complicated, so we leave a more precise description of the equilibrium strategy to the next section.

The first example, described above, constitutes the main subject of our analysis in this paper. We also sketch a simple model with two types of players (Players of different types are ones who make different marginal contributions to at least one group.) For both models the same characterization of the equilibrium payoffs for small waiting costs holds: The expected equilibrium payoffs are "close" to cooperative equilibrium (f-core) payoffs of a static game with a continuum of players (and, in the 2-type case, with the relative numbers of new participants entering the game each period reflected by the measures of players of each type in the static game).

Our models provide an example of the achievement of a cooperative outcome from a noncooperative process. Intuitively, this result rests on a few modelling assumptions: in large finite games relatively small groups (in this paper, ones of bounded size) and in games with an atomless measure space of players, groups of measure zero, can realize all gains to group formation;[7] there are many participants, albeit spread over time; and waiting is costly. In these circumstances subgame perfection and the recursive structure of our models have the effect of making future participants resemble current competitors in limiting possibilities for effective strategic (i.e. "non-payoff-taking") behavior in the current period, and optimizing strategic behavior approaches "payoff-taking", competitive behavior as waiting costs are allowed to become small.[8] Later (in Section 4) we further discuss our motivation from the viewpoint of cooperative game theory and some relationships to the literature.

Although our model involves infinite time, it is naturally truncated to form a finite-time-period model, and one which is experimentally testable. It is plausible to expect that, for small values of c, significant deviations from the equilibrium path, for example an oversupply of participants of some type, would lead to adaptive behavior, and in the case of an oversupply, a willingness to accept a lower wage. The game seems to exhibit the right sort of negative feedback to stabilize the equilibrium path by processes of learning and adaptation of aspiration level (we use the term "aspiration level" in the sense of Simon (1959), and Sauermann and Selten (1962), for example). It is our intent to investigate a truncated model to see if, in the laboratory, actual payoffs approach the infinite game equilibrium payoffs. We comment further on this in Section 5.

Our model also has similarities with bargaining models of exchange economies (cf., Rubinstein-Wolinsky (1985), Gale (1986a, 1986b), and Peters (1986)). Since in this paper we consider only two examples, neither including situations of pairwise matching of players of two different types, buyer-seller pairs, for example, a discussion of the bargaining literature mentioned above is appropriately deferred until our study is more advanced. We note that an important difference between our

---

[7]It is this aspect of large games which ensures their competitive mark t-like properties. See especially Wooders (1988) and Kaneko-Wooders (1985, 1986).

[8]We use the term "payoff-taking" as an analogue to "price-taking". The basic Hypothesis of Perfect Competition is that agents/players take pric s as given, instead of taking account of the effects of their actions on prices.

models and the exchange-economy bargaining models is that in ours, the objective of the participants is to form a profitable group -- it is of the essence that the size and composition of groups is determined endogenously. This also means, for example, that we do not restrict attention to steady-states -- there must be sufficient time for enough participants to enter the marketplace to form optimal groups; so exit from the marketplace does not necessarily occur at the same rate as entry. A static description of the dynamic game is given by a game in coalitional (ie. characteristic function) form rather than by an exchange economy. Since the game is not derived from exchange economy data, we do not compare outcomes to Walrasian outcomes, as in the bargaining literature, but instead take the f-core of a continuum game as our (static) competitive/cooperative equilibrium concept. We note, however, that especially in the case of two-types of players where profitable groups consist of one member of each type, our model would have much of the flavour of a bargaining model, for example, Rubinstein-Wolinsky (1985). Since we are interested in the impact that small numbers of players can have on outcomes, our model is less closely related to bargaining models with a continuum of players in each period, as in Gale (1986a, 1986b), for example.

Finally, our model has many similarities to the model of n-person coalitional bargaining of Chatterjee et al (1990). While again a complete discussion of the relationship of their work to ours is more appropriately deferred, we note that for their finite games strong superadditivity (so the only optimal coalition structure is the one where all players are in one coalition) plays an important role in obtaining equilibrium payoffs, to bargaining, in the core. In contrast in our model, as noted above, an important feature behind the results is that small coalitions can realize all gains to group formation.

2. THE GAME

In this and the following section we develop a model which is sufficiently tractable to enable us to characterize an equilibrium point. The model is restrictive in that, at the beginning of a period, all participants are identical.[9]

The game we consider consists of a mapping Γ from I_+, the nonegative integers, into population games, Γ^k for each $k \in I_+$. The integer k represent a population size. We think to the games Γ^k, which are games in extensive form, as population games since, as in Selten (1980), we do not keep track of individual players over time, but instead, only of the participants in the population during any one time period. Before describing the game Γ and the games Γ^k, we describe some auxiliary concepts.

---

[9] A more general model, which might be called a "recursiv population game," allows for several types of players and is a subject of our work in progress.

<u>The population</u>:

An integer $k \in I_+$, where I denotes the integers and I_+, the nonnegative integers, represents the number of participants in the game Γ^k. We interpret the game Γ as being played over a possibly infinite number of periods, and Γ^k as the gam played in any period with k participants.

At the beginning of a period with k participants, the participants are randomly ordered $< 1,\dots,j,\dots,k >$ and every participant has equal probability of being j^{th} in the ordering. The jth participant in the ordering is called the jth <u>decision-maker</u>. The decision makers act successively within the period.

<u>Money units M</u>:

We assume that there is a smallest money unit, given by $1/u!$ where u is a positive integer and that all payments must be made in integer multiples of $1/u!$ We define $M(u)$ as the set of <u>payable amounts</u> so

$$M(u) = \{\ell/u! : \ell \in I\}$$

(The purpose of taking $u!$ is to ensure that for any rational number r these exists a u_0 such that for all $u \geq u_0$, $r \in M(u)$.)

<u>Firms</u>:

A fixed number $n \in I_+$ of participants are required to fill (or operate) a firm. A full firm--one with n members including a firm-former and $n - 1$ workers--has total revenue of n.

[We remark that in principle we could have allowed a more general technology, say $f(n)$ as the total revenue of a firm with n member for each $n \in I_+$, and required only that $f(n)/n$ is bounded. This, however would not be a significant generalization since, with positive waiting costs, in equilibrium only firms which are of "optimal size", say the smallest n' such that $\frac{f(n')}{n'} \geq \frac{f(n)}{n} - c$ for all n, will form.]

<u>Alt rnatives</u>:

The jth decision maker has the following four types of alternatives, all written as ordered pairs:

(w,-), <u>making a wage offer</u>, where $w \in M(u)$, $0 \leq w \leq \frac{n}{n-1}$. An alternative of this kind is to be interpreted as forming a firm and offering to pay a wage of w to $(n-1)$ participants. (The firm former is himself viewed as a participant in the firm so, by hiring $n - 1$ other participants, he "fills" the firm.)

(-,i), accepting the wage offer of the ith participant, $i = 1,\dots,j-1$. This alternative is interpreted as accepting the wage offer of a previous decision maker i in the current population game.

(-,-), passing. In this case, the participant decides to neither form nor join a firm; he simply waits until the next period.

(0,j), choosing self-employment. The participant leaves the game Γ at the end of the current population game.

We remark that each participant is the decision maker only once within a period.

Situations:

A situation x describes a history beginning with the opening of a new population game and lists an alternative for each participant who has, up to the current time, been a decision maker. The situation $x = \langle k;a_1,\dots,a_j \rangle$ can be viewed as the history facing the current decision-maker, the $j+1^{th}$, in a population game with k participants. Formally a situation $x = \langle k;a_1,\dots,a_j \rangle$ is an ordered collection of alternatives, $a_1,\dots,a_j$, preceeded by the population parameter k, with the properties that

(a) $j \le k$ (there are no more alternatives than participants);

(b) for each $m = 1,\dots,j$, if for some positive integer i we have $a_m = (-,i)$ then $i < m$ and $a_i = (w_i,-)$ for some $w_i \in M(u)$ (only offers made by previous decision makers can be accepted); and,

(c) for any $m = 1,\dots,j$ with $a_m = (w_m,-)$, we have $|\{i\colon a_i = (-,m),\ i \le j\}| \le n - 1$ (a decision maker can only accept outstanding wage offers -- wage offers of firms with vacancies).

A situation $x = \langle k;a_1,\dots,a_j \rangle$ is a decision situation (for the $j+1^{th}$ decision maker) if $j < k$. It is a terminal situation if $j = k$, and an initial situation if $j = 0$, in which case we write simply $x = \langle k \rangle$.

For given k, let X^k denote the set of situations, Z^k, the set of terminal situations, and $X^k \backslash Z^k$, the set of decision situations. We note that X^k is finite since k is finite, each participant can choose one of only 4 kinds of alternatives, each of 3 of these kinds consists of only 1 alternative, and the 4th kind contains only a finite number of alternatives since the number of different wage offers which can be made by a participant who chooses to form a firm is finite (since $M(u) \cap \mathbb{R}_+ \cap \{w\colon w(n-1) \le n\}$ is finite).

Let $X = \bigcup_k X^k$, $Z = \bigcup_k Z^k$, and $X\backslash Z = \bigcup_k X^k\backslash Z^k$.

Choices:

The choice function C is defined on the set of decision situations $X\backslash Z$ by the condition that $y \in C(x)$, where $x = < k;a_1,\ldots,a_j >$, iff $y = < k;a_1,\ldots,a_j,a_{j+1} >$ is a situation.

(We remark that we could have defined choices as alternatives rather than as situations, since the requirements that x and y are situations determines a set of "permissible" alternatives.)

As indicated previously, participants make decisions successively with the ordering of their decision making determined by the random ordering of the participants. We note that we will introduce waiting costs, which prevent participants from forever passing (choosing $a = (-,-)$) rather than attempting to conclude their market-place activities.

The Transition to the Next Period:

The number of participants in a period depends on the terminal situation reached in the previous period. Let z be a terminal situation, $z = < k;a_1,\ldots,a_k > \in Z$. We describe the set of participants who leave the game Γ at the situation z. These participants include: the set of firm-formers whose offers have been accepted by

n - 1 workers, denoted by

$$\theta(z) = \{j\colon \text{the } j^{th} \text{ alternative in } z \text{ is } a_j = (w_j,-) \text{ for some } w_j \text{ and } |\{i\colon a_i = (-,j)\}| = n - 1\};$$

and the set of employees of full firms,

$$\Phi(z) = \bigcup_{j\in\theta(z)} \Phi_j(z)$$

where

$$\Phi_j(z) = \{i\colon a_i = (-,j),\ j \in \theta(z)\}$$

is the set of employees of the full-firm $j \in \theta(z)$. The set of all participants who leave the game at the terminal situation z, denoted by $\varsigma(z)$, consists of th full-firm formers $\theta(z)$, their employees $\Phi(z)$, and those who chose self-employment:

$$\varsigma(z) = \{i\colon a_i = (0,i)\} \cup \theta(z) \cup \Phi(z).$$

Those participants in the game Γ^k who are not in $\varsigma(z)$ are "continu d" into the next period, and a new participant joins th game at the beginning of the next period.[10] This determines a population for the next population game, defined by the population transition function.

$$K(z) = k - |\zeta(z)| + 1$$

for each $z \in Z^k$ and for each k.

Terminal Situation Payoffs:

At a terminal situation z workers in full-firms receive the wages they accepted and successful firm-formers receive the net revenues of their firms. Unsuccessful firm-formers and those workers who accepted their wage offers each pay $c + \gamma$, exogenously given payable amounts, with c interpreted as a "waiting cost" and γ as a penalty for taking an unfruitful action. The unfilled firms are then dissolved, and their members become participants in the next population game (i.e., the game in the subsequent period) and indistinguishable from other participants in that game. Participants who passed pay the cost c and also become participants in the next population game. Finally, those who chose self-employment leave the game at z and receive the payoff of zero.

For each k and each $z \in Z^k$, a (terminal situation) payoff is given by $h = (h_1, \ldots, h_k)$ with, for each $j = 1, \ldots, k$,

$$h_j(z) = \begin{cases} w_i & \text{if } i \in \Theta(z) \text{ and } j \in \Phi_i(z) & \text{(a worker in a full firm receives the wage he accepted);} \\ n-(n-1)w_j & \text{if } j \in \Theta(z) \text{ and } a_j = (w_j,-) & \text{(a full-firm-former receives the net revenue of his firm);} \\ -c-\gamma & \text{if } j \notin \Theta(z) \text{ and } a_j = (w_j,-) & \text{(an unsuccessful-firm-former former pays a fee of } c+\gamma\text{);} \\ -c-\gamma & \text{if } j \notin \Phi(z) \text{ and } a_j = (-,i) & \text{(an accepter of an offer of an unsuccessful firm pays } c+\gamma\text{);} \\ -c & \text{if } a_j = (-,-) & \text{(a passer pays } c\text{);} \\ 0 & \text{if } a_j = (0,j) & \text{(a self-employed participant receives zero, and pays no waiting costs since he leaves the game).} \end{cases}$$

where c and γ are positive numbers in $M(u)$ (and therefore rational).

The wages and net revenues of workers in full firms and the firm-formers might be best interpreted as discounted future earnings for the entire future of these

[10] We could, in principle, easily modify the model so that some fixed number enter the activity of the game at the beginning of a p riod, rather than just 1, but for this first example, we choose not to do so.

participants. We imagine these participants as exiting the market place, going into production, and no longer participating in the activity of the game. (To the extent that our analogy of a "group" to a firm suggests markets for the output of a firm and prices for output depending on the number of firms and determined endogenously, it may be misleading.)

Also, we remark that imposing a cost of γ to making a mistake -- forming a firm that does not fill, or accepting the offer of such a firm -- is expedient in that, in characterizing equilibria we can rule out many, otherwise irrelevant, decisions (for example, accepting offers of firms which cannot be filled).

We assume that $1 \geq (n-1)c + 1/u!$ so, if there is fewer than n active participants, it will be optimal to wait for more entrants rather than choose self-employment. Also, we assume that $c + \gamma > \frac{2}{u!}(n-1)$; this will be used later to show that a particular strategy is a symmetric equilibrium point.

<u>The Game Γ</u>:

The game Γ assigns to every $k \in I_+$ an extensive form component game $\Gamma^k = (X^k, Z^k, C^k, K^k, h^k)$. The constituents of Γ^k satisfying the properties above, where C^k is the restriction of C to X^k, K^k the restriction of K to $z \in Z^k$, and h^k the restriction of h to Z^k.

A few remarks may be warranted. First, during a play of the game Γ, a component game Γ^k may occur several times. (It may even occur an infinite number of tim s). Second, the component games might be regarded as "stage" games where the curr nt stage is determined by the number of participants "continued" from the previous stage (and the number of new entrants, which we assume is 1). Third, our model does not permit any "memory" of previous periods (i.e., previous component games) except that implicit in the number of participants in the current period.

<u>Strategies</u>:

We do not define strategies for individual players, but instead, global behavior strategies, which specify the behavior of a participant in every conceivabl situation. Implicit in the definition is the property that all participants use the same strat gy.

A <u>local strategy</u> b_x at the decision situation $x \in X\backslash Z$ is a probability distribution over $C(x)$. (Since $C(x)$ is finite, b_x has a finite carri r i. ., is greater than zero at only a finite number of situations.) The probability assign d to $y \in C(x)$ by b_x is denoted by $b_x(y)$. For each $x \in X$, let B_x be the s t of all local strategies.

A <u>global strategy</u> b is a function which assigns a local strategy b_x to v ry $x \in X\backslash Z$. Let B be the set of all global strategies.

Expected Payoffs:

We define expected payoffs to participants relative to a strategy choice (the same one) for all participants in all situations.

First, we introduce several additional terms.

A situation $x' \in X^k$ is reachable from $x \in X^k$ if there are situations $x_1, x_2, \ldots, x_m$ with $x_1 = x$, $x_m = x'$, and $x_2 \in C^k(x_1)$, ..., $x_m \in C^k(x_{m-1})$.

A round of the game Γ^k is an ordered list of situations, $R = (x_1, x_2, \ldots, x_m)$, where $x_1 = \langle k \rangle$ and x_m is a terminal situation and where each situation $x_{q+1} \in C^k(x_q)$. We note that a round is a list of situations occurring within one time period and a round is completely described by a terminal situation.

A chain is a list of rounds, $R_1, R_2, \ldots, R_m$ with the number of participants in the qth round equal to $K(z^{q-1})$, where z^{q-1} is the terminal situation of the q-1th round, for each $q = 2, \ldots, m$. A chain of length m occurs over m time periods.

We note that we can easily "truncate" the game after t rounds by setting the payoff in all possible t^{th}-round terminal situations z simply equal to that given by $h(z)$ and setting $K(z) = 0$ for every terminal situation of the tth-round (stopping the game).

The expected payoffs to initial situations in rounds are recursively determined because of the recursive structure of situations within rounds and of rounds within chains. Therefore we define expected payoffs by defining payoffs to chains of length t and then letting t go to infinity. Let $E(b,k,t)$ denote the expected payoff, relative to the strategy b, to a participant in Γ^k when only t rounds are to be played. Define.

$$E(b,k,0) = 0 \quad \text{and}$$

$$E(b,k,t) = \frac{1}{k} \sum_{j=1}^{k} \sum_{z \in Z^k} \pi(z|k,b)[h_j(z) + \delta_j E(b,K(z),t-1)] \quad \text{for} \quad t = 1,2,\ldots$$

where $\pi(z|k,b)$ denotes the probability that the terminal situation z is reached from the initial situation $\langle k \rangle \in X^k$ when all participants choose according to b and where $\delta_j = 0$ if j is among the participants leaving the game (i.e., $j \in \zeta(z)$) and $\delta_j = 1$ if j is continued into the next period.

We remark that this expectation is prior to the random ordering of participants. Given an ordering, $\sum_{z \in Z^k} \pi(z|k,b)[h_j(z) + \delta_j E(b,K(z),t-1)]$ is the expected payoff of the jth decision maker, and we average over all decision makers $j = 1, \ldots, k$ to obtain $E(b,k,t)$.

The expected payoff relative to b to a participant in Γ^k is defined by

$$E(b,k) = \limsup_{t \to \infty} E(b,k,t).$$

(Note that, if positive, $E(b,k)$ is finite sinc it is bounded above by n. However, $E(b,k)$ is not bounded below and might, therefore, be minus infinity.) We can also define the expected payoff relative to b of a terminal situation $z \in \Gamma^k$ to the jth decision maker by

$$E_j(b,z) = h_j(z) + \delta_j(z)E(b,K(z))$$

and the expected payoff relative to b at the decision situation x of the jth decision maker by

$$E_j(b,x) = \sum_{z \in Z_x} \pi(z|x,b)E_j(b,z)$$

where Z_x denotes the set of terminal situations reachable from x and $\pi(z|x,b)$ denotes the probability that z is reached, given x and b.

We have now defined an expected payoff for all participants at every $x \in X$ when all participants play b.

Stationary Equilibrium Points:

First, we define an optimal choice at a decision situation x, relative to a strategy b to be used by all subsequent participants in this and all subsequent rounds. Our definition is simply that a choice is optimal if it maximizes the expected payoff to the decision maker in the new situation created from x by his choice.

At a decision situation $x = < k;a_1,\ldots,a_{j-1} >$, let $\rho(x)$ denote the jth decision maker, called the decision maker at x. A choice $y \in C(x)$ is optimal at x with respect to a strategy b iff

$$E_{\rho(x)}(b,y) = \max_{w \in C(x)} E_{\rho(x)}(b,w)$$

A global strategy b is a stationary equilibrium point if, for every decision situation $x \in X\backslash Z$ and every decision maker $\rho(x)$, letting b_x be the strategy b restricted to the decision situation x, the following holds: Every $y \in C(x)$ with $b_x(y) > 0$ is optimal at x.

Local and Global Optimality:

Our definition above of a stationary equilibrium point is based on local optimization. We will show that local optimization implies global optimization. For this purpose we introduce the notion of a strategy d, used by one participant, which deviates for at most t rounds from a global strategy b used by all other participants. We denote the expected payoff to the deviation d by $E(d,b,k,t)$, where k is the population in the first round and t is the number of rounds. We then let t go to infinity. The definition of $E(d,b,k,t)$ is recursive:

$E(d,b,k,0) = E(b,k)$ (when d differs from b for 0 rounds, the expected payoff to the deviation d is the same as that to b) and

$$E(d,b,k,t) = \frac{1}{k}\sum_{j=1}^{k}\sum_{z\in Z^k}\pi(z|k,j,d,b)[h_j(z)+\delta_j E(d,b,K(z),t-1)]$$

(when d differs from b for t rounds, after t rounds, the expectation to the deviation is the same as the expectation to b)

where $\pi(z|k,j,d,b)$ is the probability that z is reached given the jth decision maker uses the strategy d and all other participants use b.

Let $E(d,b,k) = \limsup_{t\to\infty} E(d,b,k,t)$.

<u>Proposition 1</u>:

Let b be a stationary equilibrium for Γ. Then, for any population k and any deviation d we have

$$E(d,b,k) \leq E(b,k)$$

Proof: Suppose not. Then there is a smallest t, say $\tilde{t}$, for which

$$E(d,b,k,\tilde{t}) > E(b,k).$$

We construct a new game $\Gamma^{k,\tilde{t}}$ which begins at <k> and has the same rules as Γ with the exception that every play is cut off after $\tilde{t}$ rounds and the terminal situations at the end of the $\tilde{t}^{th}$ rounds become endpoints. The payoff to an endpoint z (of the $\tilde{t}^{th}$ round) to the jth participant is given by

$$h_j(z) + \delta_j E(b,K(z))$$

Since $y \in C(x)$ and $b_x(y) > 0$ implies that y is optimal at x with respect to b, for every situation x, the restriction of b to $\Gamma^{k,\tilde{t}}$ is an equilibrium point of $\Gamma^{k,\tilde{t}}$. (It is well-known that for games of finite length, local optimality. implies global optimality.) Since a participant in <k> has an equilibrium expected payoff of $E(b,k)$ for the restriction of Γ to $\Gamma^{k,\tilde{t}}$, we cannot have $E(b,k) < E(d,b,k,\tilde{t})$.

□

For completeness, we note that we can define the <u>expected payoff relative to a strategy b and a deviation d used by one participant to a terminal situation</u> z by

$$E_j(d,b,z) = h_j(z) + \delta_j(z)E(d,b,K(z))$$

and <u>to a decision situation x</u> by

$$E_j(d,b,x) = \sum_{z\in Z_x} \pi(z|x,j,d,b)\, E_j(d,b,z)$$

where $\pi(z|x,j,d,b)$ d notes the probability that z is reached from x, given that the jth decision mak r uses d and all oth r decision makers use b.

It follows from Proposition 1 that when b is a stationary equilibrium point then

$$E_j(d,b,x) \leq E_j(b,x)$$

for any deviation d and any decision situation x.

The above observation justifies our definition of a stationary equilibrium point using local optimization. Even though a stationary strategy is defined with respect to local optimization, such a strategy is stable against deviations with global strategies.

<u>Subgame Perfectness</u>:

We remark that a stationary equilibrium point b is a subgame perfect equilibrium point since b induces an equilibrium point on every subgame (Selten, 1965, 1975).

<u>Expect d Payoffs to Efficient Strategies</u>:

In this subsection and the next, we prepare some auxiliary concepts which will facilitate the description of an equilibrium strategy.

To show existence of a stationary equilibrium, in view of Proposition 1 we need only show existence of a global strategy b^* which is optimal at x for any decision situation $x \in X\backslash Z$. We will construct such a strategy, thus proving existence. The strategy will have the additional property of efficiency, which facilitates the description of expected payoffs. In this subsection, we describe expected payoffs to efficient strategies.

A terminal situation $z \in Z^k$ is <u>efficient</u> if there is no other terminal situation $z' \in Z^k$ with

$$\sum_{j-1}^{k} h_j(z') > \sum_{j-1}^{k} h_j(z)$$

Since $1 > (n-1)c$, it is clear that z is efficient if and only if $\sum_{j-1}^{k} h_j(z) - rn - \ell c$ where r is the largest integer such that $rn \leq k$ and $k - rn + \ell$. The integer r is the efficient number of firms and ℓ is the "left-over" participants. (Since there is a cost to unsuccessful firm-formation efficiency requires that no firms that cannot be filled will form.)

Recall that, at the terminal situation z, $\Theta(z)$ denotes those participants who have formed full firms and $\Phi(z)$ denotes the participants who have accepted the offers of now full firms. The expected payoff to the jth decision maker at a terminal situation z depends on the number of players in the next round. Suppose the strategy to be used in all subsequent rounds, say b^*, leads to ffici nt t rminal situations.

If there are ℓ participants in the next p riod, $\ell < n$, then the expected payoff in the next period is the expected payoff to waiting $n-\ell$ more periods until there are n participants in the market and it becomes feasible for another firm to form. Since all participants are equally likely to be the firm former, if there are n participants the expected payoff is 1 for each participant. Therefore in this case the expected payoff at the end of the current period for a participant who is continued into the next period is $1 - (n-\ell+1)c$. If there are $rn+\ell$ participants in the next period, $\ell < n$, then with probability $\frac{rn}{rn+\ell}$ a participant next period will be in a filled firm by the end of the period, and with probability $\frac{\ell}{rn+\ell}$ he will be continued into into the following period. This yields the following expected payoff to the jth participant in a terminal situation z , when in all subsequent periods efficiency is assumed:

$$E_j(b^*,z) = \begin{cases} w_i & \text{if } i \in \Theta(z) \text{ and } a_j = (-,i) \\ n-(n-1)w_j & \text{if } j \in \Theta(z) \text{ and } a_j = (w_j,-) \\ 1-g(k')-\gamma & \text{if } j \notin \Theta(z) \cup \Phi(z) \text{ and } a_j = (w_j,-) \text{ or } a_j = (-,i) \\ 1-g(k') & \text{if } a_j = (-,-) \\ 0 & \text{if } a_j = (0,j) \end{cases}$$

where $k' = K(z)$, where the number of participants next period, and

$$g(k') = \begin{cases} \frac{(n-\ell)\ell c}{rn+\ell} + c & \text{if } k' = rn+\ell,\ r > 0,\ 0 \le \ell < n \\ (n-\ell+1)c & \text{if } k' = \ell,\ 0\ \ \ell \le n \end{cases}$$

and r and ℓ are nonnegative integers.

[The function $g(\cdot)$ can be determined by solving recursively for $E(b^*,k)$. Suppose $K(z) = k'$ and $k' = rn + \ell$ with $\ell < n$.

To obtain $g(k')$ for $r = 0$ note that

$$\begin{aligned} & E(b^*,\ell) - c \\ &= E(b^*,\ell+1) - 2c = \dots \\ &= E(b^*,n) - (n-\ell)c - c \\ &= 1 - (n-\ell+1)c. \end{aligned}$$

Now suppose $r > 0$. Then

$$\begin{aligned} & E(b,k') - c \\ &= \frac{rn}{rn+\ell} + \frac{\ell}{rn+\ell}[E(b,\ell+1)-c] - c \end{aligned}$$

$$= \frac{rn}{rn+\ell} + \frac{\ell}{rn+\ell}[1-(n-(\ell+1))c-c] - c$$

$$= 1 - \frac{(n-\ell)\ell c}{rn+\ell} - c$$

$$= 1 - g(k') \;].$$

<u>Competitive Firms and Benchmarks</u>:

At a decision situation $x = < a_1,\ldots,a_{j-1} >$, whether or not the decision maker at x, $\rho(x) = j$, can optimally deviate from a symmetric strategy b depends on the configuration of vacancies and wages of unfilled firms at x, and the number of participants remaining to choose in the current period. In this subsection we define certain sets of firms and a "benchmark wage" that will be used in the next section to describe our equilibrium strategy. We first define a set of "competitors", $F^c(x)$, of $\rho(x)$ if he chooses to form a firm. Then we define the set $F^*(x)$ of the "best acceptable" firms for situations where it may be optimal for the decision maker to accept an outstanding wage offer.

Competitors

Let $x \in X^k$ be a situation, $x = < k;a_1,\ldots,a_{j-1} >$. Let

$M_m(x) = \{i: a_i = (-,m)\}$ (workers signed up by firm m),

$V_m(x) = (n-1) - |M_m(x)|$ (vacancies of firm m),

$F(x) = \{m: a_m = (w_m,-)$ for some w_m and $V_m(x) \neq 0\}$ (the firms with vacancies),

$Q(x) = |F(x)|$ (the number of firms with vacancies),

$P(x) = |\{m: a_m = (-,-)\}|$ (the passers).

Consider an ordered list of descriptions of triples

$$<(f^1(x),v^1(x),w^1(x)),---(f^q(x),v^q(x),w^q(x)),---,(f^{Q(x)}(x),v^{Q(x)}(x),w^{Q(x)}(x))>$$

where $f^q(x) \in F(x)$ and, when $f^q(x) = m$ for some $m \in \{1,\ldots,j-1\}$, the number $v^q(x)$ is given by $v^q(x) = V_m(x)$, and the number $w^q(x)$ by $w^q(x) = w_m$, and

(1) if $m \in F(x)$, $V_m(x) \neq 0$, then for some $q = 1, \ldots, Q(x)$, we have $m = f^q(x)$ (all firms with vacancies are included in the listing);

(2) $w^{q-1}(x) \geq w^q(x)$ for all $q = 2,\ldots,Q(x)$ (wage offers are nonincreasing as q increases);

(3) if $w^{q-1}(x) = w^q(x)$, then $v^{q-1}(x) \leq v^q(x)$ for all $q = 2,\ldots,Q(x)$.

The ordered list $<(f^1(x),v^1(x),w^1(x)),\ldots, (f^{Q(x)}(x),v^{Q(x)}(x),w^{Q(x)}(x))>$ describes the firms with outstanding offers and orders them according to their wage offers and, among firms with the same wage offers, according to their vacancies. We note that the ordering of the firms $f^1(x),\ldots, f^{Q(x)}(x)$ is not uniquely determined by the situation x since two different firms may have th same vacancies and wages; however, the orderings $v^1(x),\ldots, v^{Q(x)}(x)$ and $w^1(x),\ldots, w^{Q(x)}(x)$ are uniquely determined since interchanging two equal numbers effects no change in a list of numbers.

Let $\tilde{q}(x)$ be the largest number in $\{1,\ldots,Q(x)\}$ such that

(1) $\sum_{q=1}^{\tilde{q}(x)} v^q(x) \leq k - (j-1)$ (the vacancies of the firms $f^1(x)\ldots,f^{\tilde{q}(x)}(x)$ can feasibly be filled); and

(2) $w^q(x) \geq 1$ for all $q = 1,\ldots,\tilde{q}(x) - 1$.

The set of firms $\{f^1(x),\ldots, f^{\tilde{q}(x)}(x)\}$ is simply the "best" firms that can feasibly be filled, subject to the condition that all, except perhaps one, have wage offers greater than or equal to 1. We define

$$F^c(x) = \{f^q(x)\colon q = 1,\ldots,\tilde{q}(x)\};$$

this will be a description of competitors of the decision maker at x, if he chooses to form a firm. The wages and the vacancies of the firms in $F^c(x)$ are crucial factors in determining the optimal choice of wage of the decision maker if he forms a firm. The firms $\tilde{q}(x) + 1,\ldots,Q(x)$ are "noncompetitive"--their wages are too low to directly affect the wage offered by the decision maker if he forms a firm. They only affect decision making in that those already "signed-up" with these firms will, according to our equilibrium strategy, be continued into the next period. The wage $w^{\tilde{q}(x)}(x)$ is a sort of benchmark and in certain situations will constrain the wage offer of $\rho(x)$, if he forms a firm. Observe that if $F^c(x) = \phi$, then either there are no outstanding wage offers at x (i.e. no unfilled firms) or no firms with vacancies can feasibly be filled. Also observe that if $k - (j-1) < n$, then $F^c(x)$ may be nonempty but the jth participant cannot optimally form a firm since it cannot fill; in this case $F^c(x)$, although nonempty, will be irrelevant and only $F^*(x)$, defined next, need be considered.

The Best Job Offers

We next define the set of firms which are making "acceptable" offers. This set consists of the firms which the decision maker j might, in equilibrium, optimally choose to join. Whether or not an offer is acceptable depends on whether or not the decision maker has the possibility to form a firm which can feasibly be filled. Therefore we consider cases; one where $k - (j-1) < n$, so the jth decision maker cannot form a firm which will fill, and the other, where $k - (j-1) \geq n$.

Assume $k - (j-1) < n$. Then define $F^*(x)$ as the set of firms $m \in F(x)$ with properties that for each m in $F^*(x)$,

(1) $0 < V_m(x) \leq k - (j-1)$ (the firm m can feasibly be filled);

(2) $w_m \geq 1 - g(P(x) + \sum_{m' \epsilon F(x)} (n-V_{m'}(x)) + k - (j-1) + 1)$

where w_m satisfies $a_m = (w_m,-)$ (since the number in parentheses is the number of passers plus the number who are in firms that will not fill if no outstanding offers are accepted, plus the number of participants remaining to choose, including j, plus the new entrant into the market place next period, the condition is that the wage offered, w_m, is at least as big as the expected payoff to passing if no more offers are accepted this round); and

(3) there does not exist another firm $a_{m'} = (w_{m'},-)$ with $0 < V_{m'}(x) \leq k - (j-1)$, and with either: (a) $w_{m'} > w_m$, or (b) if $w_{m'} = w_m$, with $V_{m'}(x) < V_m(x)$ (among those outstanding offers, those of the firms in $F^*(x)$ are "best" in that, among the firms which can feasibly be filled, the firms in $F^*(x)$ offer the highest wage, and among the firms offering this wage, have the least vacancies).

We note that $F^*(x)$ consists of <u>all</u> of the acceptable firms, whose wage offers and vacancy status make their wage offers "best" and preferable to waiting until the next period, and $F^*(x)$ is uniquely determined.

For $k - j \geq n - 1$, we define

$F^*(x) = \{m: a_m = (w_m,-), w_m \geq 1, w_m = w^1(x), \text{ and } V_m(x) = v^1(x)\}.$

In our equilibrium, if an option open to the decision maker $\rho(x)$ is to form a firm which can feasibly be filled, then he will not accept a wage offer unless the offer is at least 1. Note that $F^*(x)$, both in this case and that above, may be empty.

We introduce some additional notation required in the next subsection. We let $f^*(x)$ denote a randomly selected member of $F^*(x)$, according to the uniform distribution on $F^*(x)$. Also, it is convenient to define, for $q = 0,\ldots,Q(x)-1$, the number of participants, $N^q(x)$, there will be in the next round if, in the current round, no further participants accept offers of the firms $f^{q+1}(x),\ldots, f^{Q(x)}(x)$, and as many as possible new firms form and fill. Formally,

$$N^q(x) = P(x) + \sum_{q'=q+1}^{Q(x)} (n-v^{q'}(x)) + \ell^q(x) + 1$$

where $\ell^q(x)$ is a nonnegative integer, $\ell^q(x) \leq n$, given by

$$k - j + 1 - \sum_{q'=1}^{q} v^{q'}(x) = \ell^q(x) + r'n$$

for some nonnegative integer r'. We note that $(n-v^{q'}(x))$ is the number of decision makers who've already accepted the offer of the firm $f^{q'}(x)$ and the firm former himself and $\ell^q(x)$ is the number of decision makers remaining after all firms $f^1(x),\ldots,f^q(x)$ are filled and, before the decision maker $\rho(x)$ has made his choice, as many as possible new firms form and fill.

A Stationary Equilibrium Point e:

We construct a stationary equilibrium point e by describing e_x at $x = \langle k;a_1,\ldots,a_{j-1} \rangle$. For simplicity of exposition, instead of writing out e_x as a probability distribution over situations $y \in C(x)$, we describe e_x in terms of alternatives. The alternative chosen will either be selected with probability 1 or chosen from $F^*(x)$ according to the uniform distribution, in which case it is described simply by $a_j = (-,f^*(x))$. We denote the equilibrium alternative chosen by $a_j^*(x)$. We make informal parenthetical remarks intended to motivate the choice.

Case I. $F^c(x) = \phi$.

(a) Suppose $k - j = rn - 1$ for some positive integer r. Then

$$a_j^*(x) = (w_j^*(x),-)$$

where $w_j^*(x)$ is the smallest payable amount (in $M(u)$) greater than or equal to

$$1 - g(N^0(x)+\ell) \quad \text{for } \ell = 2,\ldots,n.$$

[When $F^c(x) = \phi$ and $k - j = rn - 1$ (so there are no outstanding wage offers), the decision maker $\rho(x)$ forms a firm paying a wage just large enough to make the last $n - 1$ participants to be decision-makers in the current round at least as well off by accepting as by passing.]

(b) Suppose $k - j > n - 1$ and $k - j \neq rn - 1$ for some positive integer r. Then

$$a_j^*(x) = (1,-).$$

[When $F^c(x) = \phi$ and $k - j > n - 1$, $k - j \neq rn - 1$, the decision maker cannot use his position in the ordering to take advantage of the last $n - 1$ decision makers in the round and therefore forms a firm offering a wage of 1.]

(c) Suppose $k - j < n - 1$. Then

$$a_j^*(x) = (-,-).$$

[Since $F^c(x) = \phi$, and $k - j < n - 1$ so no firm can feasibly be filled and there is no alternative preferable to passing.]

Case II. $F^c(x) \neq \phi$.

IIA. Suppose $w^{\tilde{q}(x)} \geq 1$.

[Note that when $w^{\tilde{q}(x)} \geq 1$ either there are no outstanding wage offers less than 1 or no firm having made such an offer could feasibly be filled with the remaining participants, after the firms $f^1(x),\ldots,f^{\tilde{q}(x)}(x)$ are filled.]

(a) If $k - j - \sum_{q=1}^{\tilde{q}(x)} v^q(x) = rn - 1$ for some positive integer r let $w_j^*(x)$ be the smallest payable amount greater than or equal to

$$1 - g(N^{\tilde{q}(x)}(x)+\ell) \quad \text{for} \quad \ell = 2,\ldots,n.$$

(i) If $w^1(x) \geq n - (n-1)w_j^*(x)$ then

$a_j^*(x) = (-, f^*(x))$, and

(ii) If $w^1(x) < n - (n-1)w_j^*(x)$, then

$a_j^*(x) = (w_j^*(x), -)$.

[The wage $w_j^*(x)$ is just large enough to be acceptable -- at least as good as passing -- for the last $n - 1$ decision makers. The decision maker $\rho(x)$ acc pts one of the best offers outstanding if this has a higher payoff than forming a firm offering $w_j^*(x)$; otherwise he forms a firm.]

(b) If $k - j - \sum_{q=1}^{\tilde{q}(x)} v^q(x) \neq rn - 1$ for some positive integer r, then

$a_j^*(x) = (-, f^*(x))$.

[If he can not take advantage of the position of the last $n - 1$ decision makers, the decision maker $\rho(x)$ accepts a best outstanding wage offer.]

IIB. Suppose $w^{\tilde{q}(x)}(x) < 1$.

(a) If $k - j - \sum_{q'=1}^{\tilde{q}(x)-1} v^{q'}(x) = rn - 1$ for some positive integer r, let

$$\tilde{w}_j(x) = \max \{1-g(N^{\tilde{q}(x)-1}(x)+\ell),\ w^{\tilde{q}(x)}(x) + 1/u!: \ell = 2,\ldots,n\}.$$

and let $w_j^*(x)$ be the smallest payable amount greater than or equal to $\tilde{w}_j(x)$. Then

(i) If $w^1(x) \geq n - (n-1)\, w_j^*(x)$,

$a_j^*(x) = (-, f^*(x))$, and

(ii) If $w^1(x) < n - (n-1)\, w_j^*(x)$,

$a_j^*(x) = (w_j^*(x), -)$.

[The wage $w_j^*(x)$ is just large enough to be acceptable to the last $n - 1$ decision makers and also higher than the highest wage offer outstanding l ss than 1. The decision maker $\rho(x)$ either accepts one of the best offers outstanding or forms a firm offering $w_j^*(x)$, whichever is better.]

(b) If $k - j - \sum_{q'=1}^{\tilde{q}(x)-1} v^{q'}(x) > n - 1$ and not equal to $rn - 1$ for any positive integer r, and

(i) If $w^1(x) < 1$ (or, equival ntly, $\tilde{q}(x) = 1$), then

$$a_j^*(x) = (1,-),$$

(ii) If $w^1(x) \geq 1$ (or, equivalently, $\tilde{q}(x) > 1$), then

$$a_j^*(x) = (-,f^*(x)).$$

[If the decision maker $\rho(x)$ cannot take advantage of the position of the last $n - 1$ decision makers he simply accepts the best offer outstanding, if it is greater than or equal to 1; otherwise he forms a firm offering a wage of 1.]

(c) If $k - j - \sum_{q'=1}^{\tilde{q}(x)-1} v^{q'}(x) \leq n - 1$, then

(i) If $F^*(x) \neq \phi$,

$$a_j^*(x) = (-,f^*(x)) \text{ and}$$

(ii) If $F^*(x) = \phi$

$$a_j^*(x) = (-,-).$$

[If he cannot form a firm which will fill, the decision maker accepts a best offer outstanding or passes -- whichever is better. Note that $F^*(x) = \phi$ and $F^c(x) \neq \phi$ implies that $\tilde{q}(x) = 1$ and that $w^{\tilde{q}(x)}(x)$ is too small to be acceptable.]

__Proposition 2__:

The strategy e described above is a stationary equilibrium point.

Proof:

For ease in notation, if $y \in C(x)$ for any situation $x = \langle k;a_1,\ldots,a_j \rangle$, we write $y = \langle x,a_{j+1} \rangle$ when $y = \langle k;a_1,\ldots,a_j,a_{j+1} \rangle$.

We take as given a situation $\hat{x} = \langle k, \bar{a}_1,\ldots,\bar{a}_{J-1} \rangle$ and ask if $y = \langle \hat{x},a_J^*(x) \rangle$ is an optimal choice for the Jth decision maker, where $a_J^*(x)$ is chosen in accordance with e.

We divide the proof into sections, following the numbering in the description of e.

Case I.

(a) Suppose the situation $\hat{x} = \langle k;\bar{a}_1,\ldots,\bar{a}_{J-1} \rangle$ satisfies the conditions of Case I(a) in the description of e and the Jth decision maker chooses $a_J^*(\hat{x}) = (w_J^*(\hat{x}),-)$, where $w_J^*(\hat{x})$ satisfies the description in I(a), (specifically, $w_J^*(\hat{x})$ is the smallest payable amount greater than or equal to $1 - g(N^0(\hat{x})+\ell)$ for $\ell = 2,\ldots,n$). We note that $\ell = 2,\ldots,n$ because, if the Jth decision maker forms a firm, subsequent decision makers will be looking at situations wh re at l ast the firm-form r J is already in the firm. Note also that since $F^c(x) = \phi$ and $k - (J-1) \geq n$ ther are no outstanding wage offers.

Now, we ask if, when $a_J^*(\hat{x}) = (w_J^*(\hat{x}),-)$, according to Case I(a), will th strategy e prescribe acceptance of the wage offer of the Jth participant by $n - 1$ subsequent decision makers? First, consider the simplest case, $k - J = n - 1$. For e to prescribe acceptance by the $J + 1^{th}$ participant he must be facing a situation $y = < \hat{x}, a_J^*(\hat{x}) >$ with $F^*(y) = \{J\}$ and $k - (J+1) \leq n - 1$; this is the sort of situation described by IIB(c). We need only show now that $F^*(y) = \{J\}$. This is true because $w_J^*(\hat{x})$ is the highest outstanding offer (there are no others), $V_J(y) = n - 1$, and we have

$$w_J^*(\hat{x}) \geq 1 - g(N^0(x)+\ell), \qquad \ell = 2,\ldots,n$$

and

$$P(y) + \sum_{m' \epsilon F(y)} (n-V_{m'}(y)) + k - J + 1$$

(from (2) in the definition of $F^*(\cdot)$) equals

$$P(x) + (n-(n-1)) + (n-1) + 1 = N^0(x) + \ell \quad \text{for} \quad \ell = n.$$

Consequently, according to e, the $J + 1^{th}$ participant will accept the offer of the Jth participant. Now suppose the decision makers, $J + 1,\ldots, J + i - 1$ hav all accepted the offer of the Jth participant. Then, for e to prescribe acc ptanc by the $J + i^{th}$ participant, letting y^i denote the situation facing the $J + i^{th}$ participant, we must have $F^*(y^i) = \{J\}$; this follows immediately from the obs rvation that

$$w_J^*(x) \geq 1 - g(N^0(x)+\ell), \quad \ell=2,\ldots,n$$

implies that

$$w_J^*(x) \geq 1 - g(P(y^i) + n - (n-i+1) + k - (J+i) + 2) = 1 - g(N^0(x) + n).$$

Since i was an arbitrary integer between 1 and $n - 1$ we can conclude that all decision makers $J + i$, $i = 1,\ldots,n - 1$, each confronting a situation describ d by IIB(c)(i), will accept the offer of the Jth participant, i.e. choose $a_{J+i}^*(y^i) = (-,J)$.

Now suppose $r > 1$ so $k - J = rn - 1$. Let $y^1 = < \hat{x},a_J^*(\hat{x}) >$ denote the situation facing the $J + 1^{th}$ decision maker. This situation satisfies th prop rti s of IIB(b)(i), since, arguing as above, we have $F^c(y^1) = \{J\}$ (there are no wage off rs outstanding greater than of equal to 1), $w^1(y^1) = w_J^*(\hat{x}) < 1$, $k - (J+1) > n - 1$ and $k - (J+1) \neq r'n - 1$ for any $r' > 0$ (since $k - J \geq 2n - 1$ and $k - J = rn - 1$, and, by assumption, $n > 1$). According to e, the $J + 1^{th}$ participant chooses $a_{J+1}^*(y^1) = (1,-)$. Let $y^2 = < \hat{x},a_J^*(\hat{x}),a_{J+1}^*(y^1) >$.

Now the d cision mak r $J + 2$ faces th situation y^2, which, w claim, is described by IIB(b)(ii):

(1) The firms of th J^{th} and $J + 1^{th}$ participants can both feasibly be filled, so $F^c(y^2) \neq \phi$ and $f^1(y^2) = J + 1$, $f^2(y^2) = J$, $w^1(y^2) = 1$ and $w^2(y^2) = w_J^*(x) < 1$;

(2) Since $v^1(y^1) = n - 1$, we have $k - (J+2) - v^1(y^2) = k - J - 2 - (n-1) = k - J - n - 1 = rn - 1 - n - 1 = (r-1)n - 2 \neq r'n - 1$ for any positive integer r'.

Therefore according to e the $J + 2^{th}$ participant chooses $a_{J+2}^*(y^2) = (-,J+1)$. Using similar arguments we can obtain $a_{J+i}^*(y^i) = (-,J+1)$ for all $i = 2,\ldots,n + 1$, where $y^i = < \hat{x}, a_{J+1}^*(y^1), a_{J+2}^*(y^2), \ldots, a_{J+i}^*(y^i) >$ and therefore the firm of the $J + 1^{th}$ decision maker fills.

More generally, the $J + r'n + 1^{th}$ decision maker, for any positive r', faces a situation analogous to y^1 and described by IIB(b)(i); therefore according to e he chooses (1,-). The next $n - 1$ participants all choose $(-,J+r'n+1)$ according to IIB(b)(ii). Forming of firms offering wages of 1, and filling of these firms continues until there are only $n - 1$ participants yet to make their choices. Let y^* denote the situation facing the 1st of these participants, i.e. the $k - (n-2)^{th}$ decision maker. Then $F^*(y^*) = \{J\}$ and the situation y^* satisfies the properties of II(B)(c) so the $k - (n-2)^{th}$ decision maker, and all subsequent decision makers i, according to e, choose $(-,J)$, thus filling the firm of the Jth decision maker.

It is clear that the choice of a wage offer w with $w > w_J^*(\hat{x})$ would decrease the expected payoff of the Jth decision maker, when all subsequent decision makers use the strategy e. Suppose the Jth decision maker chooses $w < w_J^*(\hat{x})$. Then for some ℓ_0, $2 \leq \ell_0 \leq n$, we have $w < 1 - g(N^0(x)+\ell_0)$. In this case the choices prescribed by e for the participants $J + 1,\ldots,k - (n-2)$ remain the same as the case where the wage $w_J^*(\hat{x})$ is chosen by the Jth participant, for $r' > 1$. Let y^* denote the situation facing the $k - (n-2)^{th}$ participant. Now $J \notin F^*(y^*)$ since the wage of the Jth participant is too low and, according to e (specifically IIB(c)) the last remaining $n - 1$ participants choose to pass, leaving the firm of the Jth participant unfilled.

We conclude that $a_J^*(\hat{x}) = (w_J^*(\hat{x}),-)$ is optimal at $\hat{x}$ for the decision maker J.

(b) At a situation $\hat{x}$ described by Case I(b), the local strategy prescribed by e is $a_J^*(\hat{x}) = (1,-)$. As above, we can argue that there are no higher outstanding wage offers, since $F^c(\hat{x}) = \phi$. Let y^1 denote the situation $< \hat{x}, a_J^*(\hat{x}) >$. Since $1 > 1 - g(k')$ for all $k' = 1,2,\ldots$, and since $k - J > n - 1$, we have $F^*(y^1) = \{J\}$. The situation y^1 is covered by Case IIA. Since $k - J \neq rn - 1$ for any positive integer r, $k - (J+1) - (n-1) \neq r'n - 1$ for any positive integer r', and y^1 is described by IIA(b). Thus, e prescribes acceptance of the offer of the Jth firm by the $J + 1^{th}$, participant. For the same reasons, e prescribes acceptance of J's offer by the ith decision maker, $i = J + 2,\ldots,J + (n-1)$. Therefore, if $a_J^*(\hat{x}) = (1,-)$ is not optimal at $\hat{x}$ for the Jth decision maker, he must be able to increase his exp cted payoff by setting a lower wage or by accepting an outstanding wag off r or by passing.

The expected payoff to J of using $a_J^*(\hat{x}) = (1,-)$ is 1. Passing has an expected payoff of less than one. Therefore, if the Jth decision maker can profitably deviate it must be by selling a wage lower than w. Suppose the Jth decision maker sets a lower wage, say $w < 1$. However, according to e, the $J + 1^{th}$ decision maker is then either a situation described by I(b) (since $k-J > n - 1$ and $k - J \neq rn - 1$ for any positive integer r implies $k - (J+1) > n-1$) or by IIB(b)(i). In ither case, according to e he chooses (1,-). The next (n-1) participants accept the offer of the $J + 1^{th}$ decision maker. The procedure of forming firms offering 1, and the filling of these firms continues until there are fewer than $n - 1$ participants remaining (since $k - J > n - 1$, $k - J \neq rn - 1$ for any positive integer r implies $k - J - r'n < n - 1$ for the largest r' such that $k - J - r'n \geq 0$). Thus, according to e the firm of the Jth decision maker is not filled if he offers a wage less than 1, and his expected payoff is less than 1.

We conclude that the choice of $a_J^*(\hat{x}) = (1,-)$ is optimal at x.

(c) This is the easiest case. Since $F^c(\hat{x}) = \phi$ there are no outstanding offers from firms which can feasibly fill. Since $k - J < n - 1$ the Jth decision maker cannot successfully form a firm which will fill. Therefore he passes.

Case IIA:

(a) Suppose the situation $\hat{x}$ is described by IIA(ii), and the Jth participant chooses $a_J^*(\hat{x}) = (w_J^*(\hat{x}),-)$ where $w_J^*(\hat{x})$ is the smallest payable amount greater than or equal to

$$1 - g(N^{\tilde{q}(\hat{x})}(\hat{x})+\ell) \quad \text{for } \ell = 2,\dots, n.$$

According to e, the next $\sum_{q=1}^{\tilde{q}(\hat{x})} v^q(\hat{x})$ decision makers will accept outstanding offers of firms $f^1(\hat{x}),\dots,f^{\tilde{q}(\hat{x})}(\hat{x})$, unless some participant, say the mth, faces a situation satisfying the conditions of IIA(a)(ii), or IIB(a)(ii). We can immediately rul of IIA(a) since $J \in F^c(<\hat{x},a_J^*>)$, which is routine to confirm, and $w_J^*(\hat{x}) < 1$. We argue that IIB(a) is also impossible. Let $\ell = (m-J)$. For IIB(a) to occur when all subsequent decision makers use the strategy e we would have to have

$$\sum_{q=1}^{q'} v^q(\hat{x}) = \ell - 1 \quad \text{for some } q' \text{ and}$$

$$k - (J+\ell) - \sum_{q=q'+1}^{\tilde{q}(\hat{x})-1} v^q(\hat{x}) = r'n - 1$$

for some integer $r' \geq 1$. However, substituting for ℓ from the first into the s cond expression gives us

$$k - J - \sum_{q=1}^{q'} v^q(\hat{x}) - 1 - \sum_{q=q'+1}^{\tilde{q}(\hat{x})-1} v^q(\hat{x})$$

$$= k - J - \sum_{q=1}^{\tilde{q}(\hat{x})-1} v^q(\hat{x}) - 1 = rn,$$

which is a contradiction. We can conclude that the next $\sum_{q=1}^{\tilde{q}(\hat{x})} v^q(\hat{x})$ participants will accept the offers of the firms $f^1(\hat{x}),\ldots, f^{\tilde{q}(\hat{x})}(\hat{x})$. Note that these firms must be the only ones with wage offers greater than or equal to 1 outstanding at x since more could feasibly be filled.

The next question to be answered is whether or not the firm of the Jth participant will be filled. According to e, after the outstanding vacancies at $\hat{x}$ and new firms offering wages of 1 are all filled, there will be n - 1 participants left to make their choices. By construction of $w_J^*(\hat{x})$, according to e we have $F^*(\hat{x}) = \{J\}$ and all these participants will choose $(-,J)$.

As in previous arguements, the Jth participant cannot profitably set a lower wage.

Suppose the situation $\hat{x}$ is described by IIA(a)(i), and the Jth participant chooses $a_J^*(\hat{x}) = (-,f^*(\hat{x}))$. From the condition that $w^1(\hat{x}) \geq n - (n-1)w_J^*(\hat{x})$, the Jth decision maker cannot profitably deviate by forming a firm. By construction of $F^*(\hat{x})$ and e, the firm joined by the Jth participant will fill. Thus $a_J^*(\hat{x}) = (-,f^*(\hat{x}))$ is optimal at $\hat{x}$.

(b) At a situation $\hat{x}$ satisfying the conditions of IIA(b), the participant cannot form a firm which, according to e, might fill (depending on the number of participants remaining), unless he offers a wage of 1. But he can expect a payoff of at least 1 by choosing $a_J^*(\hat{x}) = (-,f^*(\hat{x}))$ and this is therefore optimal at $\hat{x}$.

Case IIB.

When the situation $\hat{x}$ is described by IIB, the optimal choices of the Jth participant are to accept one of the best outstanding wage offers, or to "compete" with $f^{q(\tilde{x})}(\hat{x})$ for (some or all of) the last n - 1 participants to be decision-mak rs, or to form a firm offering 1.

(a) If the Jth decision maker competes for the last n - 1 participants, i.e., $\hat{x}$ is described by IIB(a)(ii) he must offer a wage greater than $w^{\tilde{q}(\hat{x})}(\hat{x})$, as w will argue below. Also, his wage offer must be sufficiently large so that the last n - 1 participants all optimize by choosing $(-,J)$. If all firms $f^{\tilde{q}(\hat{x})}(\hat{x}),\ldots,f^{Q(\hat{x})}(\hat{x})$ are unfilled, $f^1(\hat{x}),\ldots,f^{\tilde{q}(\hat{x})-1}(\hat{x})$ (for som appropriate listing of the firms) are filled, and as many as possible "new" firms, offering wages of 1 ar filled then the $k - \ell^{th}$ participant, $\ell \in \{1,\ldots,n-2\}$, can expect the payoff

$$1 - g(N^{\tilde{q}(\hat{x})-1}(\hat{x})+n).$$

if he passes. Since $\tilde{w}_J(\hat{x}) \geq 1 - g(N^{\tilde{q}(\hat{x})-1}(\hat{x})+\ell)$ for all $\ell = 2,\ldots,n$, each participant can optimally accept the offer of the Jth participant, and the firm of the Jth participant will be filled. Therefore, it is clear that IIB(a)(i) and (ii) describe optimal choices.

We must still argue that for $\hat{x}$ described by IIB(a)(ii) the Jth participant cannot profitably deviate by setting a wage equal to $w^{\tilde{q}(\hat{x})}(\hat{x})$. If he did, then according to e, the first of the last $n - 1$ decision makers would choose the firm $f^{\tilde{q}(\hat{x})}(\hat{x})$ with probability 1 if $v^{\tilde{q}(\hat{x})}(x) < n - 1$, and the firm of the Jth participant would not be filled. Therefore suppose $v^{\tilde{q}(\hat{x})}(\hat{x}) = n - 1$. Now suppose the first of the last $n - 1$ decision makers chooses one firm, from all those offering wages equal to $w^{\tilde{q}(\hat{x})}(\hat{x})$ and including $(\cdot,J)$, with equal probability. If his actual choice is not $(\cdot,J)$, then the firm of the Jth participant will not be filled, and his terminal situation payoff will be $-c - \gamma$. We need to ensure that the following condition holds:

$$\pi_1[n-(w^{\tilde{q}(\hat{x})}(\hat{x}))(n-1)] - \pi_2(c+\gamma)$$

$$< n - (w^{\tilde{q}(\hat{x})}(\hat{x})+1/u!)\ (n-1)$$

where π_1 is the probability that the first of the last $n - 1$ decision makers chooses the firm of the Jth participant and π_2 is the probability that he does not. Indeed, we have $\pi_1 = 1/2$ and $\pi_2 = 1/2$, and substituting 1/2 for π_1 and π_2 into the above inequality (which will not affect the unequality sign), yields

$$c + \gamma > \frac{2}{u!}\ (n-1)$$

which is insured by our conditions on c, γ, and u.

(b)(i) When there are no outstanding offers greater than or equal to 1 the Jth decision maker forms a firm offering a wage of 1. Note that since

$$k - J - \sum_{q=1}^{\tilde{q}(\hat{x})-1} v^q(\hat{x}) = k - J > n - 1, \neq rn - 1,$$

according to e no subsequent decision maker will be in a position described by IIB(a) or IIA(a). Therefore the next $n - 1$ decision makers accept the offer of J.

(b)(ii) The Jth decision maker can do no better than accept one of the b st offers outstanding greater than or equal to 1.

(c) Finally, if $F^*(\hat{x}) = \phi$, then no offer dominates passing -- either th re are no outstanding offers or no firms with vacancies can feasibly be filled, or for any other firm m with $V_m(\hat{x}) > 0$, we have

$$w_m < 1 - g(P(\hat{x}) + \sum_{m'\varepsilon F(\hat{x})} (n-V_{m'}(\hat{x}))+k-J+2),$$

the right-hand side of which is the expected payoff to passing given the Jth participant passes. [Note that if the Jth participant passes then, according to e, so will $J + i$, $i = 1,\ldots,k - J$.] If $F^*(\hat{x}) \neq \phi$, we have the expected payoff of passing less (or no greater) than the highest wage offer of a firm which can feasibly be filled; according to e the Jth participant accepts one such offer and it is clear that he can do no better. [Note that if the Jth participant accepts the offer of $m \varepsilon F^*(\hat{x})$, then the equilibrium choice of the $J + i^{th}$ participant is to accept the offer of firm m for $i = 1,\ldots,V_m(\hat{x})$, so firm m will fill.]

□

3. SOME ADDITIONAL ASPECTS OF THE EQUILIBRIUM

Clearly e is not a unique equilibrium. For example, if $n = 3$ and $k = 9$, then equilibrium alternatives are:

$$a_1 = (w_1^*(<9>),-),\quad a_2 = (1,-),\quad a_3 = (-,2)$$
$$a_4 = (-,2),\quad a_5 = (1,-),\quad a_6 = (-,5)$$
$$a_7 = (-,5),\quad a_8 = (-,1),\quad a_9 = (-,1).$$

(Recall that $w_1^*(<9>)$, in the notation of Section 2, is the smallest payabl amount greater than or equal to $1-g(4)$). Another set of alternatives consistent with equilibrium would be:

$$a_1 = (w_1^*(<9>),-),\quad a_2 = (1,-),\quad a_3 = (1,-),\quad a_4 = (-,2)$$
$$a_5 = (-,3),\quad a_6 = (-,2),\quad a_7 = (-,3),\quad a_8 = a_9 = (-,1),$$

with all participants $2,\ldots,7$ having a terminal situation payoff of 1. These sets of equilibrium choices are not in any significant way different. If $w_1^*(<9>) > 1-g(4)$ (because of the smallest money unit) stationary equilibrium must have $a_1 = (w_1^*(<9>),-)$, $a_8 = a_9 = (-,1)$, and all participants $2,\ldots 7$ having a terminal situation payoff of 1; this can be achieved consistently with equilibrium in several ways. Thus we must further restrict the equilibrium to obtain a uniqueness r sult.

In this section, we show that all stationary equilibria with two "shortness" properties are efficient. We also show that, with an additional symmetry requirement, the equilibrium is unique.

Our first "shortness" property is that, all else equal, participants prefer optimal choices which shorten their stay in the game. This property ensures, for example, that if a participant is indifferent between accepting a wage offer and passing, he accepts the wage offer.

The Participation-Shortness Property:

A stationary equilibrium point b has the participation-shortness property if, at every decision situation $x = < k; a_1, \cdots, a_{j-1} >$, for all optimal choices y' and y'', where $y' = < k; a_1, \cdots, a_j' >$, and $y'' = < k; a_1, \cdots, a_j'' >$, with $a_j' = (-, i)$ for some i and $a_j'' = (-,-)$, then the probability assigned by b_x to a_j'' is zero. (Given optimal choices of joining a firm or passing, a participant joins a firm).

The second property we impose is a "within-period" shortness property, ensuring that, within a period, if optimal choices with the participation-shortness property include forming a firm or accepting an outstanding wage offer, the participant accepts the wage offer.

Within-Period-Shortness:

A strategy b has the within-period shortness property if, at every decision situation $x = < k; a_1, \ldots, a_{j-1} >$, for all optimal choices y' and y'' and $y'' = < k; a_1, \ldots, a_j'' >$ with $a_j' = (-,i)$ for some i and $a_j'' = (w,-)$ for some w, the probability assigned by b_x to a_j'' is zero. (Given optimal choices of joining a firm or forming a firm, a participant joins a firm.)

Proposition 3:

Let b^* be a stationary equilibrium with the participation-shortness property and the within-period shortness property. Then b^* is efficient in the sense that every terminal situation reached from any initial situation $< k >$, when all participants use b^*, is efficient.

Proof:

Suppose not. Then there is an initial situation $< k >$ and a terminal situation z, reached from $< k >$ with positive probability, which is not efficient. There are several cases:

The inefficiency of z might simply be that there are unfilled firms and participants who have accepted their offers or some participants might choose self-employment or some participants who have passed but should, for efficiency, have joined, or established, firms. These are inconsistent with equilibrium, so suppose there are no such participants.

Case 1: $k < n$.

Then the only alternative consistent with equilibrium for the jth decision maker is to pass (since γ, in the definition of the terminal situation payoff function h, is positive).

Case 2: $k = n$.

In this case, the only choice for the 1st decision maker consistent with a subgame perfect equilibrium is $(w_1^*(\langle k \rangle), -)$. The participation-shortness property ensures that the remaining $n - 1$ participants accept.

Case 3: $k > n$, $k = rn + \ell$ for some positive integers r and ℓ.

For this case, the 1st decision maker can do no better than the choice $(1, -)$. The $2^{nd}, \ldots, n^{th}$ decision makers can also do no better than a payoff of 1 each and, because of the within-period-shortness of b^*, accept the offer of the 1st decision maker. Similarly the next $(r-1)n$ players form and fill firms. The last ℓ players pass, and the terminal situation z must be efficient, a contradiction.

Case 4. $k = rn$.

As in Case 2, the 1st decision maker must choose $(w_1^*(\langle k \rangle), -)$. The 2nd decision maker can do no better than the alternative $(1, -)$. By within-period shortness, the next $n - 1$ decision makers accept. This forming and filling of firms must continue until there are only $n - 1$ decision makers remaining. By participation shortness, they must accept the offer of the 1st decision maker.

□

We will next show that, if we impose an additional symmetry property, the only stationary equilibrium satisfying these and the participation-shortness property is e. We remark that another way of achieving our objective of characterizing the stationary equilibria would be to determine a class of stationary equilibrium points "equivalent" (up to some permissible randomization and orderings of firm-forming and filling) to e.

<u>A Symmetry Property</u>:

A stationary equilibrium b is <u>symmetric with respect to wage offers</u> if at every decision situation $x = \langle k; a_1, \ldots, a_{j-1} \rangle$ whenever y and y' are in the carrier of b_x, with $y = \langle k; a_1, \ldots, a_j \rangle$ $y' = \langle k; a_1, \ldots, a_j' \rangle$, and $a_j = (-, i)$ for some i, $a_j' = (-, m)$ for some m, then $b_x(y) = b_x(y')$.

<u>Proposition 4</u>:

There is one and only one stationary equilibrium with the participation-shortness property, the within-period shortness property, and that is symmetric with respect to wage offers, and it is e.

Proof:

First observe that e has the required properties. Also, rereading the construction of e and the proof of Proposition 2 should suffice to convince the reader of the validity of Proposition 4. □

4. FURTHER DISCUSSION

In this section, we consider our model and results from the viewpoint of cooperative game theory. Also, as part of this discussion and to indicate future directions of this research, we sketch another example with two types of players ("players of the same type" are ones who make the same contribution to every coalition). In brief, our research indicates that as waiting costs go to zero, equilibrium payoffs converge to core payoffs of a cooperative (static) game, with a continuum of players and finite coalitions, where the static game is to be interpreted as a limit idealization or representation of the dynamic game.

Before proceeding, it is convenient to recall some game-theoretic concepts.

Games in Characteristic Function Form

A game in characteristic function form is an ordered pair (N,v) where $N = \{1,\ldots,m\}$ is a finite set, called the set of players, and v is a function assigning a nonnegative real number to every subset G of N. The number $v(G)$ is called the worth of the coalition G and v is called the characteristic function of the game. The empty coalition has value zero:

$$v(\phi) = 0.$$

The game is called superadditive if the following is true for any two coalitions G and S with $G \cap S = \phi$:

$$v(G \cup S) \geq v(G) + v(S)$$

Two players i and j in N are substitutes if for all $G \subset N$, $i \notin G$, $j \notin G$, we have $v(G\cup\{i\}) = v(G\cup\{j\})$; substitute players contribute the same amount to any coalition they might each joint. (This implies $v(\{i\}) = v(\{j\})$).

A payoff for the game (N,v) is a vector $y = (y_1,\ldots,y_m) \in R^m_+$ with $y(N) \leq \sum_{S \in P(N)} v(S)$ for some partition $P(N)$ of N. The payoff is in the core if $y(S) \geq v(S)$ for all coalitions S.

Another auxilliary concept is a "pregame". A pregame with T types is an ordered pair (T,Ψ) where $T \in R_+$ and Ψ is a function, called the worth function, from I^T_+, the T-fold Cartesian product of the nonnegative integers, to the nonnegative reals with

$$\Psi(\underline{0}) = 0$$

Let $N = \{1,\ldots,m\}$ be a finite set and let α be a function from N to $\{1,\ldots,T\}$, called a type function. From the pregame Ψ we can construct a derived game (N,v_α) by defining

$$v_\alpha(S) = \Psi(s)$$

where $s \in I_+^T$ and

$$s_t = |S \cap \alpha^{-1}(t)|,$$

the number of players of type t in S. Note that if $\alpha(i) = \alpha(j)$, then players i and j are substitutes.

Now, we briefly describe a continuum game with finite coalitions. (We refer the reader to Kaneko-Wooders (1986) for a thorough introduction to continuum games with finite coalitions, covering nontransferable, as well as transferable, utility games.) Let (T,Ψ) be a pregame with T types, let $A = [0,1)$ be the 0-1 interval with Lebesgue measure and let α be a type function, $\alpha\colon A \to \{1,\ldots,T\}$. As above, we have a derived game (A,v_α) when we define

$$v_\alpha(S) = \Psi(s)$$

for each $S \subset A$, $|S|$ finite, where $s_t = |S \cap \alpha^{-1}(t)|$. A _payoff_ (for a continuum game) is a function h from A to R_+ with the property that for some "measurement-consistent" partition P of A into finite coalitions $h(S) \leq v_\alpha(S)$ for every $S \in P$. (We will not define measurement-consistency here since it is not essential to understanding the discussion. We note only that measurement-consistency ensures that the absolute magnitudes of finite coalitions are consistent with the proportional magnitudes described by the measure.) The payoff h is in the core, called the _f-core_, if for every finite coalition $S \subset N$, we have $h(S) \geq v_\alpha(S)$.

To illustrate, suppose we have the situation implicit in our example: any n players can realize a payoff of n. Then one particular measurement-consistent partition is given by

$$\{\{i, i+1/n, i+2/n, \ldots, i+(n-1)/n\}: i \in [0,1/n)\}$$

and the payoff h given by

$$h(i) = 1 \quad \text{for all } i \in A$$

is in the f-core of the game (A,v_α).[11]

An Example with Two Types of Players:

Observe that we can extract from our example in Sections 2 and 3 a representing pregame with 1 type, say $(\{1\},\Psi)$ with $\Psi(s)$ defined by

$$\Psi(n) = n$$

$$\Psi(s) = 0 \quad \text{for all } s \neq 0.$$

For the derived continuum game (A,v_α) the unique payoff in the f-core is $h(i) = 1$

---

[11]There are _many_ measurement consistent partitions, even many for which h is a payoff.

for all $i \in A$. Also, for any continuum game with finite coalitions (derived from the pregame) the f-core is a singleton given by $h(i) = 1$ for all players $i \in A$. When we let waiting costs, c, and the smallest money unit, $1/u!$, go to zero, the expected equilibrium payoff to any situation converges to the f-core payoff 1 for each participant. It is very interesting to ask if this property of convergence of equilibrium expected payoffs to core payoffs also holds when we have more than 1 type in the underlying pregame. At this point we can answer affirmatively for an example with two types. We describe briefly the example, which we will also use to discuss a possible experiment.

Suppose the payoff to a firm depends on the numbers of participants of each type in the firm. Let $\Psi(s_1,s_2)$ denote the payoff to a firm with s_1 members of type 1 and s_2 members of type 2. We set

$$\Psi(1,2) = 10$$
$$\Psi(2,0) = 4$$
$$\Psi(0,1) = 1$$
$$\Psi(1,1) = 0.$$

No other firms are permitted.

We can construct a dynamic game from the function Ψ analogously to the construction of the example in Section 2. Now, however, in choosing the alternative of forming a firm, a participant would have to announce a number of vacancies for each type of "worker" and a wage for each type. Of course, a participant can only accept offers for workers of his type. Instead of having only 1 player join the game at the beginning of a period, we could have k_1 of type 1 and k_2 of type 2 join. We take $k_1 = k_2 = 1$, for simplicity. Also, we define a symmetric stationary equilibrium point as one where all players of the same type use the same strategy and the strategy depends only on the situation. We leave the remaining details to the imagination of the reader.[12]

With some additional restrictions it can be shown that, as $c \to 0$, the expected payoffs to stationary equilibrium strategies converge to 2 for a player of type 1 and 4 for a player of type 2. These payoffs 2 and 4 are core payoffs for a static game with 4ℓ players of each type, where ℓ is any positive integer. To see this, observe that when there are 4ℓ players of each type, we can form 2ℓ coalitions with profiles (1,2) and ℓ with profiles (2,0). For this coalition structure the payoffs of 2 for a player of type 1 and 4 for a player of type 2 are feasible, and clearly cannot be improved upon. Also, since for this case every payoff in the core has the equal-treatment property (treats players of the same type equally) this payoff is the only one in the core. To develop our ideas for comparing the dynamic game with an infinite static game, we turn to continuum games with finite coalitions.

To compare the limiting stationary equilibrium payoffs of the dynamic game to f-

---

[12] We anticipate that this example will be a special case of more general results in progress, so do not specify more detail here.

core payoffs, first we note that for the static game to represent the limit dynamic game (as $c \to 0$), we require that the ratio of the measures of each type, $\frac{k_1}{k_2}$, equals 1. Therefore, we let $A = [0,1)$ be the player set, and let $\alpha: A \to \{1,2\}$, be a function from the player set A to the set of types $\{1,2\}$, with $\alpha(i) = 1$ if $i \in [0,\frac{1}{2})$ and $\alpha(i) = 2$ if $i \in [\frac{1}{2},1)$; the players in $[0,\frac{1}{2})$ are of type 1 and those in $[\frac{1}{2},1)$ of type 2. Consider the partition P given by

$$\{\{i,\tfrac{1}{2}+i,\tfrac{3}{4}+i\}: i \in [0,\tfrac{1}{4})\} \cup \{\{i,\tfrac{1}{8}+i\}: i \in [\tfrac{1}{4},\tfrac{3}{8})\}$$

and let h be the function from A to $\mathbf{R}_+$ defined by

$$h(i) = 2 \quad \text{if} \quad i \in [0,\tfrac{1}{2})$$

$$h(i) = 4 \quad \text{if} \quad i \in [\tfrac{1}{2},1).$$

Then for each S in P, $h(S) \leq v_\alpha(S)$, there does not exist a (finite) coalition S for which $h(S) > v_\alpha(S)$ so the payoff h is in the f-core, and it is the only such payoff.

Our conjecture is that our results will hold for any dynamic game of the sort we construct in our examples, with an underlying pregame with T types, for any finite T. We remark that the case with 3 or more types seems more subtle than those with only 1 or 2 types.

<u>Relationships to the Literature</u>:

The cooperative game concept, the core (called the f-core) of a continuum game with finite coalitions, used to describe equilibrium payoffs to the dynamic games as waiting costs go to zero, arises out of a long line of research exploring the relationship between cooperative and price-taking behavior. Since this paper initiates a new approach to large games in coalitional form and the f-core, and is meant to "set the stage" for research in progress we conclude this section with a more complete discussion of the previous research, its relationship to the current work, and the motivation arising from the previous work for the current work.

The hypothesis of perfect competition, price-taking, is central to economic theory. The first attempt to explain price-taking by the possibility of individual agents in an economy making trades among themselves is due to Edgeworth (1881). He defined an allocation as in the contract curve if no group of agents, using their own endowments, could improve upon the allocation. He conjectured that, when the number of agents becomes large, the contract curve converges to the perfectly competitive allocations. In (1959) Shubik phrased Edgeworth's bargaining problem as an n-person game and showed that for such "Edgeworth market games", the core, a game-theoretic analogue of the contract curve, coincides with the contract curve and shrinks in a well-defined way as the number of traders increases. In (1963), Debreu - Scarf showed that the core[13] converges to the competitive equilibrium allocations in a framework using traditional neoclassical assumptions about agents and their preferences. In (1964), Aumann introduced his model of an economy with an atomless measure space of agents and argued that a "mathematical model appropriate to the intuitive notion of perfect competition must contain infinitely many participants ... one can integrate over a continuum, and changing the integral at a single point does not affect the value of the integral; that is the actions of a single agent are negligible" (Aumann, 1964, p. 39). In this way, Aumann beautifully and elegantly captured the notion of perfect competition in the context of exchange economies. He then showed that when coalitions are required to be sets of positive measure, the Aumann-core concept coincides with the competitive equilibrium. (We note that in Aumann's framework, just as single players are negligible relative to the entire economy, single players are negligible relative to any admissible coalition.) Aumann's model rules out trade by "small" groups of agents, for example, ones containing a bounded (or even countable) number of players, since these are groups of measure zero in the continuum.[14]

The importance of the effectiveness of small coalitions is clearly recognized by Mas-Colell (1979), who shows that approximate core-equilibrium equivalence obtains in large finite economies with "improvement" only by small coalitions (but without any restriction on cooperative outcomes for the total player set). (See also Khan (1974)). Kaneko-Wooders (1985, 1986) introduce the notion of a continuum game with finite coalitions (ones containing only a finite number of members), where outcomes of cooperation must be achievable by cooperation within members of partitions[15] of the player set into finite coalitions, and define the f-core as the set of payoffs that cannot be improved upon by any finite coalition.[16] Hammond-Kaneko-Wooders (1989) and Kaneko-Wooders (1989) adapt the approach of Kaneko-Wooders (1985, 1986) to model a continuum economy

---

[13]Debreu - Scarf, and also other writers on economic models use the concept of the "core of an economy", -- a representation of the core in allocation space. The core, in these models, (like the contract curve) refers to allocations of goods which have the property that no group of agents could achieve unanimously preferred allocations for the group membership using only their own endowments.

[14]See Kaneko-Wooders (1986, 1989, 1990) and Hammond-Kaneko-Wooders (1989) for further discussion of the nature of coalitions in Aumann's model.

[15]The partitions must satisfy some measure-theoretic conditions.

[16]We note that the f-core is especially appealing when coalition sizes are not only finite, but bounded; to illustrate, in a marriage model where players are to be paired, it is natural to allow 2-person coalitions.

Under very mild conditions, Kaneko-Wooders (1985, 1986) show nonemptiness of the f-core of continuum games. Hammond-Kaneko-Wooders (1989) show equivalence of the f-core of an economy with a continuum of agents and the set of competitive allocations and nonequivalence of the Aumann-core. Kaneko-Wooders (1989) show that the continuum with finite coalitions is the limit of large games/economies with relatively small effective coalitions.

where coalitions are finite so individual agents can trade one with another or in any finite group, thus modelling the individual player as effective in the pursuit of his own self-interests through trading, while simultaneously ineffective in influencing any broad economic totals. We note that in the important context of exchange economies without externalities the Aumann-core equals the f-core while in other contexts, the Aumann-core is not necessarily equal to the f-core and the competitive allocations (see especially Hammond-Kaneko-Wooders (1989) and Kaneko-Wooders (1990)).

Rather than an exchange-economy framework as considered above, we use a framework where, underlying the dynamic game is a game in coalitional form. This is because we are interested in the "competitive" properties of large games which may be generated by diverse economic models, including, for example, ones with "clubs", local public goods, and coalition production, besides exchange economies. It has long been conjectured that economies where small groups can realize all, or almost all, gains to collective activity are "competitive" or "market-like" (See especially Tiebout (1956) and Buchanan (1965)). It has been shown in several contexts that for such situations asymptotic core-equilibrium equivalence obtains, where the equilibrium concepts are modifications of the Arrow-Debreu competitive equilibrium. (See, for example, Bohm (1974), Wooders (1980, 1989), and Tomasiunas (1990)). Another way of showing large games satisfying inessentiality of large coalitions are "competitive" is to show that they share properties of large exchange economies:[17] Approximate cores of large finite games and cores of continuum limit games are nonempty (see Wooders (1988) for several references) and; the Shapley value is in approximate cores of large games (Wooders-Zame (1987)). Yet another approach is to show that a large game satisfying inessentiality of large coalitions is equivalent to a uniquely determined market games (Wooders 1988). In this paper we take a noncooperative approach to studying competitive properties of large economies with small effective groups. Rather than beginning with a game in coalitional form and studying cooperative equilibria we begin with a strategic game and a process of price formation. Also, we go beyond the previous work in that we are looking at situations where the number of players active (in the market) is finite and may be small.

In this paper, it is essential that, given a positive waiting cost c, the "most profitable" group or firm sizes are bounded. If it were worthwhile to wait for an arbitrarily large number of agents to enter the marketplace,[18] then our results would not obtain. Thus it is natural to use the continuum with finite coalitions as our limiting static game. That model, and its f-core, is based on the ideas that, even in

---

[17] Several of the papers referenced assume "boundedness of marginal contributions to coalitions"; this, however, implies inessentiality of large coalitions (Wooders (1990)). Another assumption previously used, that of "boundedness of average payoffs to coalitions", while less restrictive, implies a weak form of inessentiality of large coalitions.

[18] For example, if the technology were $f(n) = n^2$ describing the total revenue of a firm with n members, $n \in I_+$.

cooperative theory, it is appropriate to model the individual as effective in the pursuit of his own self-interests and nonnegligible relative to other players with whom he might form a group. Moreover, in continuum games, finite coalitions cannot affect aggregate outcomes (outcomes for positive proportions of the total player set). It is such considerations that led Kaneko-Wooders (1986) to conjecture that the continuum game with finite coalitions framework may be useful in investigating the noncooperative foundations of cooperation, in particular carrying out Nash's program. The current paper can be viewed, in part, as initiating one such investigation.

5. MOTIVATION FROM EXPERIMENTAL GAME THEORY

While a fundamental hypothesis of economic theory is that in large economies/games (one with "many" players), players behave as "price" or "payoff" takers, this idea has not been subject to experimental testing. Clearly, it is very expensive to run experiments with many subjects. However, when we think of "large" as relative to an optimal group size, of 2 or 3 players for example, then "large" may be, in absolute terms, still quite small, 16-20 perhaps, and experimental testing of behavior in large economies becomes feasible.

In describing the dynamic game in Section 2, we defined expected payoffs by truncating the game after t periods and then letting t go to infinity. Thus, the game can be viewed as a limit of large finite games. We conjecture that, for our 2-type example, if there are enough players (and thus enough periods), in experimental testing the payoffs of approximately 2 each for "most" players of type 1 and 4 each for "most" players of type 2 would emerge. This conjecture is based on the idea that, in complex situations, subjects do not appear to use backward induction, but instead in a variety of situations, reach approximately "competitive" outcomes, and the core is a competitive solution concept in that it equates demands and supplies for players (given their core payoffs/prices).

As noted in the introduction, it is plausible to expect that significant deviations from the equilibrium path would lead to adaptive behavior. For example, if type 1 players refuse offers of less than or equal to 2, eventually it will become obvious that type 1's are accumulating in the market place, leading type 2's to offer lower wages and type 1's to accept them. The excess supplies provide a clearly visible indication that type 1's might accept lower wage offers. Alternatively, if any type 1's form firms with two type 1 members, this would also serve to alert participants to the possibility that type 1 players might accept offers of 2.

We have undertaken an initial pilot study, which suggests that after subjects learn the rules of the game, the payoffs for most players of type 1 tend to \$2.00 and for most players of type 2, to \$4.00. This study consisted of three plays of the game, with 16 to 19 players in each game. Initially, wage offers and acceptances tended to be "equal split", but then payoffs of type 1 players decreased until, during the last play, 2 for a player of type 1 was the norm. The tendency towards the cooperative outcomes was not uniform. It seemed that the aspirations of type 1 players adjusted downwards during plays of the game, as supplies of type 1 relative to type 2 increased, and between plays, but new plays started with higher aspiration levels for type 1 than those at the end of the preceeding play. These studies, however, were only undertaken as pilots, and we cannot draw any firm conclusions at this time.

6. CONCLUDING REMARKS

1. Bargaining: In the models mentioned in the introduction, on bargaining and competition, matching of pairs of traders occurs via a random, exogenous process. We endogenize the process of group formation and, for our purposes, we view group formation and payoff distribution as inextricably linked. It appears however, that some sort of random group formation would yield similar results.

2. The Procedure: Part of the purpose of the cooperative game-theoretic background of this paper (Wooders (1983, 1988), and other related papers on large games in coalitional form) was to identify and capture some "essential" properties of competitive economies in a broad framework of large games, enabling us to ask whether or not economic structures, such as economies with local public goods or coalition production are competitive. In contrast, in this paper we consider a very specific procedure of group formation and payoff distribution. We might ask what characteristics of procedures ensure competitive (f-core) limiting outcomes within the context of our dynamic framework.

For example, we might substitute a demand-committment procedure for the current wage-offer-committment procedure. Have players announce payoff demands (instead of wage offers), and have later players in the period form groups consisting of players who have "attractive" stated payoff demands. We conjecture (in fact, it seems clear) that this procedure would yield the "same" payoffs as our equilibrium e insofar as possible, given indivisibilities of "money". (Consider the crucial case with $k = n$. The first participant could successfully, in equilibrium, demand the payoff w given by the biggest number in $M(u)$ less than or equal to $1 + (n-1)g(n+1)$, providing $n - w$ is divisible by $n - 1$. If all the relevant parameters are in $M(u)$, then the conjectured payoff to the 1st decision maker would be the same for both procedures.) It may be interesting to have a characterization of procedures which are consistent with "competitive" outcomes.

Anonymity (only a player's type is known, not his name), for example, appears to be an important property in enabling our results.

3. The truncated games: Another aspect of our model which needs further study is the truncated games so we have a better understanding of the effects of length of the game.

4. Nontransferable Utility (NTU): The models considered in this paper are ones with "transferable utility" (TU) or an idealized money. The f-core concept has been applied

to NTU games in coalitional form[19] in Kaneko-Wooders (1985, 1986). We conjecture that our model and results can be extended to the NTU framework, but also conjecture that several subtleties not present in the TU case will arise.

5. **Thin Markets**: Finally, we return to thin markets. Our model describes a situation in which, even though markets are thin, expected payoffs differ only by some bounded multiple of waiting costs from competitive payoffs. Moreover, if the number of participants in the market is large, most participants receive competitive payoffs. This is, perhaps, partially due to the fact that within each period competition has Bertrand-like aspects. As suggested in 2. above, our intuition is that other procedures, such as "random grouping" and multi-lateral bargaining would lead to the same outcome. The features driving the result seem to be that "small" groups are effective, there are "many" participants, (although spread over time), and subgame perfection.

References

Aumann, R.J. (1964). Markets with a Continuum of Traders. Econometrica 32: 39-50.

Boëhm, V. (1974). The limit of the Core of an Economy with Production. International Economic Review 15: 143-148.

Buchanan, J.M. (1965). An Economic Theory of Clubs. Economica 32: 1-14.

Chatterjee, K., B. Dutta, D. Ray, and K. Sengupta, (1990). A Noncooperative Theory of Coalitional Bargaining, (manuscript).

Debreu, G. and H. Scarf (1963). A Limit Theorem on the Core of an Economy. International Economic Review 4: 235-246.

Edgeworth, F.Y. (1881). Mathematical Psychics. Kegan Paul: London.

Everett, H. (1954). Recursive Games. Princeton: Princeton University (mimeographed).

Gale, D. (1986a). Bargaining and Competition, Part 1: Characterization. Econometrica 54: 785-806.

Gale, D. (1986b). Bargaining and Competition, Part 2: Existence. Econometrica 54: 806-818.

Hammond, P., M. Kaneko, and M.H. Wooders (1989). Continuum Economies with Finite Coalitions: Core, Equilibria, and Widespread Externalities. Journal of Economic Theory 49: 113-134.

Khan, M.A. (1974). Some Remarks on the Core of a "Large" Economy. Econometrica 42: 633-642.

Kaneko, M. and M.H. Wooders (1990). Widespread Externalities and Perfectly Competitive Markets. Virginia Polytechnic and State University Discussion Paper (Revised April 1990) Number E90-01-03.

Kaneko, M. and M.H. Wooders (1989). The Core of a Continuum Economy with Wide spread Externalities and Finite Coalitions: From Finite to Continuum Economies. Journal of Economic Theory 49: 35-168.

---

[19] And to NTU economies in Hammond-Kaneko-Wooders (1989) and Kaneko-Wooders (1989).

Kaneko, M. and M.H. Wooders (1986). The Core of a Game with a Continuum of Players and Finite Coalitions; The Model and Some Results. Mathematical Social Sciences 12: 105-137.

Kaneko, M. and M.H. Wooders (1985). Nonemptiness of the Core of a Game with a Continuum of Players and Finite Coalitions. Working Paper No. 295, Institute of Socio-Economic Planning, University of Tsukuba.

Mas-Colell, A. (1979). A Refinement of the Core Equivalence Theorem. Economic Letters 3: 307-310.

Nash, J.F. (1951). Noncooperative Games. Annals of Mathematics 54: 289-295.

Peters, M. (1986). Ex Ante Price Offers in Matching Games: Non-Steady States. Department of Economics, University of Toronto Working Paper No. 861.

Rubinstein, A. and A. Wolinsky (1985). Equilibrium in a Market with Sequential Bargaining. Econometrica 53: 1133-1150.

Sauermann, H. and R. Selten (1962). Anspruchsanpassungstheorie der Unternehmung Zeitschrift für die gesamte Staatswissenschaft 118; 577-597.

Selten, R. (1981). A Noncooperative Model of Characteristic Function Bargaining. In: Essays in Game Theory and Mathematical Economics in Honor of Oskar Morgenstern, V. Böhm and H. Nachtkamp (eds.), in Gesellschaft, Recht, Wirtschaft, Band 4, Wissenschaftsverlag Bibliographisches Institut, Mannheim-Wien-Zürich, pp. 131-151.

Selten, R. (1980). A Note on Evolutionary Stable Strategies in Asymmetric Animal Conflicts. Journal of Theoretical Biology 83: 93-101.

Selten, R. (1975). Reexamination of the Perfectness Concept for Equilibrium Points in Extensive Games. International Journal of Game Theory 4: 25-55.

Selten, R. (1965). Spieltheoretische Behandlung eines Oligopolmodells mit Nachfrageträgheit. Zeitschrift für gesamte Staatswissenschaft **121**: 301-324.

Shubik, M. (1959). Edgeworth Market Games, in R.D. Luce and A.W. Tucker, eds., Contributions to the Theory of Games, IV, Annals of Mathematical Studies, 40, Princeton: Princeton University Press, 257-278.

Simon, H.A. (1959). Theories of Decision Making in Economics and Behavioral Science. The American Economic Review 49: 253-283.

Tamosiunas, R. (1990): Replica Economics with Finite Public Goods. Mathematical Social Sciences (forthcoming).

Wooders, M.H. (1990). Notes on Boundedness of Average and Marginal Contributions to Coalitions and Inessentiality of Large Coalitions. (Manuscript).

Wooders, M.H. (1989). A Tiebout Theorem. Mathematical Social Sciences 18: 33-55.

Wooders, M.H. (1988). Large Games are Market Games 1: Large Finite Games. CORE Discussion Paper #8842.

Wooders, M.H. (1983). The Epsilon Core of a Large Replica Game. Journal of Mathematical Economics 11: 277-300.

Wooders, M.H. (1980). The Tiebout Hypothesis: Near Optimality in Local Public Good Economies. Econometrica 48: 1467-1485.

Wooders, M.H. and W.R. Zame (1987). Large Games; Fair and Stable Outcomes. Journal of Economic Theory 42: 59-93.

GPSR Compliance

The European Union's (EU) General Product Safety Regulation (GPSR) is a set of rules that requires consumer products to be safe and our obligations to ensure this.

If you have any concerns about our products, you can contact us on ProductSafety@springernature.com

In case Publisher is established outside the EU, the EU authorized representative is:

Springer Nature Customer Service Center GmbH
Europaplatz 3
69115 Heidelberg, Germany

Batch number: 10844669

Printed by Printforce, the Netherlands